W9-AEV-977

510.9
H62

144624

MINNESO   CE

DISCARDED

DATE DUE

DISCARDED

Minnesota Studies in the
PHILOSOPHY OF SCIENCE

HERBERT FEIGL, FOUNDING EDITOR

VOLUME XI

History and Philosophy of Modern Mathematics

EDITED BY

WILLIAM ASPRAY AND PHILIP KITCHER

CARL A. RUDISILL LIBRARY
LENOIR RHYNE COLLEGE

UNIVERSITY OF MINNESOTA PRESS, MINNEAPOLIS

510.9
H62
144624
May 1988

The University of Minnesota Press
gratefully acknowledges publication assistance provided
for this book by the Babbage Institute together with
the Minnesota Center for Philosophy of Science.

Copyright © 1988 by the University of Minnesota
All rights reserved. No part of this publication may be
reproduced, stored in a retrieval system, or transmitted, in
any form or by any means, electronic, mechanical, photo-
copying, recording, or otherwise, without the prior written
permission of the publisher.

Published by the University of Minnesota Press
2037 University Avenue Southeast, Minneapolis MN 55414.
Published simultaneously in Canada
by Fitzhenry & Whiteside Limited, Markham.
Printed in the United States of America.

Library of Congress Cataloging-in-Publication Data

History and philosophy of modern mathematics.
(Minnesota studies in the philosophy of science;
v. 11)
"The volume is the outgrowth of a conference held at
the University of Minnesota, Minneapolis, 17-19 May
1985"—Pref.
Bibliography: p.
Includes index.
1. Mathematics—History—Congresses. 2. Mathematics—
Philosophy—Congresses. I. Aspray, William, 1952-. II. Kitcher,
Philip, 1947- . III. Series.
Q175.M64 vol. 11 501 a [510'.9] 86-30884
[QA26]
ISBN 0-8166-1566-7
ISBN 0-8166-1567-5 (pbk.)

The University of Minnesota is
an equal-opportunity educator and employer.

Contents

Preface

The purpose of this volume is to bring together a number of different perspectives on modern mathematics, with the aim of understanding how the work of historians, philosophers, and mathematicians can be integrated in an interdisciplinary study of mathematics. Several of the essays adopt such an interdisciplinary approach, whereas others favor an autonomous pursuit of the history of mathematics or of the philosophy of mathematics. In our judgment, this juxtaposition of old and new styles accurately represents the state of the field at this time.

The volume is the outgrowth of a conference held at the University of Minnesota, Minneapolis, 17-19 May, 1985. The conference and the publication of proceedings were supported by generous grants from the Alfred P. Sloan Foundation and the College of Liberal Arts and Institute of Technology of the University of Minnesota. We also wish to thank a number of the scholars who served as commentators at the conference: Thomas Drucker of the University of Wisconsin-Madison; Philip Ehrlich of Brown University; Emily Grosholz of Pennsylvania State University; Joelle Proust of CNRS/Nice; and Helena Pycior of the University of Wisconsin-Milwaukee.

In 1974, Garrett Birkhoff of Harvard University organized and the American Academy of Arts and Sciences sponsored a Workshop on the Evolution of Modern Mathematics. (The proceedings are published in the November 1975 issue of *Historia Mathematica*.) The purpose was to bring together a distinguished group of historians and mathematicians to review and discuss the history of mathematics since 1800, a subject that had received relatively little attention in comparison with earlier periods of mathematics history. The meeting was an unmitigated success, both in providing a forum for discussion between these two bodies of professionals, and in producing a valuable set of proceedings.

In our opinion, in the decade since the 1974 workshop there has been an upsurge of interest in an interdisciplinary approach to historical and philosophical problems connected with mathematics. There have also been recent developments internal to each discipline: historians of mathematics have recognized the need for a more sophisticated historiography and for greater attention to the social context in which the work of mathematicians is set; philosophers of mathematics have moved away from the grand foundational programs of the early twentieth century and the preoccupation with the demise of those programs.

For these reasons, we believed it was important once again to take appraisal of our fields. Looking to the 1974 workshop as our model, we brought together a distinguished group of historians, mathematicians, and for the first time philosophers actively working in this area. Indeed, as a measure of the continuity of this area, seven of the authors represented in this volume were participants in the 1974 workshop. Our authors were requested to keep in mind not only the historical, philosophical, and mathematical significance of their topics, but also to devote special attention to their methodological approach. These approaches came under considerable discussion during the conference. We can only hope that our proceedings will be as influential as those of the 1974 workshop.

W. A. and P. K.

AN OPINIONATED INTRODUCTION

An Opinionated Introduction

Each of the essays that follow could be sent forth as an orphan, with only the most perfunctory comment, to make its own solitary way in the world, and some of them would undoubtedly do well. However, we believe that there are important reasons for setting them side by side, that each gains from the company of the others. In our judgment, the present volume offers a representation of the current state of history and philosophy of mathematics. Hence we have felt encouraged—perhaps foolhardily—to offer perspectives on the past, the present, and the future. This introduction attempts to sketch the history of the philosophy of mathematics in the last century and the recent history of the history of mathematics. We then endeavor to relate the individual studies not only to one another, but also to the traditions of work in the history of mathematics and in the philosophy of mathematics. We conclude with some speculations about the present trends and possible futures. Quite plainly, any effort of this kind must embody the authors' ideas, perspectives, and, maybe, prejudices about what kinds of researches are important. So we have rejected the plain title "Introduction" and announced our essay for what it is—an opinionated introduction.

1. Philosophy of Mathematics: A Brief and Biased History

Philosophers of mathematics may argue about many things, but, until recently, there has been remarkable agreement concerning when their discipline began. Prevailing orthodoxy takes the history of the philosophy of mathematics to start with Frege. It was Frege who posed the problems with which philosophers of mathematics have struggled ever since, Frege who developed modern logic and used it to undertake rigorous exploration of the foundations of mathematics, Frege who charted the main philosophical options that his successors were to explore. To be sure, there were earlier thinkers who considered mathematics from a philosophical point of view, scholars of the stature of Kant, Mill, and Dedekind. But they belong to prehistory.

The orthodox account embodies something of Frege's own view of what he achieved, as well as reflecting his harsh judgments about some of his predecessors and contemporaries (see Frege 1884). Within the last decades there have been criticisms of the conventional wisdom, in part because philosophers have discovered merit in the work of some of Frege's predecessors (Parsons 1964, 1969 and Friedman 1985 on Kant; Kessler 1980 and Kitcher 1980 on Mill; Kitcher 1986 on Dedekind) and in part because there have been systematic and thorough studies of Frege's place in the history of philosophy (Sluga 1980; Resnik 1980). Nevertheless, because Frege has so often been seen as the *fons et origo* of real philosophy of mathematics, it is well to begin with him and his accomplishments.

Frege was originally trained as a mathematician, and it is clear from his early writings that he became interested in integrating late-nineteenth-century mathematics into an epistemological picture that he drew (at least in large measure) from Kant. During the 1870s he seems to have become convinced that Kant's ideas about arithmetic were incorrect and that the deficiency could only be made up by a thorough revision of logic, a project that he began in his brilliant monograph (1879). By 1884, he was prepared to publish his most articulate account of the philosophy of mathematics. At the outset, he adopted the view that the central problem for the philosophy of mathematics was to identify the foundations of mathematics (1884, 1-2). Holding that the task of setting arithmetic on a firm foundation was a natural extension of the enterprise of the foundations of analysis (pursued by Dedekind, Weierstrass, Heine, and others), Frege undertook to review the possibilities.

Since the goal of the foundationalist is to display the proper justification for the mathematical statements under study, Frege suggested that there were as many possible approaches as his background epistemology allowed—to wit, three. The basic division is between justificatory procedures that are a priori and those that are a posteriori. Within the category of the a priori, there are two options. Arithmetic is either ultimately derivable from logic plus definitions of the special arithmetical vocabulary, or it is founded (at least in part) on some a priori intuition (as Frege took Kant to claim). Having argue that arithmetical statements do not admit empirical justification and that no a priori intuition is available in the case of arithmetic, Frege concluded that arithmetic was simply a development of logic.

In fact, he knew of and argued against a fourth possibility. From the perspective of the *Grundlagen* (Frege 1884), arithmetic is taken to be a science with a definite content. Elsewhere, Frege responded to the proposal that, strictly speaking, arithmetical statements are meaningless and arithmetic is simply a game that mathematicians (and others) play with signs (see the essays collected in Frege/Kluge 1971). We believe that, until the end of his life, Frege operated with the view that there were only a very small number of possibilities for finding the proper justification of arithmetical statements and that the strategy of elimination of alternatives, adopted in his early writings, had a profound effect on the development of his views (see Kitcher 1979).

Frege's eliminative argument for logicism was only the first step in the project of *Grundlagen*. Having argued that arithmetic had to be a development of logic, Frege took it upon himself to exhibit the development. One important part of the task was to delineate clearly the principles of logic—an enterprise that Frege had already begun (1879) and was to continue (1893-1903). Another was to show how the concepts of arithmetic could be defined in logical terms. Most of *Grundlagen* was devoted to the latter project, and Frege argued first that attributions of number are assertions about concepts and subsequently that the natural numbers should be identified with the extension of particular concepts.

If Frege was the "onlie begetter" of the philosophy of mathematics, then it is easy to identify his offspring. First, and of prime importance in the later history of the field, is his conception of the central problems. Philosophers (or philosophically minded mathematicians) should disclose the foundations of parts of mathematics by identifying the proper justifications for the statements belonging to those parts. Second is his taxonomy of options: logicism, formalism, appeals to some kind of intuition (in the manner of Kant), and (definitely bringing up the rear) empiricism. Third is the collection of arguments that are designed to reveal logicism as the only possible position. Fourth is the analysis of arithmetical discourse and, in particular, the claim that arithmetical statements record the properties of (logical) objects. Last, but definitely not least, is the elaboration of a logical system, which, to our mind, constitutes the greatest single achievement in the history of logic—bar none.

Of course, Frege's ideas suffered a temporary setback. In 1902 Russell discovered that the logical system, so painstakingly constructed and so little appreciated, was inconsistent. Ironically, if the discovery of Russell's paradox cast doubt on Frege's substantive ideas about the nature of mathematics, it served only to underscore the importance of the task in which he had been engaged. In the last decades of the nineteenth century, professional mathematicians rarely saw Frege as tackling a genuine problem that was continuous with the difficulties that Weierstrass and his school had attempted to address. Although the disentangling of concepts of continuity and convergence had an obvious mathematical payoff, there seemed little point to the minute excavation of the foundations of arithmetic. The paradoxes of set theory, first noted by Cantor and Burali-Forti in the 1890s and sharply presented in Russell's observation about Frege's system, changed that confident assessment. They showed that the intuitive reasoning about sets (or systems, collections, or multitudes) on which mathematicians had relied, increasingly and with ever greater explicitness, during the latter part of the nineteenth century, were surprisingly hard to represent in a precise fashion. Nobody had doubted that there was a "proper justification" for the principles of arithmetic. But the first serious attempt to identify it had exposed difficulties in widely adopted modes of reasoning that were disturbingly reminiscent of those that had been recently resolved in the foundations of real analysis.

For the first three decades of the twentieth century, the philosophy of mathematics was dominated by the rival claims of three main foundational programs, each of which pursued the Fregean task of exposing the proper justification for arithmetic (and analysis), and each of which embodied a response to the paradoxes. This is not to deny that there were critics, most prominently Poincaré, who offered alternative visions of what philosophy of mathematics ought to be (see Goldfarb, this volume). But Frege's identification of philosophy of mathematics with foundations of mathematics won the day, and the principal differences were internal to the general Fregean framework. Closest to Frege's original approach was the program undertaken by Russell and Whitehead in *Principia Mathematica*, although it was apparent, almost from the beginning, that some of the principles that Russell and Whitehead took as their starting assumptions were not uncontroversially *logical*. The theory of types was able to circumvent the paradoxes, but the accommodation of arithmetic it offered was purchased at the cost of axioms, most notoriously the axioms of infinity and reducibility, that sat uneasily with Frege's characterization of laws of logic—laws without which no thinking is possible.

A second proposal, deeply at odds with Frege's doctrine that mathematics is a doctrine all of whose statements have an identifiable content, was Hilbert's suggestion that a foundation for mathematics should be provided by constructing finitary consistency proofs for mathematical systems. As Hilbert's program was articulated in its mature form (Hilbert 1926), there is a body of contentful mathematical statements whose credentials are above suspicion. This body of contentful statements is to be used in proving the consistency of formal systems in which contentful mathematics is itself embedded and which also include ideal statements. The role of these ideal statements is to facilitate transitions among the contentful statements, and the set-theoretic paradoxes show us that, if we are careless in adjoining such ideal auxiliary statements to our mathematical systems, we generate contradictions. Consistency proofs will guard us against such possibilities—so long, of course, as they are given within a mathematical framework (contentful mathematics) that cannot itself be questioned. Hence, for Hilbert, the important task is to give finitary consistency proofs for formal systems that formalize classical mathematics. Gödel's theorems seemed to show that the task cannot be completed, even when the part of classical mathematics in question is first-order arithmetic. (For illuminating discussions of Hilbert's program and of the significance of Gödel's incompleteness theorems, see Kreisel 1958; Resnik 1974; Detlefsen 1979; Tait 1981; Simpson forthcoming.)

The third program, developed by Brouwer, sought a foundation for mathematics by returning to Kant. Brouwer claimed that we construct the objects of mathematics and that our knowledge of their fundamental properties is based on an a priori intuition. Through this intuition, we are able to recognize a potential infinity of mathematical entities (the natural numbers), and these can form the basis for further constructions. But classical mathematics goes astray both in its positing of objects that are beyond the reach of human constructive powers and it its use of nonconstructive existence proofs, achieved through the employment of the law of excluded middle. Brouwer, and later Heyting (1956), proposed to substitute for classical mathematics an alternative that would be free of these mistakes and that would be true to the constructive/intuitive foundation of mathematics. To the response that the result was a mutilation of mathematics, in which long, ugly proofs of intuitionistic analogs replaced the simple and elegant classical proofs and theorems, they responded that such alterations were needed to avoid the profligate and incoherent metaphysics of the classical mathematicians. (For seminal works in classical intuitionism, see Brouwer 1949, Heyting 1956; a recent development of constructivist mathematics is Bishop 1967; Dummett 1977 gives a profound interpretation of the intuitionist program; for an interesting alternative view, see Shapiro 1985).

The three programs we have mentioned correspond to three of the four possibilities that occur in Frege's eliminative argument for logicism. Russell and Whitehead carry on with the logicist strategy, Hilbert develops a sophisticated version of the idea that (at least part of) mathematics is a formal game, and Brouwer suggests that the foundation of mathematical knowledge lies in nonlogical, a priori, intuition. The one of Frege's options that does not issue in a developed program is the suggestion of which he was most scornful: there is no empiricist tradition in the philosophy of mathematics in the early twentieth century (although Curry's version of formalism has some connections with empiricism; see his 1951).

In noting the relationship between the dominant themes in philosophy of mathematics from 1900 to 1930 and Frege's basic epistemological categories, we should not overlook the fact that each of the programs also has roots in nineteenth-century mathematics (for different amplification of the theme, see Resnik 1980 and Stein, this volume). However, insofar as they continue the Fregean identification of philosophy of mathematics with the foundations of mathematics and insofar as they see the task of laying the foundations as finding a priori justifications for parts of mathematics, they are naturally seen as internal modifications of a very general research program that Frege set for his successors. We shall now suggest that the next phase of developments in the philosophy of mathematics remained deeply Fregean in spirit.

During the 1920s and early 1930, logicism underwent a fundamental change, a change that would affect the credibility of each of the alternative views of the foundations of mathematics. Inspired by their reading of Wittgenstein's *Tractatus*, the logical positivists suggested that the idea of tracing arithmetic back to logic-plus-definitions gave the false appearance that two different types of principles were fundamental to mathematics. Instead, they recommended, we should think of logic itself as tacitly embodying the definitions that we have fixed for the logical vocabulary, the connectives and quantifiers. Thus the entire corpus of logic and mathematics can be conceived as elaborating the conventions that underlie our language; the statements that make up this corpus are true by virtue of meaning (see Carnap 1939; Ayer 1936; Hempel 1945). From this perspective it does not matter, in the end, whether the axioms of various bits of mathematics are (or are derivable from) laws of logic. What counts is that they should be (or be derivable from) principles that embody the conventions that govern the primitive vocabulary.

The significance of this change for the credibility of the other programs is that it provides a way to accommodate certain ideas of formalism and to remove the sting of the intuitionist critique of classical mathematics. One of the chief points of dispute between Hilbert and Frege had surrounded Hilbert's early proposal that the primitive vocabulary of an axiomatic system is implicitly defined by the axioms that are set down. On Frege's conception of logic (which held no place for the notion of an uninterpreted statement), this claim of Hilbert was absurd (see Resnik 1980 for a lucid discussion of the controversy.) In the wake of the development of the semantics of logic (see Goldfarb 1979 and Moore, this volume), Frege's conception of logic was abandoned and logicists came to adopt Hilbert's proposal. Thus, at a time when the mature Hilbert program was in difficulties, it appeared possible to salvage part of Hilbert's early motivations. Furthermore, instead of conceiving intuitionism to be a rival mathematics, the emphasis on the conventionality of logic made it possible to see intuitionistic logic and mathematics as embodying a different set of conventions. The system based on those conventions could be tolerated as an interesting curiosity, something that could be placed alongside classical mathematics but that could not pose any threat of displacement (see Carnap 1937).

The new logicism represented both a return to Frege and a departure. Because of Frege's opposition to the founding of mathematics on intuition, his scorn for empiricism, and his emphasis that mathematical statements are meaningful, the logical positivists could legitimately cite him as a predecessor. However, their ideas about logic and semantics were importantly different from those that Frege had espoused (see Dummett 1973; Sluga 1980; Ricketts 1985). Hence there was a genuine transition between one version of logicism, on which the principles of logic were seen as fundamental to all thought, to another, on which logic was true by convention (Friedman, this volume, provides an illuminating account of part of the transition).

The new form of logicism avoided awkward questions about how we know the principles that prescribe to all thought, but it quickly became vulnerable to a different challenge. In 1936, Quine published a seminal article containing two distinct arguments against the idea that logic and mathematics are true by convention. The first of these arguments contends that any part of our knowledge can be replaced by a system whose principles are true by convention. Thus it appears (although Quine does not hammer home the point) that the thesis that logic and mathematics are true by convention cannot explain the epistemological status of these disciplines, unless their epistemological status is the same as that of other branches of inquiry. The second argument, adapted from an idea of Lewis Carroll, is that logic could not be true by convention for the simple reason that logic is required to extract the consequences of any conventions we may lay down.

Quine's second argument seems to have been more immediately influential than his first, for it prompted a reformulation of the canonical doctrines of the new logicism. Although logicists continued to describe logic and mathematics as conventional, they preferred to speak of the truth of these disciplines as resting on semantical rules—rules that were explicitly formulated in the construction of formal systems aimed at explicating prior mathematical usage and that were taken to underlie that prior usage— rather than thinking of truth as resulting from an explicit convention (see Carnap 1939; Hempel 1945). Quine's first argument unfolded into a critique of the notion of semantical rule (and the related notion of truth by virtue of meaning) designed to circumvent these reformulations. Thus in (1953), in (1962), and in a host of later writings, Quine argued at length that there are no statements true by virtue of meaning. One consequence of these arguments was a muted (and underdeveloped) empiricism.

As logicism suffered under Quine's attack on its fundamental conception of logical truth, a different Fregean theme underwent a revival. Since he did not believe the truths of logic to be true in virtue of meaning *and therefore as devoid of factual content* (as some of the positivists maintained), Frege saw no tension between claiming that arithmetic is disguised logic and declaring that numbers are objects. Logical positivism downplayed the latter declaration, and the positivists sometimes maintained that, as statements that are true in virtue of meaning, truths of logic and mathematics are devoid of any reference. Quine's scrutiny of the concept of truth by virtue of meaning led him to pose the ontological questions: what, if anything, is mathematics about? Accepting Frege's analysis of the logical forms of arithmetical statements and adopting the semantics for first-order languages developed by Tarski, Quine came to Frege's own Platonistic conclusion: mathematics is about abstract objects, numbers and sets. (Earlier, Goodman and Quine [1947] had explored the possibility of managing without any abstract objects at all, but they concluded that nominalism was inadequate to allow for the truth of those mathematical statements that are required in empirical science.) Quine's later writings have attempted to show that the only abstract objects needed by mathematics and science are sets and to integrate this Platonistic position with his epistemology.

Quine was not alone in fostering a renaissance of Frege's doctrine that numbers are objects. In investigating one of the great outstanding problems in set theory in the 1940s, the truth of Cantor's continuum hypothesis (which states that the power of the continuum is $\aleph_1$), Gödel pondered the possibility that the continuum hypothesis is independent of the axioms of set theory (that is, the axioms of a standard set theory such as ZF or NBG). A possible response to the discovery of independence (which Gödel envisaged and which was established in part by his own work and in part by subsequent work of Paul Cohen) would be to claim that the continuum hypothesis has no truth value until we specify the system in which it is to be embedded. Thus, just as there are Euclidean and non-Euclidean geometries, and just as the parallel postulate holds in the former but not the latter, so, perhaps, we shall have to talk about Cantorian and non-Cantorian set theories. Gödel took this response to be profoundly misguided. Even if the continuum hypothesis were found to be independent of the rest of set theory (as in fact it was), we should continue to ask whether it is true and to seek principles that give an adequate and complete characterization of the universe of sets. Pursuing the point, Gödel contended that there is a definite world of mathematical entities, that these entities are independent of human thought, and that human beings have a way of apprehending the fundamental properties of these entities. In a phrase that was to resonate in subsequent philosophy of mathematics, he wrote that the truth of the axioms of set theory "forces itself upon us" (Gödel 1964).

Like Quine, Gödel departed from the positivist conception that, in mathematics and logics, truth is dependent not on reference to objects but on meaning (or, in the more primitive versions, on conventions). Frege's old conception of mathematics as a body of statements about abstract objects came to occupy center stage in the philosophy of mathematics, appearing as the only serious possible view about mathematical ontology and mathematical truth. But some of the connections within Fregean epistemology had now been broken. Frege's own wedding of the ontological thesis to the claim that mathematics is logic seemed no longer available, for, according to the orthodox view of logic, logic has no special content, not even objects so ethereal as those apparently needed for mathematics. Quine's attacks on the notion of truth by virtue by meaning had apparently led him to a version of empiricism about mathematics that was not vulnerable in the ways that Mill's empiricism had been, but it was far from clear what positive claims about mathematical knowledge were embodied in Quine's writings. Finally, Gödel's ontological Platonism was accompanied by a clear and straightforward epistemological thesis that fractured Frege's old (Kantian) link between intuition and the construction of mathematical objects. According to Gödel, we have the ability to intuit the properties of *mind-independent* objects, and this accounts for our knowledge of the basic principles of set theory (and, in the future, it may account for the principles that we add to complete our knowledge of the universe of sets). Yet, although it may be relatively easy to understand, Gödel's view is not so easy to believe, for the notion of intuition on which it relies appears somewhat nebulous. Frege's old warning is apt: "We are all too ready to invoke inner intuition, whenever we cannot produce any other ground of knowledge" (1884, 19).

During the 1950s and 1960s, the philosophy of mathematics centered around efforts to displace the versions of logicism, formalism, and intuitionism that had been current in previous decades and to articulate a *neo-Fregean* view of mathematics whose distinctive claim was that set theory (specifically some extension of ZF) provided the foundations for mathematics. To a large extent, epistemological issues were neglected in favor either of technical explorations (concerning the import of the Gödel-Cohen discoveries or the significance of the Löwenheim-Skolem theorem, for example) or of attempts to show the superiority of neo-Fregeanism to one or the other of the (now unfashionable) earlier foundational programs. However, in the 1960s and early 1970s, a number of important criticisms of the Platonist orthodoxy began to emerge, and these criticisms have played a major role in shaping current work in the philosophy of mathematics.

On the neo-Fregean conception, all mathematics is embedded in set theory and all the entities of mathematics are sets. In particular, numbers are sets. But, it appears, if numbers are sets then they must be particular sets, and the philosopher should be able to identify *which* ones they are. However, mathematicians have long recognized that arithmetic can be reduced to set theory in many different ways: the natural numbers can be identified with the von Neumann numbers, the Zermelo numbers, or with any of an infinite number of other sequences of sets. Quine (1960, 1970) used the point to argue for his own ideas about reference and reduction, but the most forceful posing of the problem occurred in an article by Benacerraf (1965). In effect, Benacerraf's discussion delineated three options for the philosopher of mathematics: we can either try to find some principle for choosing one of the set-theoretic identifications of the natural numbers as privileged (and Benacerraf carefully explored some main possibilities, with negative results), we can try to maintain the thesis of ontological Platonism without identifying numbers with any particular sets (as Quine had argued, and as subsequent writers—White 1974; Field 1974; Resnik 1980, 1981; Maddy 1981—were to do), or we can abandon ontological Platonism. In the 1960s and 1970s, few philosophers were prepared to consider the last option.

However, there were occasional expressions of dissent. In 1967, Putnam contrasted the "mathematics of set theory" picture (our neo-Fregeanism) with an alternative that he called "mathematics as modal logic" (Putnam 1967). His suggestion was that mathematical statements can be formulated as necessary conditionals whose antecedents are the conjunctions of axioms of standard mathematical theories. In this way, we may capture parts of the idea that mathematics is concerned with the properties of structures, while avoiding the conventionalist notions that accompanied that idea during the heyday of positivism. Putnam also proposed that neither the neo-Fregean picture nor the modal logic picture should be viewed as offering the fundamental correct view of mathematics. Both should be recognized as "equivalent descriptions."

A more direct attack on ontological Platonism was offered in 1973 by Chihara, who developed a detailed account of a nominalist version of mathematics. Chihara proposed to reevaluate the Quinean argument for Platonism by showing how the mathematics needed for science could be articulated without commitment to abstract entities. Chihara's work may fairly be taken as the beginning of a revival of interest in nominalism, which has been pursued in his own subsequent work and (in different ways) in Gottlieb (1980) and Field (1980, 1982, 1984, 1985).

However, the most influential study of the new orthodoxy in philosophy of mathematics was another article by Benacerraf (1973). Benacerraf constructed a dilemma, designed to show that our best views about mathematical truth do not fit with our ideas about mathematical knowledge. Start from the relatively unproblematic theses that there are some mathematical statements that are true and that some of these are known by some people. Then a task of philosophy is to provide an account of the truth of those statements that are true and an account of how we know the mathematics that we know. On our best accounts of truth in general and of the form of mathematical statements (accounts that derive from Frege and Tarski), true statements of mathematics owe their truth to the properties of and relations among mathematical objects (paradigmatically sets). It appears that such objects are outside space and time—for there seem to be too few spatiotemporal objects to go around, and, in any case, the truths of mathematics seem independent of the fates of particular spatiotemporal entities. According to Benacerraf, on our best account of human knowledge, knowledge requires a causal connection between the knower and the objects about which the knower knows. Since mathematical objects are outside space and time, there can be no such connection between them and human beings. Hence, on the best accounts of truth and knowledge, mathematical knowledge turns out to be impossible after all.

Since the early 1970s, much of the research in the philosophy of mathematics has been devoted to evaluating the merits of neo-Fregeanism in the light of the studies that I have briefly reviewed. Particularly important has been the task of finding an appropriate response to Benacerraf's dilemma. Some writers (notably Steiner 1975; Maddy 1980; Kim 1982) have argued that neo-Fregeanism is unthreatened by the dilemma: they hold that a proper understanding of the conditions on human knowledge will allow for us to have knowledge of objects that are outside space and time. Others (Lear 1977; Jubien 1977) have deepened the dilemma and have proposed revisions of neo-Fregeanism to accommodate it. In addition, some writers have proposed approaches to the ontology of mathematics that question the thesis that mathematical objects are sets. The less radical of these is *structuralism*, the doctrine that mathematics describes the properties of mathematical structures. More radical is the nominalism of Chihara, Gottlieb, and Field.

Structuralism is a natural reaction to Benacerraf's other problem, the question of how to identify the *genuine* natural numbers among all the possibilities that set theory supplies. It is tempting to reply that all the set theoretical identifications share a common structure and that it is this structure that is the subject matter of arithmetic, not any particular ω-sequence of sets. One can remain completely within the preferred ontological framework of neo-Fregeanism by supposing that arithmetic describes the properties of a structure that is multiply realized in the universe of sets (White 1974; Field 1974; see Kitcher 1978 for criticism and Maddy 1981 for an attempt at revival). Alternatively, the structuralist may treat mathematical structures as entities in their own right, contending that set theory, like other mathematical disciplines, is concerned with a particular kind of structure (Resnik 1975, 1981, 1982; Shapiro 1983a). On either of these approaches, mathematical entities seem to be abstract objects, so that the structuralist ontology will have to be supplemented with a response to Benacerraf's dilemma. But structuralism may also be articulated with an eye to coping with the epistemological difficulty raised by Benacerraf, perhaps by using notions from modal logic (as in Putnam 1967 or in Jubien 1981a, 1981b) or by working out a constructivist version of structuralism (Kitcher 1978, 1980, 1983). We believe that it is worth noting that structuralism has some obvious kinship with the ways in which some contemporary mathematicians (especially algebraists) present their discipline—and this seems particularly true of the version of structuralism favored by Resnik and Shapiro.

Contemporary nominalism has been elaborated in three rather different ways. Chihara's version (1973, 1984) retains the traditional idea that statements appearing to bear commitment to the existence of abstract objects can be reconstrued as assertions about linguistic entities. Gottlieb (1980) proposes to reinterpret the quantifiers, making use of substitutional quantification to avoid commitment to abstract entities (see Parsons 1971 for a lucid account of substitutional quantification and Parsons 1982 for an illuminating appraisal of Gottlieb's program). Field (1981, 1982, 1984, 1985) has made the most decisive break with the tradition, arguing that physical theories can be reformulated so that they bear no commitment to numbers or to other mathematical entities. The conclusion he draws is that mathematical knowledge is simply logical knowledge, and that mathematical vocabulary functions as a device for obtaining consequences that would have been more difficult to reach without that vocabulary. It seems to us that there is an obvious similarity here to Hilbert's notion of "ideal statements," and indeed Field's program might seem closer to Hilbert's formalism rather than to traditional nominalism. (For penetrating evaluations of Field's work, see Malament 1982 and Shapiro 1983b.)

Besides these two lines of reaction to problems (actual or perceived) with neo-Fregeanism, there have been a number of other important developments in recent philosophy of mathematics. Some philosophers have been primarily concerned to explore the special issues that arise within various parts of logic and set theory, the status of conceptions of set, the credentials of second-order logic, the role of reflection principles, and so forth. Thus Parsons has produced an important series of essays (collected in his 1984) that examines questions of ontology as they arise locally in various versions of set theory and in other parts of mathematics. Boolos (1975, 1984, 1985) has offered a fresh approach to second-order logic, which resists the orthodox account (originally defended by Quine 1963 and 1970) that second-order logic is a cloudy way of doing set theory. Both these investigations have provided glimpses of fresh approaches to the more general problems discussed above, approaches that preserve insights from traditions that had previously seemed outmoded.

Because the influence of Frege's ideas on contemporary philosophy of mathematics is so evident, it should hardly be surprising that some current work in the philosophy of mathematics is centered around reappraisals of Frege. Thus, there have been detailed explorations of parts of Frege's philosophy of mathematics (Resnik 1980; Wright 1983), attempts to articulate Fregean themes (Hodes 1984; Boolos 1985), and an important series of investigations of the conceptions of logic and knowledge in which Frege's technical work was set (Goldfarb, this volume; Ricketts 1985). Although philosophical exploration of the origins of the main categories of analytic philosophy may initially appear rather remote from questions about mathematics, it is surely appropriate to remind ourselves that the problems whose history we have been tracing are continuous with the central problems of epistemology and metaphysics in the twentieth century. Understanding how the latter problems emerged will, perhaps, enable us to avoid errors and confusions in formulating them, mistakes that may lead us to favor nonsolutions or may prevent us from finding any solutions at all.

We have attempted to give a whirlwind tour of present work in *mainstream* philosophy of mathematics and to show where that work comes from. We now want to emphasize a point that has, we hope, been obvious to the reader who is following the story for the first time. Although our tale begins with the researches of mathematicians, originating with questions that seem to arise from the mathematics of the late nineteenth century, and although, with the discovery of the paradoxes, the task of providing foundations for mathematics seems to assume some mathematical importance, the distance between the philosophical mainstream and the practice of mathematics seems to grow throughout the twentieth century. Philosophy of mathematics appears to become a microcosm for the most general and central issues in philosophy—issues in epistemology, metaphysics, and philosophy of language—and the study of those parts of mathematics to which philosophers most often attend (logic, set theory, arithmetic) seems designed to test the merits of large philosophical views about the existence of abstract entities or the tenability of a certain picture of human knowledge. There is surely nothing wrong with the pursuit of such investigations, irrelevant though they may be to the concerns of mathematicians and historians of mathematics. Yet it is pertinent to ask whether there are not also other tasks for the philosophy of mathematics, tasks that arise either from the current practice of mathematics or from the history of the subject.

A small number of philosophers (including one of us) believe that the answer is yes. Despite large disagreements among the members of this group, proponents of the minority tradition share the view that philosophy of mathematics ought to concern itself with the kinds of issues that occupy those who study other branches of human knowledge (most obviously the natural sciences). Philosophers should pose such questions as: How does mathematical knowledge grow? What is mathematical progress? What makes some mathematical ideas (or theories) better than others? What is mathematical explanation? Ideally, such questions should be addressed from the perspective of many areas of mathematics, past and present. But, because the tradition is so recent, it now consists of a small number of scattered studies, studies that may not address the problems that are of most concern to mathematicians and historians or explore episodes or areas within mathematics that most require illumination.

If the mainstream began with Frege, then the origin of the maverick tradition is a series of four papers by Lakatos, published in 1963-64 and later collected into a book (1976). Echoing Popper, Lakatos chose the title *Proofs and Refutations*, and the choice is, in part, apt. One obvious theme of the essays is the discussion of a segment of the history of mathematics in terms of categories that are adapted from Popper's methodology for the natural sciences. We believe that the significance of the papers is best appreciated by looking past the suggested Popperian solutions to the problems that Lakatos was seeking to address. Selecting the sequence of attempts to formulate precisely and to prove the Euler-Cauchy conjecture about the relation among the faces, edges, and vertices of polyhedra as his historical example, Lakatos posed such questions as: How do mathematical definitions get revised? How are methods of proof modified? Are there methodological rules that mathematicians follow in pursuing these changes? If so, what are they? Now we take it that questions like these are important both for historians of mathematics and, perhaps, for practising mathematicians. Any serious investigation of the history of mathematics must embody some ideas about how the subject proceeds when it is done properly, and Lakatos's questions promise to move historiography beyond a set of tacit ideas, distilled from a (possibly incompatible) collection of sources, toward the explicit formulation of methodological principles. If we had such a canon, then historians could use it to investigate the match between actual history and the ideal, perhaps finding intriguing cases in which some extrinsic factor prompted a departure from the recommendations of methodology. Moreover, mathematicians might find it illuminating both to see how their chosen field of investigation had emerged from the mathematics of the past and how certain kinds of methodological considerations were paramount in fashioning its central concepts. It is not even out of the question that the answers to Lakatos's questions might help illuminate disputes among mathematicians about the legitimacy of various approaches or the significance of certain ideas.

Although Lakatos's discussion was the first extended philosophical treatment of methodological issues in mathematics, there were similar themes in the earlier work of Wilder, who subsequently developed his approach further (1975). Steiner (1975) and Putnam (1975) argued for the importance of nondeductive arguments in mathematical *justification* (as opposed to mathematical discovery, which had been explored with great thoroughness and elegance by Polya [1954]). In subsequent work, Steiner has discussed the character of mathematical explanation (1978) and has posed again Wigner's famous question about the "unreasonable effectiveness" of mathematics in the scientific investigation of the world.

Grosholz (1980, 1985) has investigated the manner in which new mathematical fields emerge from the synthesis of prior disciplines. The general task of articulating an account of mathematical methodology, begun by Lakatos, has been tackled by writers from several different perspectives (see, for example, Hallett 1979; Kitcher 1983).

The work that we have mentioned focuses on a few episodes in the history of mathematics and on a scattering of fairly elementary examples from contemporary mathematics. Many mathematicians are inclined to take this as symptomatic of the inevitable irrelevance of the general approach, rather than as a side effect of the fact that responsible studies begin with examples that are best suited for developing the central ideas of the new field. We hope that the enterprise of articulating the methodology of mathematics will attract scholars who are able to investigate examples from all phases of the development of mathematics, including the most exciting current contributions, and that, as this is done, the skepticism of parts of the mathematical community will be overcome.

The current state of the philosophy of mathematics thus reveals two general programs, one that is continuous with Frege's pioneering endeavors, and which conceives of the philosophy of mathematics as centered on the problem of the foundations of mathematics, and another that is new and less well developed, and which takes the central problem to be that of articulating the methodology of mathematics. It is natural to wonder whether the two programs are compatible. Both Lakatos and Kitcher proceed to the enterprise of exploring the methodology of mathematics after arguing that there is no a priori foundation for mathematics. Parsons (1986) has questioned the idea that there is a genuine inconsistency between the search for a priori foundations and the contention that there is a serious problem of the growth of mathematical knowledge. Moreover, the work of some writers who have pondered methodological questions (Steiner 1975, 1983; Maddy 1982) seems to presuppose that the two programs are ultimately compatible.

The present volume brings together contributors to the Fregean tradition and enthusiasts for new departures. We shall now take a look at the recent history of the history of mathematics, before offering our views about how exactly the spectrum of historical and philosophical approaches is distributed among the essays.

2. History of Modern Mathematics: A Brief and Biased History

We have seen that contemporary philosophy of mathematics has grown out of two traditions: one studying the foundations of mathematics in the analytic style prevalent in twentieth-century philosophy, and the other a recent approach drawing on both history and philosophy to investigate the methodology of mathematics. Contemporary history of mathematics also has two flourishing traditions. One is an older tradition, affiliated with professional mathematics and mathematics education, that studies the history of great men and ideas from their published papers. The other tradition, affiliated most closely with postwar professional history of science, focuses on conceptual and social issues and employs a wide range of source materials.

One of our objectives is to identify trends in the history and philosophy of modern mathematics since the last assessment of the field in 1974 at the American Academy of Arts and Sciences Conference on the Evolution of Modern Mathematics (see the November 1975 issue of *Historia Mathematica* for details). The "great ideas" tradition has flourished for hundreds of years (see Jayawardene 1983) with no recent signs of change. We only hastily recount its history before turning our attention to newer trends that have been stimulated in great part by the historical community. This focus on recent trends may seem to disparage the continuing contributions made in the old style of mathematics history, although it is not our intention to do so. Two other caveats should be kept in mind. First, our attention is restricted to the history of modern mathematics; thus, we do not consider the ample fine scholarship of mathematics before the nineteenth century. Second, the reader may detect a bias toward English language literature and to research conducted by mathematicians and historians of the United States. This shortcoming is a simple reflection of our knowledge of the state of the field.

Although historians of mathematics have traced their discipline back to antiquity, our interest is in more recent historical work. The first important modern turn came with the Göttingen encyclopedia historian Kastner, who introduced a higher standard of scholarship at the end of the eighteenth century. In the nineteenth century a number of editions of manuscripts, bibliographies, and general histories of mathematics were published in Europe, especially in Germany (e.g., Cantor 1880-1908; Todhunter 1861).

In the first two decades of the twentieth century, science educators gained an appreciation for the value of the history of science in teaching science because of the human dimension it gives to science as well as its demonstration of the importance of science in western civilization. This recognition stimulated the preparation of elementary histories of the sciences and shorter historical articles useful to undergraduate teachers of science. Smith (1906) was among the first to produce these materials for mathematics. It is evident from the number and duration of these publications that history has proved to be an effective tool in introducing students to the culture of mathematics. General histories by Cajori (1968), Smith (1923-25), Archibald (1932), Bell (1940), Boyer (1968), and Eves (1976), as well as the journal *Scripta Mathematica*, founded by Archibald, Smith, and others in 1932 for the philosophical, historical, and expository study of mathematics, all had this educational purpose in mind.

This pedagogic movement stimulated more scholarly research. The results are seen in the historical dissertations written in the 1920s and 1930s at Columbia under Smith and at Michigan under Louis Karpinski, the increased production of historical articles, and the new bibliographies (e.g., Loria 1946) and source books (e.g., Smith 1929) that appeared.

Meanwhile, the mathematical community came to appreciate the value of technical historical surveys of mathematics to their research—both to reappropriate methods and facts lost over time and to understand the origins of research areas. The work that established history's research value beyond any doubt was the *Encyklopädie der mathematischen Wissenschaften* (1898-1935), in which Felix Klein and other distinguished mathematicians attempted a comprehensive review of mathematics as it existed at the turn of the century. Many other historical studies (e.g., Dickson 1919-23) and volumes of collected papers were prepared by and for mathematicians between the world wars. After 1945 these internalist histories appeared at an accelerated rate.

The seeds of a new tradition in history of mathematics were sewn in the 1930s by George Sarton through his efforts to professionalize history of science in the United States. Sarton's own contributions to the history of mathematics (Sarton 1936) are best viewed as a continuation and reinforcement of the older tradition. Indeed, the orientation of historians in the 1930s and 1940s was decidedly internalist. Their work focused on the content of science, and they were skeptical of history of science or mathematics conducted by those without advanced scientific training. This led to a natural alliance between the fledgling history of science and the scientific disciplines and to a unified approach that "explored the filiation and intellectual contexts of successful ideas" (Thackray 1983, 17).

In the 1950s the first professional historians of science entered the universities, bringing with them an increase in the number and sophistication of historical studies of science. Most of these studies were internalist, but a separate "externalist" tradition developed, inspired by the work of Robert Merton. The externalist studies traced the social structure of the scientific community and the relation of this community to the external world. Internalist and externalist approaches were seldom employed in the same study.

In this period the rise of the history of science profession had little impact on the history of modern mathematics, except perhaps to publicize the importance and legitimacy of historical study and to stimulate the mathematical community to produce greater numbers of collected works, monographs, and smaller studies. Modern mathematics received almost no attention from historians of science in this period for at least two reasons. First, the number of practitioners was small and they tended to be scientific generalists, or at least to focus on a wide range of scientific activities in a single chronological period. Mathematics, especially after the mid-eighteenth century, was considered to have a content and method distinct from the other sciences, and even today this belief creates a gulf between historians of modern mathematics and historians of other modern sciences. Second, following Sarton and Merton, early historians of science were attracted to the great revolutions in science—particularly to the Scientific Revolution. Thus, the little energy of early historians of science for mathematics was consumed in the study of topics from the sixteenth through the early eighteenth centuries, particularly the rise of algebra, the beginnings of analytic geometry, and, most of all, the calculus.

In the 1960s and early 1970s the history of science profession grew at a rapid pace, stimulated in part by the substantial government support to science and its cultural study. As the number of practitioners increased, subspecialties in histories of the individual sciences emerged. History of science programs—Harvard and Wisconsin in particular—produced their first historians of modern mathematics: Michael Crowe, Joseph Dauben, Judith Grabiner, Thomas Hawkins, Uta Merzbach, and Helena Pycior. Since then, a small but continuous stream of young historians with training in both history of science and mathematics has entered the field.

The line drawn between internal and external history of science was too tenuous to resist fading. In the 1970s and 1980s, historians of science have balanced their internalist studies with social studies and have advanced a more sophisticated historiography incorporating both internal and external factors. These studies increasingly consider social factors when examining not only the institutions of science, but also the form and content of scientific ideas.

Social histories of modern mathematics are relatively uncommon, probably because in comparison with other sciences mathematics is regarded as least affected by factors beyond its intellectual content. Yet mathematicians have long recognized the importance of communities such as those in Göttingen, Paris, Berlin, and Cambridge in sponsoring particular styles of research and producing certain kinds of research mathematicians; a number of studies of institutions, their educational programs (Biermann 1973), and the individuals who shaped them (Reid 1970, 1976) have appeared.

For Sarton, the study of national science made no sense because history of science was the study of scientific ideas, which knew no national boundaries. But mathematicians have long understood that there are national differences, both in the subjects studied and the way in which they are approached. One famous example is the geometric approach to calculus in vogue in eighteenth- and early-nineteenth-century Britain in contrast to the analytic approach of Leibniz favored on the Continent. But contrasts can be drawn in nineteenth-century mathematics, as well, for example, between German and British approaches to applied mathematics, between the Italian and French work in projective geometry and the German work in transformational geometry, and in the German dominance of the arithmetization of analysis. The appreciation of national styles of mathematics has resulted in a few studies, for example of the introduction of Continental methods into British analysis (Enros 1979). American mathematics has come under close scrutiny, partly through the interests of American mathematical societies and further encouraged by the nation's bicentennial celebration in 1976 (Tarwater 1977). In the 1980s, American mathematics has become an active area of research.

Although mathematicians and historians have come to understand the value of studying professional societies, journals, prizes, institutions, funding agencies, and curricula, they have considerably less appreciation for the study of the social roots of the form and content of mathematics. This is evidence of the firmly seated belief that mathematicians but not their ideas may be affected by external factors. This attitude is slowly beginning to change, as Daston (this volume) and others are able to demonstrate the interplay between social factors and mathematical ideas. Another group assaulting this belief are the feminist historians (J. LePage, E. Fee, E. F. Keller), who variously are trying to establish cognitive differences between male and female mathematicians and to explain why mathematics has long been considered a male vocation.

Historical studies conducted before the 1960s sometimes projected contemporary standards of proof, rigor, problem definition, and discipline boundary onto their mathematical subject. The effect was to "explain" the subject in anachronistic concepts and terminology, to select topics for study only insofar as they had a connection to more recent developments, and to praise or damn these efforts on the basis of whether they anticipated (took a step toward) the current state of mathematical knowledge. Thus, the focus was on the great "successes" and sometimes the "blunders" of the greatest mathematicians. To accomplish this they studied the great men (there were almost no women mathematicians) without considering the lesser, but able practitioners that comprised the wider mathematical community. These studies of the great ideas were accompanied by anecdotal biographies of the great men, in the worst cases amounting to no more than hagiographic tributes. These authors relied principally upon their mathematical acumen and personal experiences to evaluate the published corpus of the great mathematicians, instead of examining a wider range of published and unpublished documents.

One historian of science has noticed a similar trend in historical writing about the sciences (Thackray 1983, 33):

> Discipline history by scientists has usually been based on an individualistic epistemology, in keeping with the image of the scientist as one voyaging through strange seas of thought, alone. There are also individualistic property relations in science, giving importance to the adjudication of rival claims. One result has been an historical interest in questions of priority—of who first exposed "error" and established "right" answers, or who developed successful instruments and techniques.

With the advent of a professional history of science, a new and more sophisticated historiography has arisen and is being put into practice in the history of mathematics. This historiography measures events of the past against the standards of their time, not against the mathematical practices of today. The focus is on understanding the thought of the period, independent of whether it is right or wrong by today's account. The historiography is more philosophically sensitive in its understanding of the nature of mathematical truth and rigor, and it recognizes that these concepts have not remained invariant over time. This new historiography requires an investigation of a richer body of published and unpublished sources. It does not focus so exclusively on the great mathematicians of an era, but considers the work produced by the journeymen of mathematics and related scientific disciplines. It also investigates the social roots of mathematics: the research programs of institutions and nations; the impact of mathematical patronage; professionalization through societies, journals, education, and employment; and how these and other social factors shape the form and content of mathematical ideas.

The new historiography has not been universally adopted. Historical works of the older style continue to be written by mathematicians and some historians. Perhaps because of the perceived differences in method and content between mathematics and the other sciences and because of the slow rate at which the history of science profession has produced scholars interested in modern mathematics, historiographic change has been relatively slow in coming. However, one area of considerable activity is the preservation and use of archival materials. The mathematics community has a long tradition of publishing collected papers of eminent mathematicians. This tradition continues today, but attention is also being given to the publication of collections of unpublished manuscripts and correspondence such as those of Wiener, Gödel, and Russell that are now in production. Many fine European archival collections relating to nineteenth-century mathematics (e.g., at Institute Mittag-Leffler, the Berlin Akademie der Wissenschaften der DDR, the West Berlin Staatsbibliothek Preussischen Kulturbesitz, and others in Cambridge, Freiburg, and Göttingen) have been in existence for many years, but they have received little attention over the years. Recently, historians have used these materials to great effect in preparing new interpretations and more accurate accounts of classic events—for example, Dauben (1979) on Cantor, Moore (1982) on the set-theoretic paradoxes, and Hawkins (1984) on the *Erlanger Programm*. Others have used these sources to pioneer new areas—for example, the work of Cooke (1984), Koblitz (1983), and Kochina (1981) on Kovalevskaya.

Institutions have been founded in the last decade with a major objective of collecting and preserving important archival materials on mathematics. These include the Archives of American Mathematics at the Humanities Research Center of the University of Texas, the Charles Babbage Institute for the History of Information Processing Archives at the University of Minnesota, the Bertrand Russell Archives at McMaster University, and the Contemporary Scientific Archives Center at Oxford University. For more information on archival resources in repositories in the United States, see Merzbach (1985).

The new professionalism in history of mathematics is reflected in the formation and growth of specialist societies and journals in the 1970s and 1980s. These include the Canadian and British societies for the history of mathematics and the journals *Historia Mathematica, Archive for History of Exact Science, History and Philosophy of Logic, Annals of the History of Computing, Mathematical Intelligencer,* and *Bolletino di storia della scienze matematiche.* Joseph Dauben, Ivor Grattan-Guinness, and Kenneth May have made noteworthy contributions to professionalization.

The professionalization of the history of mathematics has stimulated the production of a rich set of publications. The remainder of this section presents a brief survey of the literature of the 1970s and 1980s. For reasons of space and manageability, the survey is restricted to full-length studies. This does not do justice to the contributions of some scholars, like Thomas Hawkins or Helena Pycior, who contribute primarily through journal articles. For a fuller discussion of both book and journal literature see Grattan-Guinness (1977), Jayawardene (1983), and Dauben (1985).

Historians of mathematics have long appreciated the value of bibliographies. Dauben (1985, p. xxii) lists nine bibliographies produced in the nineteenth century, and his list does not include important ones appearing in the journal literature, notably the bibliographies produced between 1877 and 1900 by Moritz Cantor in *Zeitschrift für Mathematik und Physik.* In the twentieth century there have been four major bibliographies of the history of mathematics: Sarton (1936), Loria (1946), May (1973), and Dauben (1985). The two of greatest research value today are the last two. May's bibliography is intended to be comprehensive, and therefore lists everything of which he knew, whereas Dauben's is selective, critical, and annotated. Sarton and May in particular saw their bibliographic work as important to discipline building, as a way to define the field and to guide further research.

Mathematicians have also long been interested in general histories of their field. May (1973) lists approximately 150 general histories written between 1742 and 1968, and Dauben (1985) lists 29 such works of current value. Some (e.g., Struik 1967) provide a synoptic overview suitable for introducing students to the field, whereas others (e.g., Cantor, 4 vols., 1880-1908) with more comprehensive coverage serve as reference works to the mathematically educated. These two needs have been met in recent years by Boyer (1968) and Kline (1972), respectively. Kline's work is particularly impressive, generally accurate and far in advance of its predecessors in interpretation. There is growing opportunity, however, for a new general history of mathematics able to synthesize the more detailed, focused studies being produced today. Others (Bourbaki 1974, Dieudonné 1978) have written general histories of mathematics intended to advance their position on the "correct" approach to mathematics.

Sourcebooks of mathematics have been published throughout the twentieth century. Their function is to unite in one volume primary source material on a single topic, so as to make these materials more widely available and easier to compare. Sourcebooks have found frequent use in the classroom, and it is no surprise to learn that the first ones appeared (Smith 1929) at the time when it was first recognized that history can be an effective tool in introducing students to the culture of science and mathematics. Nor is it surprising that the number of sourcebooks to appear has increased in proportion to the professionalization of the history of mathematics. Some of these recent sourcebooks (Calinger 1982) are intended primarily for educational use, whereas others (van Heijenoort 1967; Birkhoff 1973) fulfill an additional need of the practicing mathematician for access to classic papers.

Since the late nineteenth century, mathematicians have regularly paid tribute to the most eminent members of their profession by publishing their collected works or selected editions of their papers. These volumes are not merely honorific, for they serve a useful research function. Dauben (1985) has identified about 300 mathematicians active before World War I whose collected works have been published since 1880. A survey of a subset of Dauben's list consisting of mathematicians from the nineteenth or twentieth century indicates steady publication of their collected works over the last ten decades, with little activity in the 1940s (because of the war, presumably) and a slight increase in the 1960s and 1970s. Volumes published in recent years seem to be more sensitive to historical concerns, for example in being less likely to introduce anachronistic modernization in notation. In fact, many of the recent collections have simply photo-reproduced the articles as they originally appeared in print. In the last ten years a number of the older of these collected works have been re-printed, demonstrating their enduring value. Although collected works are primarily published to meet mathematical research needs—at least in the case of modern mathematicians, whose collected works are published in far greater (absolute) numbers than those of "mathematicians" of the sixteenth through eighteenth centuries—they also serve the historical com-munity. Thus the number published and kept in print is likely to increase.

Dauben (1985) also identifies about 75 mathematicians (active before World War I) whose correspondence has been published. Of these a disproportionately high percentage (in comparison to the number of mathematical practitioners at different periods of time) flourished prior to the nineteenth century, perhaps indicating the unavailability of more recent correspondence. It may also result from a lessening in the number and quality of letters over the last century with the improved opportunities for mathematical communication through journals and professional meetings, easier travel, and more recently the widespread use of the telephone and the computer. Or perhaps an explanation is provided by the fact that early mathematicians were often involved in a range of scien-tific activities, that history of science (especially of the Scientific Revolu-tion) matured more rapidly than history of modern mathematics, and that historians place greater value on correspondence than scientists do. If the history of modern mathematics follows the trend in the history of science, the number of volumes of collected correspondence will grow and will have increasingly sophisticated annotation and analysis.

Biographies of eminent mathematicians have been popular among mathematicians and mathematics educators since early in the century, when history was first perceived as an effective introduction to the culture of mathematics. Before that time, mathematical biographies had appeared only occasionally, often written by younger contemporaries of the biographical subject, such as Koenigsberger (1904) on Jacobi. Although full-length biographies were written in the early twentieth century, most biographical writings were semipopular sketches emphasizing the personal and the anecdotal. Some of these were very popular, serving pedagogic function and also offering "intellectual challenge, inspiration, and recreation to mature practitioners" (Thackray 1983, 39). But as the field has matured, these works have outlived their historical utility. One immensely popular example (Bell 1937) is castigated by historians today for its gross historical inaccuracies.

In recent years many full-length biographies have appeared. Biographical subjects since 1970 include Babbage (Hyman 1982), Cantor (Dauben 1979), Lazare Carnot (Gillispie 1979), Courant (Reid 1976), Dedekind (Dugac 1976), Fisher (Box 1978), Fourier (Grattan-Guinness and Ravetz 1972), Gauss (Hall 1970; Wussing 1974), Hamilton (Hankins 1980), Hilbert (Reid 1970), Kovalevskaya (Cooke 1984; Koblitz 1983; Kochina 1981), Neyman (Reid 1982), Turing (Hodges 1983), and Ulam (Ulam 1976). Some of these (Reid 1970, 1976, 1982; Ulam 1976) are semipopular and continue the older biographical tradition. Others (Dauben 1979, Gillispie 1979, Hankins 1980) have employed a rich set of source materials to examine in detail the mathematical contributions of their subject in his or her social and intellectual context. There is considerable variability in the extent to which these works emphasize the mathematics (Dugac 1976; Cooke 1984) or the personalities and social context (Reid 1970; Koblitz 1983). Even the semipopular biographies are historiographically mature in comparison to biographies written earlier in the century. Constance Reid, for example, is attuned to institutional context in her studies of Courant, Hilbert, and Neyman. And as one historian (Thackray 1983, 40) has noted, biographies result in "a fresh awareness of the subtle, elusive quality of the [scientists'] ideas and of the persistence and intricacy of their patterns of thought. That awareness has done much to challenge stereotypes of science as impersonal, value-free inquiry." Popular and scholarly biographies are likely to appear in increasing numbers in the coming decades.

In the last fifteen years, mathematicians and historians have begun to publish monographs on the historical development of subdisciplines within mathematics. These studies either survey the history of a subdiscipline from its modern roots in the nineteenth century, or they examine in detail some particular problem or episode and demonstrate how it contributed to subdiscipline formation. Three studies are models of this kind of research because of their mathematical and historical acumen: Moore (1982) on Zermelo and the origins of axiomatic set theory, Hawkins (1970) on measure and integration theory, and Wussing (1969) on the abstract group concept. Others include Mehrtens (1979) and Novy (1973) on algebra;

Dugac (1980), Grattan-Guinness (1970), Goldstine (1980), Grabiner (1981), Medvedev (1976), and Monna (1975) on analysis; Bashe (1985), Goldstine (1972, 1977), and Williams (1985) on computing; Dieudonné (1974), Gray (1979), Pont (1974), and Scholz (1980) on geometry and topology; Biggs, Lloyd, and Wilson (1976) on graph theory; Edwards (1977) and Weil (1975, 1984) on number theory; and Maistrov (1974) on probability. As this list indicates, analysis has drawn by far the greatest attention. Applied mathematics, non-Euclidean and projective geometry, operations research, probability, and statistics have received little attention. Undoubtedly, the number of historical studies of subdisciplines will multiply in the coming years. Authors will come increasingly to appreciate the importance of examining correspondence, unpublished manuscripts, records of professional societies, and the work of lesser mathematicians in understanding the contributions of the leading mathematicians. These writers will also be more historically sensitive than their predecessors to projection of modern standards of notation, rigor, problem definition, and disciplinary boundaries onto the mathematics of the past.

At least mathematicians who participated in the founding of academic programs, professional organizations, or journals have long appreciated the impact of institutions on mathematics. A good example is R. C. Archibald, not only a leading figure in the rise of mathematics in the United States between the two wars, but also a devoted student of its history. His 1938 semicentennial history of the American Mathematical Society provides useful information on the society's financial affairs, programmatic activities, and key personnel. In the last fifteen years, interest in mathematical institutions has increased among both mathematicians and historians. Biermann (1973) has produced a fine study of the Berlin school of mathematics. Enros (1979) has traced the formation and effect of the Cambridge Analytic Society. Kuratowski (1980) has provided us with reminiscences of Polish mathematical figures and institutions between the two wars.

Stimulated by the interest of such senior American mathematicians as Birkhoff, Bochner, Browder, Halmos, MacLane, Stone, and Tucker and by the United States bicentennial celebration in 1976, there has been heightened interest during the last decade in the history of American mathematics (Tarwater 1977; May 1972). Historians and mathematicians including Aspray, Cooke, Grabiner, Merzbach, Reingold, Rickey, and Rider are continuing investigation of this history, and many additional studies should appear in the coming decade.

Outside of histories of specific institutions, little has been written on the social context of mathematics. Fang and Takayama (1975) present a survey of methods and theory in the social history of mathematics. Mehrtens, Bos, and Schneider (1981) have edited a useful volume of conference proceedings on the social history of nineteenth-century mathematics, focusing mainly on professionalization and education. Interest in women in mathematics has resulted in several volumes of biographical sketches of women mathematicians (Osen 1974; Perl 1978), as well as a recent industry producing full-length biographies of famous women in mathematics, including Ada Lovelace (Stein 1985), Germain (Bucciarelli and Dworsky 1980), Kovalevskaya (Cooke 1984; Kochina 1981; Koblitz 1983), and Noether (Brewer and Smith 1981; Dick 1981). A number of other, more sociologically oriented studies of women in mathematics are now underway.

Some of the most effective recent work in the history of science has brought to bear both internal and social factors in a unified historical analysis. MacKenzie's study (1981) of British statistics and Daston's article in this volume point the way to this kind of integrated historical study of the history of modern mathematics.

3. The Essays

We have divided the essays in this volume into four sections with the aim of bringing together papers that address similar topics or share common themes or approaches. The first section consists of three studies that tackle the traditional area of concern to philosophers of mathematics— logic and the foundations of mathematics. The second contains essays that articulate the historian's enterprise in different ways, either by displaying the structure of some particular episode of the history of modern mathematics or by using a historical example to comment on how that enterprise should be conducted. In the third section, historians and philosophers of mathematics explore the two fields in attempts to find illumination of one by the other. Finally, the fourth section comprises two studies of the interactions between mathematics and the broader social context.

In "Poincaré Against the Logicists," Warren Goldfarb considers Poincaré's criticisms of the logicist program. On Goldfarb's account, Poincaré was not primarily guilty of missing the point of the work of Frege, Russell, Couturat, and others. His objections were founded in a quite different conception of the philosophy of mathematics. The root of the difference is Poincaré's refusal to emancipate logic from psychology in the way for which Frege had campaigned.

Poincaré criticized the logicist definitions of the numerals on the grounds that they were ultimately circular, and he contended that the proper resolution of the set-theoretic paradoxes should proceed by honoring the vicious circle principle. Goldfarb argues that the former criticism is not an "elementary logical blunder," but the product of Poincaré's insistence that legitimate definitions must trace the obscure to the clear, where the notions of clarity and obscurity are understood psychologically. Poincaré rejected the framework elaborated by Frege, within which what we actually think of in connection with a given mathematical notion becomes irrelevant and all that is pertinent is the logical issue of what concepts are rationally presupposed. Similarly, Goldfarb contends that the full force of Poincaré's vicious circle principle and the notion of predicativity to which it gives rise can only be appreciated by recognizing Poincaré's concerns about the mutability of mathematical definitions. Here again, he was refusing to adopt a central principle of the logicist view, that "logic applies to a realm of fixed content."

Ultimately, then, the difference between Poincaré and his opponents comes down to a deep divergence in agendas for the philosophy of mathematics. Where Frege and later logicists saw the task of finding foundations as one of showing how mathematics results from the most general conditions on rational thought, Poincaré saw mathematics as the product of natural objects—human beings—so that the task of finding foundations is intimately linked to bringing clarity (judged by the standards appropriate for such beings) to areas that are currently obscure (again, judged by the standards appropriate for such beings). As Goldfarb hints, this contrast between Poincaré and the defenders of the logicist program is not only useful for throwing into relief the central tenets of logicism, but it also enables us to see interesting parallels between the early criticisms of logicism and contemporary naturalistic approaches to the philosophy of mathematics.

Michael Friedman also explores an episode from the history of logicism. "Logical Truth and Analyticity in Carnap's *Logical Syntax of Language*" illuminates the differences among different phases of the logicist program by seeing Carnap's endeavors in *Syntax* as an ingenious—but unsuccessful—attempt to find an intermediate position between the early logicism of Frege and the later conventionalism that would come under fire from

Quine. On Friedman's interpretation, the Fregean influence on Carnap runs very deep. Although it may appear that Carnap preserved little of any significance from Frege, and though the claim to reconcile Frege and Hilbert may seem spurious, Friedman argues that Carnap's own view of his connections to the tradition was correct.

Like Goldfarb, Friedman sees the Fregean enterprise as an attempt to show how mathematics is "built in to the most general conditions of thought itself." Carnap endorsed Wittgenstein's interpretation of Frege's view of logic: logic sets forth the conditions on rational thinking by elaborating those features that make any system of representation possible. But he transformed the Wittgensteinian insight, suggesting that logic can be formulated exactly as "the syntax either of some particular language or of languages in general." The disjunction prepares us for Carnap's new version of logicism. Recognizing that the principles that emerge from the syntax of "languages in general" will not suffice to generate classical mathematics, Carnap proposed to relativize logic. Thus, he offered the Principle of Tolerance, designed to permit freedom of choice with respect to languages and syntax—and thus freedom of choice with respect to logic. The task of foundations of mathematics remains in its Fregean form, for we are to show how mathematics is "built in" to the structure of thought and language, but the new relativization is supposed to accommodate the insights and achievements of Hilbert and Gödel.

Friedman goes on to argue that the new program was unstable. Carnap's sensitivity to the Gödel results forced him to complicate his account of analyticity. Indeed, Friedman claims, the full import of the Gödel theorems, appreciated only belatedly by Carnap, prevented him from characterizing analyticity along the lines favored by Frege and the early Wittgenstein, so that Carnap was pushed toward the *non*-Fregean construal of analyticity as truth-in-virtue-of-meaning that makes him vulnerable to Quinean criticisms. Hence, through its focus on an important transitional episode, Friedman's essay brings before us an interpretation of the entire history of logicism. The gap between Frege's original attempt to show how mathematics is built in to the general conditions of rational thought and the conventionalist idea that mathematical propositions are true in virtue of the concepts they contain is bridged by Carnap's heroic effort to honor Frege's view of foundations and yet discover a foundation for classical mathematics.

Gregory Moore considers the history of logic in the late nineteenth and early twentieth centuries from the different perspective of the historian. Moore's project is to investigate the way in which first-order logic became the canonical framework for the logical formulation of mathematics. His main thesis presents a striking irony: the canonization of first-order logic was achieved through the work of Skolem, specifically work that was designed to show the limitations on first-order formalism and to argue on this basis for the relativity of important mathematical concepts.

Moore reviews the history of modern formal logic from the mid-nineteenth century, beginning with Boole's algebra of logic. The connection with Boole is important for his argument, for part of the novelty of his story is the influence of the Boole-Peirce-Schröder tradition on Löwenheim (and, indirectly, on Skolem). Moore proceeds to examine the contributions of Peirce and Frege, emphasizing how different the Fregean picture of logic is from that which became accepted in the 1920s. As Moore notes, Frege did not separate the first-order part of his system from the rest, so that, in Frege's conception of the subject, there would have been no thought of responding to the difficulties with the notorious Basic Law V (the source of the Russell paradox) by banishing the second-order machinery from the province of logic. Instead, Moore proposes, the influence of Peirce and Schröder was essential for this separation to occur.

To see how this influence worked on Löwenheim and Skolem, it is necessary to achieve a clearer view of Schröder's accomplishments than has been available. Because we have seen Schröder through the lenses of Frege's criticisms, we have not understood either how he advanced the approach to logic favored by Boole and Peirce or how he influenced Hilbert. Moore seeks to correct the distortion. He also explains how Russell and Whitehead perpetuated the Fregean conception of logic, in which there is no place for any study of logic as a system or for any conception of a metalanguage. Thus Moore offers a picture of the state of logic in the first decades of the century: because of the influence of *Principia Mathematica*, Frege's conception of logic was dominant; nonetheless, there were indirect influences from the Peirce-Schröder tradition, through the work of Peano and Hilbert, and the work of Peirce and Schröder continued to be studied in its own right. (For lucid presentation of ideas that have some kinship with Moore's, see Goldfarb 1979.)

In the final sections of his essay, Moore argues in some detail that the important result for which Löwenheim is famous developed directly out of his research on Schröder's system of logic. He goes on to present

Skolem's further elaboration of Löwenheim's result, the explicit formulation of it as a theorem about first-order logic, and Skolem's connection of the resultant theorem with the relativity of notions of cardinality. A final, surprising, twist to the story is that even such talented mathematical logicians as von Neumann and Gödel were apparently unaware that the Skolem relativity thesis does not hold in second-order systems. Thus, in Moore's historical reconstruction, the naive idea that the canonization of first-order logic was inevitable once it was recognized that Frege's original system leads to paradox gives way to a complex and subtle account, in which now unfashionable historical traditions exert an influence and in which substantive philosophical doctrines have an important role to play.

Like Moore, Harold Edwards is concerned to alter our common vision of a historical episode. There is a familiar tale that reports the baneful influence of Kronecker on late-nineteenth-century mathematics. In this tale, Kronecker plays the part of the wicked and powerful gnome, whose evil schemes almost prevent the heroic prince (Cantor) from unlocking the gate of paradise (the transfinite). Fortunately, despite nervous breakdowns induced by Kronecker's machinations, despite his enforced banishment to Halle, the hero succeeds, and, in the light of the set-theoretic universe, the dark incantations of the villain are almost forgotten.

On Edwards's account, the usual historical assessment of Kronecker is almost as far from reality as most fairy stories. Edwards concedes that Kronecker's constructivist views have been a minority tradition in the history of the philosophy of mathematics. However, he finds no evidence for the allegations that Kronecker was personally hostile and aggressive, or for the charge that he was more guilty than other German professors of manipulating the social context of late-nineteenth-century mathematics to advance his own philosophical ideas. Edwards suggests that we should free ourselves from the stereotypes and consider Kronecker's work—both the actual mathematics that he produced and the philosophical viewpoints that motivated it—on its own merits. For the latter, we can find little that is explicit in Kronecker, except for a few scattered statements, and Edwards briefly suggests that we think of contemporary constructivists, such as Bishop, as pursuing Kronecker's main themes. On the other hand, Kronecker's mathematical papers are a treasure trove for historical investigation.

Edwards concedes that identifying the treasures is not easy. Kronecker's mathematical style is difficult. Edwards attempts to convey the ideas that he finds valuable by drawing a contrast between the work of Dedekind and Kronecker on divisibility in algebraic number theory. For Kronecker, it was important that the divisors be defined in such a way that the statements about them continue to hold under extensions or contractions of the field. Even more significantly, Kronecker's way of treating division was directed toward *computing* with divisors. Edwards claims that not only do these features mark a *difference* between Kronecker's approach and the familiar Dedekindian treatment in terms of ideals, but they constitute *advantages* in Kronecker's point of view. He concludes by speculating that a detailed exposition of the contributions of Kronecker may enable us to achieve a more sympathetic appreciation of constructivist ideas about mathematics.

Garrett Birkhoff and M. K. Bennett reevaluate the historical assessments of the influence of Felix Klein and his *Erlanger Progamm*. They view their work as a companion piece to a 1984 article by Thomas Hawkins that used a rich collection of published and unpublished sources to call into question the reception of Klein's work in the decades following its publication, as well as the authorship of the fundamental ideas. Relying mainly on published writings and letters by leading mathematicians (Lie, Engel, E. Cartan, and Weyl), and on their own familiarity with the role of continuous groups in geometry, Birkhoff and Bennett emphasize Klein's contributions in developing and disseminating group-theoretic concepts in geometry and geometric function theory.

Joseph Dauben shares with Edwards the concern that contemporary ideas and fashions may all too easily limit our appreciation of the content and development of past mathematics. Dauben chooses for his study an episode that has received great attention from historians and philosophers of mathematics. In Cauchy's *Cours d'Analyse*, there is a celebrated "theorem" to the effect that the sum of a convergent series of continuous functions is continuous. We know that the "theorem" is false, and historians have shown how the efforts to refine the theorem in the light of counterexamples gave rise, at the hands of Seidel and Weierstrass, to the modern distinction of convergence and uniform convergence.

Cauchy had an argument for his claim, a "proof" of the "theorem," and any satisfactory account of the development of concepts of convergence should explain how Cauchy's argument works, showing how a mathematician of his stature could have been led to advance the statements he did. Some recent developments in model theory and analysis allow us to consider a radical possibility. Perhaps Cauchy's "theorem" is actually *correct*, and it is we who are mistaken in interpreting it. Robinson's non-standard analysis permits us to consider the possibility of nonstandard continua—continua in which Cauchy's result would hold and in which his argument would work. In a provocative essay, Lakatos proposed that this is indeed the case, and suggested that we view the transition from Cauchy to Weierstrass not in terms of the disambiguation of concepts of convergence but as the replacement of the nonstandard continuum with the standard one.

Dauben presents Lakatos's approach to the history of analysis, contrasting it with the more traditional attempts to diagnose the subtle fallacy in Cauchy's reasoning. Dauben argues that the Lakatosian reinterpretation cannot be sustained, and he uses this conclusion to explore the methodological constraints that a historian of mathematics should honor. The danger that he sees, one that has become the most commonplace for historians of science during the past thirty years, is that contemporary notions will be read back into the mathematics of the past, so that the history of mathematics will consist, not in a sequence of efforts to fathom the mathematics of our predecessors on its own terms, but in "discoveries" of "anticipations" of the latest interesting ideas. We believe that Dauben's resistance to this "Whig history" is salutary for the history of mathematics, and that it is of a piece with the attempts of Moore and Edwards to reclaim those figures of the past whose ideas seem at odds with contemporary fashions.

Dauben goes on to consider the influence of historical researches on Robinson's own presentation of his ideas. He points out that, despite the formal equivalence of nonstandard and standard analysis, Robinson was still able to contend that his approach enjoyed a special intuitive evidence and that he could base this claim on the history of analysis. Moreover, in opposition to those (like Bishop) who deny the meaningfulness of nonstandard analysis, Dauben describes the findings of studies that show how beginning students are able to solve calculus problems with greater skill if they are trained in the Robinsonian approach. Thus, like the great mathematicians of the eighteenth century, it seems that the neophytes of today can often make good use of infinitesimalist reasoning.

Richard Askey offers a perspective on the practice of history of mathematics that may at first appear to run counter to that taken by Dauben. Askey suggests that historians of mathematics ought to know a lot of mathematics—in fact, he hints, far more than they do. Unless they are trained beyond "the current undergraduate curriculum and first-year graduate courses," there is a danger that they will overlook those episodes in the history of mathematics that are really significant and concentrate on peripheral issues. However, Askey does not simply suggest that history of mathematics should be turned over to professional mathematicians, as a recreation in which they can indulge when they take themselves to be on the verge of their dotage. History has its own standards and methodological canons, and Askey, despite his keen interest in history, is quite modest about his knowledge of these. The heart of his paper is thus a plea for cooperation and development. Mathematicians can contribute "protohistory" (our term, not his) by drawing the attention of historians to problems and episodes that are mathematically significant. The mathematician, playing protohistorian, will not attempt any coherent treatment of these episodes. The goal will simply be to assemble "mathematical facts" whose normal form may be the attribution of a relation of kinship between the writings of a past mathematician and some (perhaps sophisticated) piece of contemporary mathematics. Once confronted with these suggestions of kinship, the historian must go to work, following the canons and standards of the discipline. However, part of Askey's message is that doing the work properly may involve a great deal of further study in mathematics.

Historians are likely to view protohistory as incomplete in two respects. The more obvious deficiency, touched on in the last paragraph, is that the mathematician's recognition of kinship needs to be scrutinized from the perspective of an understanding of the concepts and standards in force in the historical epoch under study. In addition, we should not assume that the most significant historical problems concern work that has any straightforward connection with (or offers any anticipation of) problems and methods of current interest. Askey seems to us tacitly to appreciate the point when he couches his discussion in terms of the writing of history that mathematicians will find interesting.

Askey illustrates his general proposal by recounting some of his own historical research on series identities. After contending that some of the historical attributions commonly made by professional mathematicians seem erroneous, Askey recounts the reactions of historians to his work. His tale seems clearly to be one of missed communication. The technical discussions of series identities and hypergeometric series are dismissed as not amounting to "real history." Nonetheless, we are sympathetic to Askey's contention that his research—and work like it—brings before historians things that they ought to know. Hence, we accept *half* of his general thesis: mathematicians can help historians of mathematics by assembling "mathematical facts," an activity that Askey explicitly recognizes as not the same as full history.

But this is only part of Askey's thesis, for he argues that historical investigations may be of value in the development of mathematics. Askey relates how ongoing work on the Bierberbach conjecture led to a problem about series identities. His own historical interests had led Askey to a perspective from which he could assist in the solution of this problem. Thus, the excavation of ideas from past mathematics may contribute to current research. Here we find a familiar theme, but one which Askey illustrates in a dramatic way. Perhaps there are hints of something similar in Edwards's suggestions for resurrecting the main ideas of Kronecker's number-theoretic work.

We believe that the juxtaposition of Dauben's essay with Askey's is especially happy, for both can be seen as warning against a particular sort of danger. To do adequate history of mathematics one must avoid both dangers, relinquishing both the chauvinism of the professional historian (as Dauben surely would) and the chauvinism of the professional mathematician (as Askey clearly does). If we may be allowed yet another variation on Kant's celebrated dictum: For the purposes of doing history of mathematics, knowledge of contemporary mathematics without historiographical sensitivity is empty, historiographical sensitivity without knowledge of contemporary mathematics is blind.

Lorraine Daston is also concerned to find ways in which the mathematics of the past can be treated accurately without introducing anachronistic distinctions. "Fitting Numbers to the World: The Case of Probability Theory" considers the contemporary distinction between pure and applied mathematics and its relation to the eighteenth-century idea of "mixed mathematics." Daston focuses her discussion by considering three branches of eighteenth- and early nineteenth-century probability theory, three fields of inquiry that we might naturally call applications of probability theory. She endeavors to show how differently the mathematicians who pursued these fields conceived of them when they placed them under the rubric of mixed mathematics.

Daston begins with a review of the historical roots of ideas about the relation of mathematics to the external world. She concludes that, for the mathematicians of the Enlightenment, "abstract" mathematics occupied one end "of a continuum along which mathematics was mixed with sensible properties in varying proportions." "Mixed" mathematics occupied much of the energy of eighteenth-century mathematicians, and yet, as Daston notes, there was a sense that, in introducing a greater "mixture" of sensible ideas, mathematicians ran the risk of error and "retrogression."

Daston's first illustration concerns the art of conjecture, and her discussion centers on one celebrated problem and its impact. The problem is the St. Petersburg paradox—a paradox in virtue of the fact that application of the art of conjecture to a contrived game offers a recommendation that is intuitively unacceptable. Daston argues that we can only see the St. Petersburg paradox as a paradox—that is, as the eighteenth-century discussants saw it—if we recognize the status of the art of conjecture as a piece of mixed mathematics. The ability of the problem to threaten the credentials of probability theory must strike us as absurd if we approach the situation from the perspective of our modern distinction between pure and applied mathematics. *We* can consign the puzzle to economics. Our eighteenth-century predecessors could not.

The second example concerns the use of probability in a legal context, the probability of judgments. Daston explains how the mathematical community set itself the task of deciding on the optimal design of a tribunal of judges, where the criterion for optimality consisted in minimizing the risk of error. The hope, bizarre as it now seems, was that, by treating judges as akin to dice, the practice of legal judgment could be "reduced to a calculus." Daston relates the fate of this discipline, showing how the understanding of legal reasoning became divorced from the calculus of probabilities, so that a branch of "mixed mathematics" came to be viewed as a faulty combination of an unassailable piece of pure mathematics (probability theory) and a misguided application.

Finally, she presents us with one of the success stories in the "application" of probability theory, the development of actuarial mathematics. Here Daston seeks to understand why the use of probability theory in insurance took so long to become established, why the early ventures in actuarial mathematics were undertaken with such extreme caution, why an antistatistical attitude was displaced so slowly. The kernel of her answer is that the phenomena to be discussed seemed insusceptible to proper mathematical analysis because it appeared that statistical treatment would blur subtle distinctions that experienced insurers would be able to use advantageously in their decisions.

Taken together, the three examples show how the eighteenth-century notion of mixed mathematics differed from our concept of applied mathematics, how the distinction of a pure discipline from its applications can be achieved, and how mathematical theories are sometimes dependent on the fates of their applications. It seems to us that Daston's essay raises interesting philosophical questions concerning the traditional topic of the relation of mathematics to reality and that her treatment of the examples she has chosen offers some fruitful suggestions for addressing those issues. Moreover, her study shows the effect of the social context on the development of a branch of mathematics. In this way, Daston touches on themes akin to those pursued by Grabiner and Aspray.

If Daston demonstrates how the detailed study of the history of mathematics can have significance for philosophical issues, then it seems to us that Howard Stein's essay shows how attention to philosophical questions can shed considerable light on episodes in the history of mathematics. Stein is explicitly concerned to trace the main foundational programs of the early twentieth century to mathematical roots in the nineteenth century. He begins from the thesis that mathematics underwent a "second birth" in the nineteenth century and that it is a primary task for philosophy to understand this transformation. After reviewing the early-nineteenth-century developments in algebra, analysis, and geometry, Stein identifies several "pivotal figures"—Dirichlet, Riemann and Dedekind—the latter two greatly influenced by Dirichlet's teaching. With the filiations to the early nineteenth century in place, he then begins a more detailed account of some late-nineteenth-century developments.

The account starts with a problem, a problem that occupies anyone who ponders Dedekind's dual status as a respected mathematician and as a figure in the "foundations of mathematics." Why did Dedekind write his monograph on the natural numbers (*Was sind und was sollen die Zahlen?*)? Stein's answer consists in a careful tracing of the connections between the project of this monograph and Dedekind's earlier work in number theory, work that shows the influence of both Gauss and Dirichlet. The account culminates in the contention that the famous supplement to Dirichlet's lectures in number theory not only served as a major source in the history of algebraic number theory, but was also the origin of some of Dedekind's deepest philosophical ideas. Stein uses this contention to provide an illuminating contrast between the project undertaken by Dedekind in his monograph and the famous enterprise that Frege began in *Begriffsschrift.*

After what he concedes may sound like a panegyric on Dedekind (and one of us agrees that panegyrics are not here misplaced), Stein turns to consider the work of Kronecker. After noting Kronecker's famous stringent requirements on mathematics, the insistence on constructivity, he points out that Kronecker excepted geometry from these requirements. Considering Kronecker's position in the light of Riemann's conception of geometry, Stein argues that the demands Kronecker made are ultimately unjustified. The concession granted to geometry ought to be allowed to other areas of mathematics as well.

Finally, Stein turns his attention to the work of Hilbert. Here he is concerned to recapture the insights of Hilbert's early foundational work—the work of the *Grundlagen der Geometrie*—and to view the later slogans about the meaninglessness of finitary mathematics as overstatements of a sensible attitude that was present in all of Hilbert's foundational thought. For Stein, Hilbert's fundamental point is that there is no formal requirement on mathematics beyond consistency, and the original, deep suggestion is that the consistency of systems is itself open to mathematical investigation. The results of Gödel and others thus constitute a "final irony" in the story of Hilbert's development.

Any brief summary inevitably misrepresents Stein's inquiry, for his major theme is the intricacy of the connections among early-twentieth-century foundational programs, long-standing philosophical issues, and achievements within nineteenth-century mathematics. The articulation of that theme offers a new view of the main traditions and problems in philosophy of mathematics. Michael Crowe's essay, "Ten Misconceptions about Mathematics and Its History," also proposes to revise commonly accepted views. But Crowe's strategy is more direct. He sees much historical writing as, tacitly or explicitly, adopting ideas about mathematics and its history that cannot be sustained on closer inspection. Instead of offering a detailed discussion of a single episode, Crowe argues by assembling counterexamples drawn from a variety of periods and subjects in the history of mathematics.

Traditional historiography is in the grip of a conception of mathematics as deductive, certain, and cumulative. Mathematics students, many mathematics teachers, and even some historians of mathematics think that mathematical statements are invariably correct, that the structure of mathematics accurately reflects its history, that mathematical proof is unproblematic, and that standards of rigor are unchanging. They also assume that the methodology of mathematics is radically different from the methodology of science, that mathematical claims admit of decisive falsification, and that the philosophical options are empiricism, formalism, intuitionism, and Platonism. Crowe's attack on this tenfold conception uses a number of recent philosophical sources, notably Lakatos, and ranges over examples from many branches of modern mathematics—algebra, geometry, analysis, vector algebra and analysis.

We believe that Crowe has issued a broad and clear challenge, and we think that there are three main lines of response. First, one may object to his treatment of the mathematical examples, holding that when the cases are elaborated more carefully, the theses that he is concerned to rebut can survive unscathed. Second, a critic might urge that the theses themselves cannot be so straightforwardly assessed and that considerable preliminary conceptual analysis is required before they can be confronted with historical analysis. (What does it mean to claim that mathematics is "certain" or that "standards of rigor" are immutable?) Finally, there is the rejoinder that Crowe has simply erected a straw man and that the actual practice of doing history of mathematics is free of the simplistic ideas that he castigates. Whatever the merits of any (or all) of these objections, we think that Crowe has raised interesting questions about the historiography of mathematics and that the theses that he has selected deserve to be analyzed and evaluated by philosophers and historians of mathematics.

As its title suggests, Felix Browder's "Mathematics and the Sciences" focuses on the relationship between mathematics and the natural sciences. Browder is concerned to illustrate a recent major trend in preconceptions about the most significant directions in mathematical research. He starts by reviewing a number of ways in which, in the last decade, developments in the natural sciences, most notably physics, have employed very sophisticated mathematical ideas. He continues by considering the relationship between mathematics and computer science, noting that those parts of mathematics that seem least applicable in physical modeling—to wit, algebraic number theory and mathematical logic—have been of enormous value in computer science. Browder suggests that the old distinction between "applicable" and "inapplicable" mathematics may be out-moded, and he opposes any attempt to separate natural science from the "artificial" sciences.

However, any picture of mathematics that ignores the autonomy of mathematical research appears to Browder to be misguided. Whether the solution of a mathematical problem is undertaken in the context of a physical theory or without any particular physical result in mind, those who tackle that problem are functioning as mathematicians. Thus Browder is led to consider historically influential attempts to specify the nature of mathematics and of mathematical problems. He goes on to note two different ways in which important ideas in recent mathematics have emerged from interactions between prior mathematics and the sciences. He concludes by relating these examples to the history of mathematics, arguing that the current dialogue between mathematics and the sciences is merely the latest exchange in a conversation that has been going on since the seventeenth century and that has been of value to all the disciplines that have been party to it.

Philip Kitcher's "Mathematical Naturalism" offers an agenda for the philosophy of mathematics that is distinct from the set of problems that have dominated the subject since Frege. Kitcher argues that the identification of philosophy of mathematics with the construction of a foundation for mathematics depends on a commitment to apriorism. The principles to which mathematics is to be reduced are supposed to be a priori, and, if they do not have this special epistemological status, then the reduction loses its point. Kitcher outlines some arguments for believing that an apriorist theory of mathematical knowledge will not succeed, and he recommends a naturalistic approach to mathematical knowledge, according to which the mathematics of one generation is built on the achievements of the previous generations. Within this framework, the problems that occupy center stage are those of understanding rational development of mathematics and of characterizing mathematical progress.

The later sections of Kitcher's essay are focused on these problems. Kitcher argues that there are various different notions of rationality that ought to be given a place in a history and philosophy of mathematics. He contrasts cases in which mathematical change is driven by factors external to the discipline and those in which the prior state of a branch of mathematics furnishes reasons for amending it in a particular way. On the basis of this distinction, he proposes that mathematical progress should be understood in terms of the advancement of ends that are ultimately external to mathematics. The connections between mathematics and practical concerns or the results of other sciences may, in some cases, be extremely remote, but, Kitcher contends, even the most "useless" parts of mathematics constitute objectively valuable accomplishments in virtue of the fact that they result from practical projects through a chain of rational transitions.

The position Kitcher sketches provides an alternative not only to apriorist epistemologies for mathematics but also to Platonist accounts of mathematical truth. Kitcher uses his treatment of the problems of mathematical rationality and mathematical progress to propose that truth in mathematics is what is achieved in the long run through the application of the principles that govern the rational development of mathematics. He points out that this proposal would have some radical consequences for the usual image of mathematics, and he concludes by suggesting some ways in which his proposals might be articulated in the history and philosophy of mathematics.

The last two essays are concerned with the interplay between mathematics and society. Judith Grabiner uses the recent controversies about the claims of artificial intelligence to illustrate a general pattern of change in the history of science. According to Grabiner, there are many episodes in the development of the sciences that go through the following phases: first, new ideas and methods are introduced; they prove successful in solving some outstanding set of problems and are taken up enthusiastically by their originators, who proclaim that vast insights are at hand; provoked by what they see as overambitious claims, detractors outside the discipline criticize the use of the new ideas and methods, finding fault even with the results originally accomplished; finally, there is a more thorough critique from within the scientific community.

Grabiner provides a number of examples of this general pattern. She discusses the methodological revolution in seventeenth-century science, the eighteenth-century "spirit of systems," the Industrial Revolution, the introduction of Darwinian ideas in mid-nineteenth-century biology. In each example she shows how the episode divides naturally into four phases. Thus, in the last case, the ideas of Darwin and Wallace achieved initial success in accounting for phenomena of biogeographical distribution, relationships among organisms past and present, adaptations, and so forth. In the hands of the social Darwinists and eugenicists, Darwinism was hailed as the key to solving all kinds of social problems. These enthusiastic extrapolations provoked continued criticism from outsiders, who objected even to the early achievements of Darwinian evolutionary theory. Finally, the more thorough internal critique from practicing biologists and social scientists trimmed away the excesses of social Darwinism while preserving the genuine biological successes.

Grabiner suggests that we are currently in the middle of the last stage in the debate about artificial intelligence. She outlines the early successes of computer science and relates the confidence with which its advocates—such as Simon, Newell, and McCarthy—predicted dramatic results both in constructing machines that would perform all kinds of intelligent tasks and in shedding light on the nature of human intelligence. Grabiner cites the philosophers Dreyfus and Searle as prominent representatives of the external critics, who announce the "triviality" of the entire venture. The role of internal critic is filled by Weizenbaum, who has attempted to chart the limits of artificial intelligence research instead of dismissing it entirely.

While Grabiner is concerned with the ways in which mathematical ideas (and scientific ideas) are elaborated, disseminated and criticized in the social context, William Aspray focuses on the impact of social structure on the development of mathematics. Aspray considers one of the most prominent success stories in the history of American mathematics, the development of mathematics at Princeton in the early decades of the twentieth century. He begins by giving a detailed account of the conditions under which professional mathematicians worked at the end of the nineteenth century. After showing how heavy teaching loads and few incentives to research were the order of the day, Aspray describes how Wilson's presidency at Princeton University initiated the building of a modern mathematics department.

A key figure in the building was Fine. Originally trained as a classicist, Fine had pursued his mathematical education in Germany and had become impressed with the high academic quality of the German universities. Returning to Princeton, first as a professor and later as dean, Fine devoted considerable energies to reorganizing the structure of appointments and the commitments to research. Veblen was also instrumental in the development of a research community in mathematics at the university, and Aspray traces the ways in which mathematical research was fostered by Fine and Veblen.

The implementation of the program would have been impossible without funding from a wealthy patron, and Fine was fortunate to secure support from a former classmate. One of the important consequences of the influx of money was the opportunity to construct a building for the mathematics department, Fine Hall, designed by Veblen to facilitate exchange of ideas among mathematicians. Aspray documents the significance of this physical facility in attracting promising young mathematicians either to come to Princeton or to return there. He also recounts the founding of the Institute for Advanced Study and shows how cooperation between the institute and the university further promoted Princeton as a center for mathematical research. The exodus of brilliant mathematicians from central Europe in the 1930s combined with the attractiveness of Princeton as a haven to create an extraordinary mathematical community.

Although Aspray does not exhibit in detail how the institutional factors he describes actually influenced the development of any particular mathematical field, his central message is abundantly clear. It is surely hard to believe that the ideas of the great mathematicians who were at Princeton in the 1930s were unaffected by their frequent exchanges with one another. Those exchanges were made possible by a number of historical contingencies: the drive of a university president, accidents in the education of a dean, the lucky business success of a classmate, an unwonted understanding of the importance of architecture, the emergence of an evil dictatorship. We do not ordinarily think of the course of mathematics as being affected by such chances. Aspray reminds us that they may easily leave their mark.

4. Common Themes and Possible Futures

We would like to conclude by pointing to some connections (and contrasts) among the essays and by offering some brief speculative comments about the possible future of history of mathematics, philosophy of mathematics, and history-and-philosophy of mathematics. Some common themes are already indicated in our division of the articles. Other connections crisscross the groupings we have imposed.

A. Reading the past through the categories of the present. Many of the writers are concerned to warn against the dangers of Whig history. This is especially obvious in the essays of Dauben and Daston, but it is implicit in the studies of Edwards, Moore, Crowe, and Kitcher as well. Askey's paper serves as an important reminder that sensitivity to the possibility that the conceptions of the past may not be those of the present needs to be tempered by an ability to pose questions about affinities between old and new ideas. We believe that Dauben, Daston, et al., have drawn an important lesson from the general history of science, a lesson that has often been neglected in the work of historians of mathematics, and that Askey's insights are complementary to their recommendations. Perhaps there may be some divergence of opinion or of emphasis implicit in the essay by Birkhoff and Bennett, which offers a more traditional exercise in the history of mathematics—to wit, the assembly of connections among the works of past mathematicians, conceived in the image of present mathematics. We believe that mathematicians will continue to find this kind of historical research interesting and illuminating.

B. Correcting common distortions of particular figures. Several essays are concerned to stress the fact that familiar ideas about some great figure or achievement of the past are quite inaccurate. Thus Edwards contends that we have a badly distorted picture of Kronecker, Moore argues that we do not understand Skolem's work, Goldfarb suggests that the debate between Poincaré and the logicists has been misunderstood, Friedman claims that Carnap's connections with Frege's logicism have been missed, and Stein proposes that we do not see how the foundational work of Frege, Hilbert, and Brouwer was connected to nineteenth-century mathematics. We see these essays as underscoring the need to be cautious in reading the past through the spectacles of the present.

C. Relating mathematics to the social context. Although only the essays by Grabiner and Aspray are explicitly devoted to studying the interactions between mathematics and social institutions, several other authors touch on this theme. Thus Daston considers the ways in which the enterprises of the eighteenth and nineteenth centuries shaped the development of probability theory, Browder examines the response of mathematics to extramathematical concerns, and Kitcher is concerned to emphasize the need to consider broader notions of rationality and to appreciate the effects of institutions on mathematical practice. None of the authors adopts the radical view that the evolution of mathematics is entirely driven by social forces, but those we have mentioned are sympathetic to the claim that the history of mathematics cannot be adequately written in a purely "internal" idiom. Thus historians of mathematics are beginning to turn to the appreciation of social factors, embracing the methodology that has affected general history of science in the past decade. Although we hold no brief for the extreme claims sometimes made on behalf of the sociology of knowledge, we believe that this is an extremely important trend that will illuminate many aspects of the historical development of mathematics.

D. *Finding a foundation for mathematics.* The question whether philosophy of mathematics should consist in identifying foundations for mathematics provokes one of the clearest divisions in the volume. Goldfarb and Friedman are both sympathetic to the Fregean project, and their subtle reconstructions of logicist ideas indicate how they believe that that project has been misunderstood. Crowe and Kitcher are, just as obviously, unsympathetic to the traditional view of philosophy as laying foundations. The character of the opposition is, in some ways, parallel to the debate between Poincaré and the logicists, and, as we have noted, Goldfarb hints at the relevance of his interpretation to current philosophical controversies.

E. *The relation between history and philosophy of mathematics.* Several of the contributors appear to envisage a fruitful relationship between historical and philosophical study of mathematics. However, we think it fair to say that there is no single view of that relationship. Moore employs the techniques of the historian to explore the genesis of a philosophically important thesis—that logic is first-order logic. Dauben uses ideas in contemporary philosophy of mathematics to illuminate the work of the past and, conversely, appeals to the history of mathematics to counter some current philosophical arguments. Daston argues at length that the historian cannot come to terms with significant features of past mathematics without philosophical analysis of major categories and distinctions. In complementary fashion, Stein contends that our understanding of contemporary programs in philosophy of mathematics will prove deficient if we do not see their roots in the history of mathematics. Crowe warns against a variety of historiographical errors that he takes to be the product of simplistic philosophical ideas about mathematics. Kitcher is also concerned to use philosophy to outline a positive historiography, and he adds the suggestion that historical research is needed to complete the philosophical project of understanding our mathematical knowledge.

If this is an adequate representation of where things stand in the present, what (if anything) can we predict about the future? In our judgment, it would be surprising if the historiographical sensitivities that many of the contributors want to stress (A) were to be forgotten in future research in history of mathematics. We suspect that there will continue to be revisions in our common understanding of those few mathematicians who have actually been studied by historians (B). We hope that the work of enhancing and deepening our appreciation of past mathematics will exemplify that cooperation between mathematicians and historians that Askey's essay identifies.

What of the interplay between mathematics and the broader social context? We think the essays of Grabiner and Aspray establish that it is sometimes necessary to locate factors in society that are efficacious in modifying the course of mathematics. Perhaps historians of mathematics will find, as general historians of science have already found, that the most illuminating studies integrate social factors with the forces recognized by those older historical traditions that focus on the "dynamics of ideas." In our judgment, Daston offers a vision of this unified approach. Kitcher's essay also offers a preliminary attempt to sketch a historiography appropriate to it.

The future course of philosophy of mathematics looks much less certain. We should note explicitly that the present collection represents only those trends in contemporary philosophy of mathematics that are concerned in some way with history. As we have emphasized in our biased history of the philosophy of mathematics, there are currently many different ways in which the post-Fregean tradition is being elaborated. Goldfarb and Friedman represent just one version of this tradition, a version that emphasizes the need for reanalyzing the philosophical roots of Anglo-American philosophy. Their contributions to the volume share the theme that the important Fregean enterprise is to show how mathematics is "built in" to the conditions of all rational thought.

Crowe, Browder, and, most explicitly, Kitcher argue for the maverick approach, originally pioneered by Lakatos, on which the project of securing foundations for mathematics is regarded as pointless. A major part of the argument is that a successful Fregean foundation for mathematics would trace mathematical theorems to axioms that have a special status (that are a priori in the sense that Kitcher ascribes to that notion). To settle the dispute between the two points of view, a number of issues need to be addressed: Is it impossible to find foundations that are a priori in this sense? Is this the sense of apriority that is crucial for the Fregean project? Would a demonstration that mathematics is reducible to axioms that are "built in" to the conditions of rationality be epistemologically significant? Is it reasonable to expect any such demonstration?

Stein's essay hints at a different image for the philosophy of mathematics, one that stresses the filiations both to the history of mathematics and to philosophical questions that have engaged thinkers since Plato. Thus, in our judgment, Stein offers a revision of the main twentieth-century tradition in philosophy of mathematics, but one that is less radical than that suggested by Lakatos, Crowe, Kitcher, et al. We think that it will be interesting to see which, if any, of the approaches we have mentioned prove influential in the philosophy of mathematics in the next decade.

Finally, what are the chances for an informative synthesis and for the development of history-and-philosophy of mathematics? We are encouraged by the number of contributors who have succeeded in bringing together historical and philosophical insights. Even if there is no uniform view of how history and philosophy should relate to one another (E), the essays that follow demonstrate that each field can help the other. To our minds, this is especially evident in the essays of Daston, Stein, and Moore, where, in very different ways, the authors have forged a genuine synthesis.

Although the communication between historians and philosophers, obvious both in the papers and in discussions at the conference, is exciting, we are less sanguine about the prospects for dialogue among historians, philosophers, and mathematicians. One disappointing feature of conference discussions was the tendency for some of the mathematicians present to dismiss the work of historians and philosophers as ignorant invasion of the mathematicians' professional turf. We believe that Askey's discussion in his essay captures what bothers these mathematicians, while formulating the point in a constructive way. Genuine progress is possible if all parties to the dialogue recognize that each of the others has professional expertise that the rest lack and that the point of the conversation is to remedy deficiencies, not to announce that one's own profession has all the answers.

It is evident that professional historians and philosophers of mathematics know less mathematics than professional mathematicians, and, even when the historian or philosopher has considerable "second-hand" knowledge, it is rare to find someone who has worked *creatively* in more than one of the disciplines. But typically the mathematician is far more ignorant of the practice and standards of history or philosophy than the historian or philosopher is ignorant of mathematics. Thus, while the mathematician may be bothered by the fact that the historian of number theory is not *au fait* with the advances of the 1960s, 1970s, and 1980s, historians and philosophers are just as irritated by crude precursor-hunting masquerading as history and by attempts at "philosophy" that consist in the mathematician's pet expositions of the parts of their subject that interest them most. We hope that it will help to make the point explicit so that more scholars will be led to pose questions of the kind that Askey takes for his title and to advance constructive suggestions of the sort presented in his essay.

In short, we counsel patience and the search for mutual understanding. The interdisciplinary study of mathematics has come a considerable distance since Birkhoff's pioneering attempt to bring historians and mathematicians together in 1974. We are optimistic that that study will continue to flourish and hope that the present collection of essays will advance that end.

References

Archibald, R. C. 1932. *Outline of the History of Mathematics*. Lancaster, Pa.: Lancaster Press.

Ayer, A. J. 1936. *Language, Truth and Logic*. London: Gollancz. Reprinted New York: Dover, 1946.

Bashe, Charles J., et al. 1985. *IBM's Early Computers*. Cambridge, Mass.: MIT Press.

Bell, Eric Temple. 1937. *Men of Mathematics*. New York: Simon and Schuster.

———1940. *The Development of Mathematics*. New York: McGraw-Hill. 2d ed., 1945.

Benacerraf, P. 1965. What Numbers Could Not Be. *Philosophical Review* 74: 47-73. Reprinted in (Benacerraf and Putnam 1983).

———1973. Mathematical Truth. *Journal of Philosophy* 70: 661-80.

Benacerraf, P., and Putnam, H., eds. 1964. *Philosophy of Mathematics: Selected Readings*. Englewood Cliffs, N.J.: Prentice-Hall. 2d ed., Cambridge University Press, 1983.

Biermann, Kurt R. 1973. *Die Mathematik und ihre Dozenten an der Berliner Universitat, 1810-1920. Stationen auf dem Wege eines mathematischen Zentrums von Weltgeltung*. Berlin: Akademie-Verlag.

Biggs, Norman L., Lloyd, E. Keith, and Wilson, Robin J., eds. 1976. *Graph Theory 1736-1936*. Oxford: Clarendon Press.

Birkhoff, Garrett, ed. 1973. *A Source Book in Classical Analysis*. Cambridge, Mass: Harvard University Press.

Bishop, E. 1967. *Foundations of Constructive Analysis*. Cambridge, Mass.: Harvard University Press.

Boolos, G. 1975. On Second-Order Logic. *Journal of Philosophy* 72: 509-27.

———1984. To Be Is to Be the Value of a Variable or among the Values of Several Variables. *Journal of Philosophy* 81:430-49.

———1985. Nominalist Platonism. *Philosophical Review* 94: 327-44.

Bourbaki, Nicolas. 1974. *Elements d' histoire des mathematiques*. Rev. ed. Paris: Hermann.

Box, J. F. 1978. *R. A. Fisher. The Life of a Scientist*. New York: Wiley.

Boyer, Carl B. 1968. *A History of Mathematics*. New York: John Wiley.

Brewer, James W., and Smith, Martha K., eds. 1981. *Emmy Noether: A Tribute to Her Life and Work*. New York: Dekker.

Brouwer, L. E. J. 1979. Consciousness, Philosophy, and Mathematics. In *Proceedings of the Tenth International Congress of Philosophy* (Amsterdam, 1948), ed. E. W. Beth, H. J. Pos, and J. H. A. Hollak. Vol. 1, pt. 2. Amsterdam: North-Holland. Reprinted in (Benacerraf and Putnam 1964).

Bucciarelli, Louis L., and Dworsky, Nancy. 1980. *Sophie Germain: An Essay in the History of the Theory of Elasticity.* Boston: D. Reidel.

Cajori, Florian. 1919. *A History of Mathematics.* New York: Macmillan.

Calinger, Ronald. 1981. *Classics of Mathematics.* Oak Park, Ill.: Moore.

Cantor, Moritz B. 1880-1908. *Vorlesungen über Geschichte der Mathematik.* 4 vols. Leipzig: Teubner. Reprinted New York: Johnson Reprint Corp., 1965.

Carnap, R. 1937. *The Logical Syntax of Language.* Trans. A. Smeaton of *Logische Syntax der Sprache.* London: K. Paaul, Trench, Trubner; New York: Harcourt, Brace.

———1939. *Foundations of Logic and Mathematics.* Chicago: University of Chicago Press (International Encyclopedia of Unified Science.

Chihara, C. 1973. *Ontology and the Vicious-Circle Principles.* Ithaca, N.Y.: Cornell University Press.

———1984. A Simple Type Theory without Platonic Domains. *Journal of Philosophical Logic* 13: 249-83.

Cooke, Roger. 1984. *The Mathematics of Sonya Kovalevskaya.* New York: Springer-Verlag.

Curry, H. B. 1951. *Outlines of a Formalist Philosophy of Mathematics.* Amsterdam: North-Holland.

Dauben, Joseph W. 1979. *George Cantor: His Mathematics and Philosophy of the Infinite.* Cambridge, Mass.: Harvard University Press.

———1985. *The History of Mathematics from Antiquity to the Present: A Selective Bibliography.* New York: Garland Publishing.

Detlefsen, M. 1979. On Interpreting Gödel's Second Theorem. *Journal of Philosophical Logic* 8: 297-313.

Dick, Auguste. 1981. *Emmy Noether, 1882-1935.* Trans. H. I. Blocher. Boston: Birkhauser.

Dickson, L. E. 1919-23. *History of the Theory of Numbers.* 3 vols. Washington, D.C.: Carnegie Institution. Reprinted New York: Stecher, 1934. Reprinted New York: Chelsea, 1952, 1971.

Dieudonné, Jean. 1974. *Cours de géometrie algébrique I.* Paris: Presses Universitaires de France.

———1978: *Abrege d'histoire des mathématiques, 1700-1900.* 2 vols. Paris: Hermann.

Dugac, Pierre. 1976. Problemes d'histoire de l'analyse mathématique au XIXeme siecle. Cas de Karl Weierstrass et de Richard Dedekind. *Historia Mathematica* 3: 5-19.

———1980. *Sur les fondements de l'analyse au XIXe siecle.* Louvain: Université Catholique de Louvain.

Dummett, M. 1973. *Frege: Philosophy of Language.* London: Duckworth; New York: Harper & Row.

———1977. *Elements of Intuitionism.* Oxford: Clarendon Press.

Edwards, Harold M. 1977. *Fermat's Last Theorem: A Genetic Introduction to Number Theory.* New York: Springer-Verlag.

Encyklopädie der mathematischen Wissenschaften mit Einschluss ihrer Anwendungen. 1898-1935. 23 vols. Herausgegehen im Auftrage der Akademien der Wissenschaften zu Berlin, Göttingen, Heidelberg, Leipzig, Munchen und Wien sowie unter Mitwirkung Zahlreicher Fachgenossen. Leipzig and Berlin: Teubner.

Enros, Philip C. 1979. *The Analytical Society: Mathematics at Cambridge University in the Early Nineteenth Century.* Dissertation, University of Toronto.

Eves, Howard. 1976. *An Introduction to the History of Mathematics.* 4th ed. New York: Holt, Rinehart and Winston.

Fang, J., and Takayama, K. P. 1975. *Sociology of Mathematics. A Prolegomenon.* Happange: Paideia Press.

Field, H. 1974. Quine and the Correspondence Theory. *Philosophical Review* 83: 200-28.

———1980. *Science without Numbers.* Princeton, N.J.: Princeton University Press.

———1982. Realism and Anti-Realism about Mathematics. *Philosophical Topics* 13: 45-69.

———1984. Is Mathematical Knowledge Just Logical Knowledge? *Philosophical Review* 93: 509-52.

——1985. On Conservativeness and Completeness. *Journal of Philosophy* 82: 239-59.

Frege, G. 1879. *Begriffsschrift, eine der arithmetischen nachgebildete Formel-sprache des reinen Denkens.* Halle, Nebert. Trans. by S. Bauer-Mengelberg as Begriffsschrift, a Formula Language, Modeled upon That of Arithmetic, for Pure Thought, in (van Heijenoort 1967).

——1884. *Die Grundlagen der Arithmetik. Eine logisch-mathematische Untersuchung uber den Begriff der Zahl.* Breslau: Koebner. Trans. as (Frege 1950).

——1893-1903. *Grundgesetze der Arithmetik, Begriffsschriftlich abgeleitet.* 2 vols. Jena: Pohle. Reprinted Hildescheim: Olms, 1962.

——1950. *The Foundations of Arithmetic: A Logico-Mathematical Enquiry into the Concept of Numbers.* Trans. by J. L. Austin, with German text, of (Frege 1884). Oxford: Blackwell.

——1971. *On the Foundations of Geometry and Formal Theories of Arithmetic.* Trans. E. H. Kluge. London and New Haven: Yale University Press.

Friedman, M. 1985. Kant's Theory of Geometry. *Philosophical Review* 94: 455-506.

Gillispie, C. C. 1979. *Lazare Carnot Savant.* Paris: Vrin.

Gödel, K. 1964. What Is Cantor's Continuum Problem? In (Benacerraf and Putnam 1965, 1983).

Goldfarb, Warren. 1979. Logic in the Twenties: The Nature of the Quantifier. *Journal of Symbolic Logic* 44: 351-68.

Goldstine, Herman H. 1972. *The Computer from Pascal to von Neumann.* Princeton, N.J.: Princeton University Press.

——1977. *A History of Numerical Analysis from the 16th through the 19th Century.* New York: Springer.

——1980. *A History of the Calculus of Variations from the 17th through the 19th Century.* New York: Springer.

Goodman, N., and Quine, W. V. 1947. Steps toward a Constructive Nominalism. *Journal of Symbolic Logic* 12: 105-22.

Gottlieb, Dale. 1980. *Ontological Economy: Substitutional Quantification and the Philosophy of Mathematics.* Oxford: Oxford University Press.

Grabiner, J. 1981. *The Origins of Cauchy's Rigorous Calculus.* Cambridge, Mass.: MIT Press.

Grattan-Guinness, Ivor. 1970. *The Development of the Foundations of Mathematical Analysis from Euler to Riemann.* Cambridge, Mass.: MIT Press.

——1977. History of Mathematics. In *Use of Mathematical Literature,* ed. A. R. Dorling. London: Butterworth.

Grattan-Guinness, Ivor, in collaboration with Ravetz, J. R. 1972. *Joseph Fourier, 1768-1830.* Cambridge, Mass.: MIT Press.

Gray, Jeremy. 1979. *Ideas of Space.* New York: Oxford University Press.

Grosholz, E. 1980. Descartes' Unification of Algebra and Geometry. In *Descartes, Mathematics and Physics,* ed. S. Gaukroger. Hassocks, England: Harvester.

——1985. Two Episodes in the Unification of Logic and Topology. *British Journal for the Philosophy of Science* 36: 147-57.

Hall, T. 1970. *Carl Friedrich Gauss: A Biography.* Trans. A. Froderberg. Cambridge, Mass.: MIT Press.

Hallett, M. 1979. Towards a Theory of Mathematical Research Programmes. *British Journal for the Philosophy of Science* 30: 1-25, 135-59.

Hankins, Thomas. 1980. *Sir William Rowan Hamilton.* Baltimore: Johns Hopkins University Press.

Hawkins, Thomas. 1970. *Lebesgue's Theory of Integration: Its Origins and Development.* Madison: University of Wisconsin Press. Reprinted New York: Chelsea, 1975.

——1984. The *Erlanger Programm* of Felix Klein. *Historia Mathematica* 2: 442-70.

Hempel, C. G. 1945. On the Nature of Mathematical Truth. *American Mathematical Monthly* 52: 543-56. Reprinted in (Benacerraf and Putnam 1964).

Heyting, A. 1956. *Intuitionism: An Introduction.* Amsterdam: North-Holland. 2d rev. ed., 1966; 3d rev. ed., 1971.

Hilbert, D. 1926. Uber das Unendliche. *Mathematische Annalen* 95. Trans. by E. Putnam and G. J. Massey as "On the Infinite" in (Benacerraf and Putnam 1964) and by S. Bauer-Mengelberg in (van Heijenoort 1967).

Hodes, Harold T. 1984. Logicism and the Ontological Commitment of Arithmetic. *Journal of Philosophy* 81: 123-49.

Hodges, Andrew. 1983. *Alan Turing: The Enigma.* New York: Simon and Schuster.

Hyman, Anthony. 1982. *Charles Babbage: Pioneer of the Computer.* Princeton, N.J.: Princeton University Press.

Jayawardene, S. A. 1983. Mathematical Sciences. In *Information Sources for the History of Science and Medicine,* ed. P. Corsi and P. Weindling. London: Butterworth, pp. 259-84.

Jubien, M. 1977. Ontology and Mathematical Truth. *Nous* 11: 133-50.

———1981a. Formal Semantics and the Existence of Sets. *Nous* 15: 165-76.

———1981b. Intensional Foundations of Mathematics. *Nous* 15: 513-28.

Kessler, G. 1980. Frege, Mill and the Foundations of Arithmetic. *Journal of Philosophy* 77: 65-79.

Kim, J. 1982. Perceiving Numbers and Numerical Relations. (Abstr.) *Nous* 16: 93-94.

Kitcher, P. S. 1978. The Plight of the Platonist. *Nous* 12: 119-36.

———1979. Frege's Epistemology. *Philosophical Review* 88: 235-62.

———1980. Arithmetic for the Millian. *Philosophical Studies* 37: 215-36.

———1983. *The Nature of Mathematical Knowledge.* New York: Oxford University Press.

———1986. Frege, Dedekind and the Philosophy of Mathematics. In *Frege Synthesized,* ed. L. Haaparanta and J. Hintikka. Dordrect: Reidel.

Kline, M. 1972. *Mathematical Though from Ancient to Modern Times.* New York: Oxford University Press.

Koblitz, Ann Hibner. 1983. *A Convergence of Lives: Sofia Kovalevskaia, Scientist, Writer, Revolutionary.* Boston: Birkhauser.

Kochina, P. Y. 1981. *Sof'ya Vasil'ievna Kovalevkaya.* Moscow: Nauka. Trans. P. Polubarinova-Kochina as *Sofia Vasilyevna Kovalevskaya. Her Life and Work.* Moscow: Foreign Languages Publishing House.

Koenigsberger, Leo. 1904. *C. G. J. Jacobi.* Leipzig.

Kreisel, G. 1958. Hilbert's Programme. Reprinted in (Benacerraf and Putnam and 1983).

Kuratowski, Kazimierz. 1980. *A Half Century of Polish Mathematics: Remembrances and Reflections.* New York: Pergamon.

Lakatos, I. 1976. *Proofs and Refutations.* Cambridge: Cambridge University Press.

Lear, J. 1977. Sets and Semantics. *Journal of Philosophy* 74: 86-102.

Loria, Gino. 1946. *Guida allo studio della storia delle matematiche.* 2d ed. Milan: Hoepli. 1st ed., 1916.

MacKenzie, Donald A. 1981. *Statistics in Britain 1865-1930. The Social Construction of Scientific Knowledge.* Edinburgh: Edinburgh University Press.

Maddy, P. 1980. Perception and Mathematical Intuition. *Philosophical Review* 89: 163-96.

———1981. Sets and Numbers. *Nous* 15:495-512.

———1982. Mathematical Epistemology: What Is the Question? *Nous* 16: 106-7.

Maistrov, L. E. 1974. *Probability Theory. A Historical Sketch.* Trans. S. Kotz. New York: Academic Press.

Malament, D. 1982. Review of (Field 1980). *Journal of Philosophy* 79: 523-34.

May, Kenneth O. 1972. *The Mathematical Association of America: Its First Fifty Years.* Washington, D.C.: The Mathematical Association of America.

———1973. *Bibliography and Research Manual of the History of Mathematics.* Toronto: University of Toronto Press.

Medvedev, F. A., ed. 1976. *The French School of the Theory of Functions and Sets at the Turn of the XIX-XX Centuries.* In Russian. Moscow: Nauka.

Mehrtens, Herbert. 1979. Das Skelett der modernen Algebra. Zur Bildung mathematischer Begriffe bei Richard Dedekind. In *Disciplinae Novae. Zur Entstehung neuer Denk- und Arbeitsrichtangen in der Naturwissenschaft. Festschrift zum 90. Geburtstag von Hans Schimank,* ed. C. J. Scriba. Göttingen: Vardenhoeck & Ruprecht, pp. 25-43.

Mehrtens, Herbert, Bos, H., and Schneider, I., eds. 1981. *Social History of Nineteenth Century Mathematics*. Boston: Birkhauser.

Merzbach, Uta. 1985. *Guide to Mathematical Papers in the United States*. Wilmington, Del.: Scholarly Resources.

Monna, Antonie Frans. 1975. *Dirichlet's Principle*. Ultrecht: Oosthoek, Scheltema en Holkema.

Moore, Gregory H. 1982. *Zermelo's Axiom of Choice: Its Origins, Development, and Influence*. Studies in the History of Mathematics and the Physical Sciences 8. New York: Springer.

Nový, Luboš. 1973. *Origins of Modern Algebra*. Prague: Academia Publishing House of the Czechoslovak Academy of Sciences.

Osen, Lynn M. 1974. *Women in Mathematics*. Cambridge, Mass.: MIT Press.

Parsons, C. 1964. Infinity and Kant's Conception of the Possibility of Experience. *Philosophical Review* 73: 182-97. Reprinted in (Parsons 1984).

———1969. Kant's Philosophy of Arithmetic. In *Philosophy, Science, and Method: Essays in Honor of Ernest Nagel*, ed. Sidney Morgenbesser, Patrick Suppes, and Morton White. New York: St. Martin's Press. Reprinted in (Parsons 1984).

———1971. A Plea for Substitutional Quantification. *Journal of Philosophy* 68: 231-37.

———1982. Review of (Gottlieb 1980). *British Journal for the Philosophy of Science* 33: 409-21.

———1984. *Mathematics in Philosophy*. Ithaca, N.Y.: Cornell University Press.

———1986. Review of (Kitcher 1983). *Philosophical Review* 95: 129-37.

Perl, Teri. 1978. *Math Equals*. Reading, Mass.: Addison-Wesley.

Polya, G. 1954. *Induction and Analogy in Mathematics*. Princeton, N.J.: Princeton University Press.

Pont, Jean Claude. 1974. *La topologie algébrique, des origines a Poincaré*. Paris: Presses Universitaires de France.

Putnam, H. 1967. Mathematics without Foundations. *Journal of Philosophy* 64: 5-22. Reprinted in (Putnam 1975), vol. 1, and in (Benacerraf and Putnam 1983).

———1975. *Philosophical Papers*. Vol. 1: *Mathematics, Matter and Method*. Vol. 2: *Mind, Language, and Reality*. Cambridge: Cambridge University Press.

Quine, W. V. 1936. Truth by Convention. In *Philosophical Essays for Alfred North Whitehead*, ed. O. H. Lee. New York: Longmans, Green. Reprinted in (Benacerraf and Putnam 1964).

———1953. *From a Logical Point of View: Nine Logico-Philosophical Essays*. Cambridge, Mass.: Harvard University Press. 2d ed., 1961.

———1960. *Word and Object*. Cambridge, Mass., and New York: Technology Press of MIT and Wiley; London: Chapman.

———1962. Carnap and Logical Truth. In *Logic and Language: Studies Dedicated to Professor Rudolf Carnap on the Occasion of His Seventieth Birthday*, ed. B. H. Kazemier and D. Vuysje. Dordrecht: Reidel. Reprinted in *The Ways of Paradox*. New York: Random House, 1966.

———1963. *Set Theory and Its Logic*. Cambridge, Mass.: Harvard University Press.

———1970. *Philosophy of Logic*. Englewood Cliffs, N.J.: Prentice-Hall.

Reid, Constance. 1970. *Hilbert*. New York: Springer.

———1976. *Courant in Göttingen and New York: The Story of An Improbable Mathematician*. New York: Springer.

———1982. *Neyman—From Life*. New York: Springer.

Resnik, M. D. 1974. On the Philosophical Significance of Consistency Proofs. *Journal of Philosophical Logic* 3: 133-48.

———1975. Mathematical Knowledge and Pattern Recognition. *Canadian Journal of Philosophy* 5: 25-39.

———1980. *Frege and the Philosophy of Mathematics*. Ithaca, N.Y.: Cornell University Press.

———1981. Mathematics as a Science of Patterns: Ontology. *Nous* 15: 529-50.

———1982. Mathematics as a Science of Patterns: Epistemology. *Nous* 16: 95-105.
Ricketts, T. G. 1985. Frege, the *Tractatus* and the Logocentric Predicament. *Nous* 19: 3-16.
Sarton, George. 1936. *The Study of the History of Mathematics.* Cambridge, Mass.: Harvard University Press. Reprinted New York: Dover, 1957.
Scholz, Erhard. 1980. *Geschichte des Mannigfaltigkeitsbegriffs von Riemann bis Poincaré.* Boston: Birkhauser.
Shapiro, S. 1983a. Mathematics and Reality. *Philosophy of Science* 50: 523-48.
———1983b. Conservativeness and Incompleteness. *Journal of Philosophy* 80: 521-31.
———1985. Epistemic and Intuitionistic Arithmetic. In *Intensional Mathematics* ed. S. Shapiro. New York: North-Holland.
Simpson, Stephen G. Partial Realizations of Hilbert's Program. Forthcoming.
Sluga, H. 1980. *Gottlob Frege.* London: Routledge & Kegan Paul.
Smith, David Eugene. 1906. *The Teaching of Elementary Mathematics.* New York: Macmillan. Reprinted 1908.
———1923-25. *History of Mathematics.* Boston: Ginn and Company. 2 vols. Reprinted New York: Dover, 1951, 1958.
———1929. *A Source Book in Mathematics.* New York: McGraw-Hill. Reprinted New York: Dover Publications, 2 vols., 1959.
Stein, Dorothy. 1985. *Ada: A Life and Legacy.* Cambridge, Mass.: MIT Press.
Steiner, M. 1975. *Mathematical Knowledge.* Ithaca, N.Y.: Cornell University Press.
———1978. Mathematical Explanation. *Philosophical Studies* 33: 135-51.
———1983. Mathematical Realism. *Nous* 17: 363-85.
Struik, Dirk, 1967. *A Concise History of Mathematics.* New York: Dover.
———1980. The Historiography of Mathematics from Proklos to Cantor. *NTM Schriftenreihe fur Geschichte der Naturwissenschaften, Technik, und Medizin* 17:1-22.
Tait, W. W. 1981. Finitism. *Journal of Philosophy* 78: 524-46.
Tarwater, Dalton, ed. 1977. *The Bicentennial Tribute to American Mathematics: 1776-1976.* Washington, D.C.: The Mathematical Association of America.
Thackray, Arnold. 1983. History of Science. In *A Guide to the Culture of Science, Technology, and Medicine,* ed. Paul T. Durbin. New York: Free Press, pp. 3-69.
Todhunter, I. M. 1861. *A History of the Calculus of Variations during the Nineteenth Century.* New York: Chelsea. Reprinted 1962.
Ulam, S. M. 1976. *Adventures of a Mathematician.* New York: Scribner's.
Van Heijenoort, Jean. 1967. *From Frege to Gödel. A Source Book in Mathematical Logic, 1879-1931.* Cambridge, Mass.: Harvard University Press.
Weil, Andre. 1975. *Essais historiques sur la theorie des nombres.* Geneva: Universite.
———1984. *Number Theory: An Approach through History from Hammurabi to Lengendre.* Boston: Birkhauser.
White, N. P. 1974. What Numbers Are. *Synthese* 27: 111-24.
Wilder, R. 1975. *Evolution of Mathematical Concepts.* New York: Wiley.
Williams, Michael R. 1985. *A History of Computing Technology.* Englewood Cliffs, N.J.: Prentice-Hall.
Wright, C. 1983. *Frege's Conception of Numbers as Objects.* Aberdeen: Aberdeen University Press.
Wussing, Hans. 1969. *Die Genesis des abstrakten Gruppenbegriffs. Ein Beitrag zur Entstehungsgeschichte der abstrakten Gruppentheorie.* Berlin: VEB Deutscher Verlag der Wissenschaften.
———1974. *C. F. Gauss.* Leipzig: Teubner.

I. LOGIC AND THE FOUNDATIONS OF MATHEMATICS

Poincaré against the Logicists

Although the great French mathematician Henri Poincaré wrote on topics in the philosophy of mathematics from as early as 1893, he did not come to consider the subject of modern logic until 1905. The attitude he then expressed toward the new logic was one of hostility. He emphatically denied that its development over the previous quarter century represented any advance whatsoever, and he dismissed as specious both the tools devised by the early logicians and the foundational programs they urged. His attack was broad: Cantor, Peano, Russell, Zermelo, and Hilbert all figure among its objects. Indeed, his first writing on the subject is extremely polemical and is laced with ridicule and derogation. Poincaré's tone subsequently became more reasonable but his opposition to logic and its foundational claims remained constant.

Poincaré's first paper on logic (1905) is a response to a series of articles published the previous year by Couturat, who was the purveyor of logicism to the French. Poincaré seems particularly outraged at the logicists' claim to have conclusively refuted Kant's philosophy of mathematics. Thus Poincaré is moved to write in the area by a purely philosophical animus.[1] No mathematical issues are on his mind; his concern is with general, philosophical claims about the nature of mathematics, and he is reacting to claims of this sort made by the partisans of the new logic. Apparently, then, to questions in the foundations of mathematics Poincaré wishes to provide answers opposed to those of the logicists. Such a description of the situation, however, is simplistic. Rather, as I hope to make clear, Poincaré's conception of what the questions in the foundations of mathematics *are* differs considerably from that held by the logicists. An examination of the differences that operate at this level can, I believe, illuminate both logicism and the rejection of logicism.

A second reason for interest in Poincaré's writings on foundations lies in the fact that two of his points continue to figure in contemporary discussions. The first is what I shall call Poincaré's *petitio* argument. Poincaré alleges that any program of grounding number theory in something else will invariably beg the question; for if that foundation is to be carried out, number theory (and in particular mathematical induction) will have to be presupposed. The second is Poincaré's *vicious circle principle*, which precludes mathematical definitions that are impredicative, that is, that involve quantified variables ranging over a universe that contains the defined entity. Indeed, Poincaré's introduction of the notion of predicativity is his most influential contribution to foundational studies.

I shall treat these two points in turn. An analysis of the *petitio* argument can provide helpful clues both about Poincaré's view of foundational questions and about the construal of logicism that is needed to turn the objection aside. Matters are less straightforward with the vicious circle principle, and my treatment will be more purely historical. I shall trace how Poincaré arrives at the principle and how, after seeing its force, he begins to argue for it more philosophically. Those arguments, together with his remarks on the technical import of the principle, provide further insight into his central motivations.

My discussion is limited to Poincaré's explicit writing on logic and foundations. It might well be most interesting and fruitful to investigate relations between his foundational views and his mathematical work in other areas, but unfortunately there is little go on. As I have mentioned, Poincaré comes to discuss foundations solely for philosophical reasons. Unlike Hilbert, Cantor, and others, no technical factors drive him on. As a result, his papers on foundations are disconnected from his positive work in mathematics. (Again, this contrasts with his philosophical writings on geometry, in which, for example, analysis situs is explicitly discussed.) Of course, it may be noted that Poincaré's style of mathematics generally was more constructive than, say, Hilbert's—Poincaré was a *French* analyst. The question is whether anything specific can be drawn from this that would illuminate his position on foundations. Such an investigation is beyond the scope of this paper. My aim below is to sort out what the foundational issues are; it is a subsequent task to find out whether and how Poincaré's stand on these issues can be clarified by attention to his mathematical work.

I

Poincaré begins his attack on the new logic with the avowed aim of showing that the new logicians have not eliminated the need for intuition in mathematics. By showing this, he says, he is vindicating Kant (1905, 815). This avowal is misleading, for in Poincaré's hands the notion of intuition has little in common with the Kantian one. The surrounding Kantian structure is completely lacking; there is no mention, for instance, of sensibility or of the categories. Indeed, in (1900), his address to the Paris Congress, Poincaré explicitly separates mathematical intuition from imagination and the sensible. Thus there is no hint of an epistemological framework in which the notion functions. For Poincaré, to assert that a mathematical truth is given to us by intuition amounts to nothing more than that we recognize its truth and do not need, or do not feel a need, to argue for it. Intuition, in this sense, is a psychological term; it might just as well be called "immediate conviction."[2] As a result of the unstructured nature of the notion, Poincaré's argument in (1905) can be purely negative. If a mathematical proposition is convincing, that is, it seems self-evident to us, and the purported logical proofs of it are insufficient, then, tautologously, intuition in Poincaré's sense is what is at work.

Such a sense of "intuition" underlies another of Poincaré's criticisms, namely, that the logicists have done nothing but rename the chapter in which certain truths are to be listed: what was called "mathematics" is now called "logic." These truths, he says, "have not changed in nature, they have changed only in place" (1905, 829). This is merely a renaming, he claims, for since the axioms the logicists use are not merely "disguised definitions," they must rest on intuition. He emphasizes this even with respect to truth-functional logic. Here Poincaré is simply ignoring a central philosophical point of logicism. Frege and Russell take logic to consist of those general principles that underlie all rational discourse and all rational inference, in every area, about matters mathematical or empirical, sensible or nonsensible. Consequently, the truth of these principles could not stem from any faculty with a specialized purview; certainly, their truth does not play any particular role in constituting the world of experience. If one persists, as Poincaré does, in claiming that intuition is needed, then one has in fact robbed "intuition" of all content.[3]

Poincaré also complains that logicism errs insofar as "in reducing mathematical thought to an empty form...one mutilates it" (1905, 817).

He sees logicists as saying something analogous to "all the art of playing chess reduces to rules for the movement of the pieces," whereas, of course, one does not understand chess just by knowing those rules. In his Paris Congress address, he stresses how intuition is needed to gain an understanding of proofs in analysis, even after the correctness of each step in the proof has been recognized; and in (1905) he talks of intuition as that which guides the choice of conventions to adopt and of routes to take toward a proof. Intuition thus becomes merged with a notion of what might be called "mathematical sagacity." For here Poincaré's concern is explicitly with the psychology of mathematical thinking, with a faculty or skill that enables a person to do mathematics, or to do it well.[4]

A similar concern is exhibited in Poincaré's discussion of the continuum in (1893), amplified in (1902). His remarks amount to a speculation about what could be called the psychogenesis of the idea of the continuum, that is, how we came to think or how we come to think of this mathematical structure. By 1903 Poincaré was acquainted with Dedekind's construction of the real numbers. He applauds it for *reflecting* the "origin of the continuum": we come to "admit as an intuitive truth that if a straight line is cut into two rays their common frontier is a point" (1902, 40). Nothing could be further from Dedekind's outlook; Dedekind, after all, explicitly offers his construction to show how to *rid* analysis of any geometrical elements.

In invoking this sense of intuition, in (1905), Poincaré raises the charge that logicism does not accurately portray the psychology of mathematical thinking. He then claims to forswear the charge: he admits that there is a distinction between context of discovery (or invention) and context of justification, and that intuition in the broader sense pertains to the former whereas the logicists are concerned with the latter (1905, 817). His more serious antilogicist arguments are not of this sort, on the surface. However, closer scrutiny reveals a continued dependence on a psychologistic conception of the foundational enterprise. This can be seen in his *petitio* argument.

There are, in fact, several forms of this argument in Poincaré's writing. The clearest is directed against the notion that mathematical induction is not a principle with content, but is just an implicit definition of the natural numbers. Poincaré notes that such a definition must be justified by showing that it does not lead to contradiction; yet any such demonstration would have to rely on mathematical induction. The argument is a good one but it has force only against a naive version of Hilbert's position. Early on, Hilbert saw that if axioms are to endow talk of their objects with warrant, a consistency proof is needed. He was unclear about the status of such a proof at first. Subsequently, he came to see that one cannot hope to ground mathematics *without residue* in this way; rather, the metamathematical reasoning employed has to be taken for granted. His formalist program rests on the idea that the residue would be finitary mathematics, a small part of number theory. Finitary mathematics includes some amount of induction—although Hilbert was never explicit about the matter—namely, induction on decidable number-theorectic predicates (that is, quantifier-free induction). There is no *petitio* since there is no claim that all number-theoretic principles are to be legitimized from scratch.

To be sure, Poincaré's arguing in this manner is explicable. Hilbert's sketch in (1904) of the application of his early ideas of implicit definition to number theory is quite unclear. The distinction between mathematics and metamathematics does not become rigorous until nearly two decades later, and the question of limitation of means available in the metatheory does not exist in 1904, even in embryo. More to the historical point, Couturat (1904), to which Poincaré is responding, does not distinguish Hilbert's notion of implicit definition from the Frege-Dedekind-Russell strategy for explicit definition of the numbers. Hence it is not surprising that Poincaré thinks of the project of reducing arithmetic to logic as being carried out by implicit definition. In point of fact, the logicists objected strongly to the idea of implicit definition. Frege and Russell (and, later on, Couturat) insist that existence is not proved by consistency; rather, consistency is vouchsafed by showing existence. Hence this form of the argument does not touch logicism.

Another form is more relevant. In a sarcastic passage (1905, 823), Poincaré examines definitions of particular numbers given in Peano-writing by Burali-Forti (1897). He notices, in the definition of zero, the use of the concept "in no case," and, in the definition of one, the concept that would be rendered in ordinary language "a class that has only one member." He concludes, "I do not see that the progress is considerable." Similarly, he cites phrases used in introducing logical machinery, such as "the logical product of two or more propositions," as showing that a knowledge of number must be presupposed. In general,

> It is impossible to give a definition without enunciating a phrase, and difficult to enunciate a phrase without putting in it a name of a number,

or at least the world "several," or at least a word in the plural. And
then the slope is slippery, and at each instant one risks falling into a
petitio principi (1905, 821).

Here Poincaré is making what is, by our lights, an elementary logical
mistake. As Frege warned, there can be an appearance of circularity; but
this appearance is dispelled when one distinguishes uses of numerical expressions that can be replaced by purely quantificational devices from the
full-blooded uses of such expressions that the formal definition is meant
to underwrite. The former involve no arithmetic, and no mathematical
induction; hence there is no *petitio*. In a reply to Poincaré, Couturat makes
this point reasonably well, if rather untechnically (Couturat 1906). He
argues, for example, that what is presupposed is not the *number* one but
rather unity (or, as Frege might prefer to say, objecthood). Poincaré never
cedes the point. In (1906), he abandons it for the sake of the discussion,
to avoid, as he puts it "the spectacle of an interminable guerrilla war."
He adds, "I continue to think that M. Courturat defines the clear by the
obscure, and one cannot speak of *x* and *y* without thinking *two*" (1906,
294).

Thus Poincaré does not accept the logical distinctions that would rebut
the charge of *petitio*. The distinction he does invoke, between "clear"
and "obscure," is a psychological one. Indeed, in this context clarity and
obscurity amount to no more than familiarity and unfamiliarity. (I take
it that *we* do not find the quantificational notions to which Poincaré is
objecting particularly obscure.) In this, Poincaré is going beyond his point
that mathematical induction is psychologically convincing enough that no
further grounding of it is needed (all we need do is cite "intuition"); he
is claiming that any attempt to ground it further *must* fail, because it will
have to reduce familiar notions to ones less familiar. Now that is not the
logicists' criterion of success; they were under no illusion that the notions
to which arithmetic is to be reduced would be more familiar to their audience. The contrast between Poincaré and Frege here is striking (there
is, however, no evidence that Poincaré was acquainted with Frege's work).
Frege explicitly mentions that familiarity poses a danger, since a familiar
inferential step will seem self-evidently correct to us, "without our ever
being conscious of the subordinate steps condensed within it" (1884, 102).
As a result, such a step will be too readily deemed intuitive, rather than
composed of smaller steps each of which is purely logical. Thus familiarity can obscure the need for analysis of our reasoning.

Poincaré's example "one cannot speak of x and y without thinking *two*" highlights the psychologistic nature of his argument. Despite Couturat's urging that *logically* the number two is not presupposed in the use of two variables, Poincaré rests on the alleged psychological fact that in this case we do always think of two. Again the contrast with Frege is stark. Frege argues emphatically that the "ideas and changes of ideas which occur during the course of mathematical thinking" are irrelevant to the foundations of arithmetic (1884, vi); and that, more generally, a sharp distinction must be made between the mental or physical conditions under which a person comes to understand, appreciate, or believe a proposition and the ultimate rational basis of the proposition. In the absence of such a distinction, Frege adeptly urges, we would have to take into account in mathematics the phosphorus content of the human brain (ibid.); this, for Frege, is a reductio. Indeed, this distinction is required even to make proper sense of the claim that mathematics is a priori. It may well be true that a person cannot do mathematics without having had experiences; but such a fact is irrelevant to the grounds for the propositions of mathematics, which is what the claim of a priori status concerns (Frege 1884, 3 and 12).

The contrast with Frege shows how Poincaré is—despite his disclaimer—construing the project of the foundations of mathematics as being concerned with matters of the psychology of mathematics and faulting logicism for getting it wrong. Now there is a subtle aspect to this difference between Poincaré and the logicists. I claimed, noncontroversially, that Poincaré's objection is logically in error: the notions used in the definition of number do not presuppose number. But to accept this is to accept the notion of presupposition that is first made available through modern logic. We all agree that the notions "in no case," "a class with one object," and so on, do not presuppose any number theory, because of the way they can be rendered using (first-order) quantificational logic. In our acceptance of this rendering as being the criterion of presupposition, we have already accepted at least part of the picture that the logicists were urging. If we take Poincaré to reject even such a use of logic, and to claim that if certain thoughts always occur to us in such cases then the content of those thoughts is "presupposed," we can make some sense of his objection. It is, of course, questionable whether such a position is ultimately coherent; but, in any case, the distance between Poincaré's stance and that of any logicist is at this point unbridgeable.

More generally, the Fregean distinctions I've been stressing—between habitually accepted inferences and fully analyzed inferences, and between empirical conditions a person must satisfy in order to arrive at certain propositions and the ultimate rational basis for the propositions—must be given content by establishing a means for saying what analysis and justification look like: that is, for saying what counts as showing that one proposition is the basis for another. Only thus can the opposition be exhibited between a Fregean project and one of describing how people actually come to hold various propositions. It is the new logic itself that provides those means. Thus the questions that logicism is asking are themselves first given sense by the framework the logicists develop for answering them. And that the new logic does this, by being explicatory of "analysis" and of "justification," is part of the reason Frege takes it to deserve the honorific "logic."

In this there is a circle of a peculiarly philosophical sort. If you don't buy the picture, you needn't buy the picture. But, in fact, entrance into the circle is hard to resist: for the notion of nonpsychologistic justification can be seen as, and was urged to be, just the fully thought-through version of the sort of rigor at which proofs in mathematics, in the ordinary, unreconstructed, sense, already aim. (This point had particular force given the achievements of the arithmetiziers of analysis.) Poincaré can then be painted, or caricatured, as holding that in pursuing mathematics we seek such justifications, but with regard to matters he wants to think of as basic (because they are familiar), the proper story is a psychological one. For Frege, to be content with this is to make the rigor of *all* proofs as illusion.

In any case, the objection I've discussed is, by our lights, simply mistaken, and Poincaré does not return to it after 1905 (although he apparently thinks well enough of it to reprint the relevant paragraphs in [1908]). He does, however, give a more sophisticated *petitio* argument in (1909b), in criticizing Russell's theory of types. Descendants of this argument are still discussed today, in large measure due to Parsons (1965). The clearest version focuses on the fact that to formulate the (formal) system of logic, to which arithmetic is to be reduced, we inductively define the formulas and the notion of derivation. The very foundations of logic, it is alleged, are thereby shown to require induction.

This version of the argument may appear to have more force than Poincaré's original one; but, I believe, the appearance is misleading. Its seeming force stems from the fact that, in contemporary logic, formal systems are automatically treated from a metatheoretical standpoint. Consider, for example, a Hilbertian view: here the justification of a mathematical proposition starts with the assertion that the proposition, as a formal object, is a theorem of a certain formal system. That assertion is, of course, a metamathematical one, requiring notions defined by recursion. Thus, even before the question of a consistency proof for the system enters, arithmetic must be invoked. (As before, this causes the Hilbertian no problem, since he does not claim to eliminate induction.) Contemporary foundational discussions, although ordinarily not committed to Hilbert's formalism, share this starting point; investigation of the status of a proposition begins with a metamathematical assertion about its provability in some system.

With respect to logicism, though, the formal system plays a markedly different role. The logical system Frege or Russell proposes is meant to be the universal language, inside of which all reasoning takes place. There is no metatheoretical stance either available or needed. Justification is just what is done inside the system. To give the ultimate basis for a proposition is to give the actual proof inside the system, starting from first principles; that is, it is to assert the proposition with its ground, not to assert the metaproposition "this sentence is a theorem." The role of formality in logicism is limited: it enables us to be sure that all the principles needed for the justification of the proposition have been made explicit, but it is neither essential to the nature of the justification nor constitutive of its being a justification. To a logicist, the charge the number theory is being invoked is based, in the end, on thinking that not judgments but the manifestations of judgments—that is, the formal symbols—are the objects of study. Frege steadfastly argued against this construal. Logic is not *about* manipulations of signs on paper, even though it may be a psychological necessity for us, in order to be sure that we are proceeding logically, to verify proofs by syntactic means.

Frege explicitly mentions a similar issue:

> A delightful example of the way in which even mathematicians can confuse the grounds of proof with the mental or physical conditions to be satisfied if the proof is to be given is to be found in E. Schröder. Under the heading "Special Axiom" he produces the following: "The principle I have in mind might well be called the Axiom of Symbolic Stability. It guarantees us that throughout all our arguments and deductions the symbols remain constant in our memory—or preferably on paper." (1884, viii)

Of course, if we are to be able to do mathematics, the signs will have to remain constant before our eyes; but that hardly says that mathematics presupposes the physics of inkblots. Similarly, if we are to be able to do logic we probably will have to be able to count, so that, for example, we can count the number of parentheses in a formula. On the logicist view, that does not show that logic presupposes counting.[5]

In sum, the logicists' conception of logic as the universal framework of rational discourse reduces the sophisticated form of Poincaré's objection to the original, naive form. That conception yields a distinction between the justification of a proposition and the means for expressing and checking putative justifications. Knowledge about the formal system of logic pertains only to the latter; possession of such knowledge is among the conditions under which a person, in fact, will be able to give correct justifications. As we have seen, it is a central tenet of antipsychologism that such conditions are irrelevant to the rational grounds for a proposition. Thus the objection is defeated.

Of course, Poincaré does not accept any such distinction; if the formal system is "comprehensible" only to one who already knows arithmetic, then the theorems of the formal system presuppose arithmetic (see 1909b, 469). Here again, "presupposition" is for Poincaré a notion based in psychology, not logic; the dispute between Poincaré and the logicists amounts, at bottom, to a difference in the conception of the foundational enterprise.

Thus in the end none of the forms of the *petitio* argument has force against logicism. Poincaré's ultimate reliance in them on a psychologistic view of the questions in the foundations of mathematics underlines what is involved in the logicists' establishment of their radically antipsychologistic view. One can reject that view, but, as Poincaré himself exhibits, such a rejection leaves little foundational role for logic at all. The rejection pervasively affects the basic notions of justification, presupposition, and proof; as a result, the general questions one asks about mathematics are considerably transformed. On this score, some current approaches that seek naturalistic accounts in the philosophy of mathematics are in the same position as psychologism.[6] Consequently, the extent to which such approaches are capable of addressing the problems that gave rise to the subject of foundations of mathematics is highly unclear.

II

Poincaré first enunciates the vicious circle principle in (1906), and his subsequent foundational writings focus almost exclusively on it. By 1906, Poincaré had gained a far more extensive acquaintance with the new logic. In particular, he had read Whitehead (1902), in which the logistic construction of the numbers is outlined, and Russell (1905), in which Russell discusses various paradoxes and gives the famous delineation of routes to avoid them—namely, the zigzag theory, the theory of limitation of size, and the no-classes theory.

Poincaré introduces the vicious circle principle in a discussion of Richard's paradox, the paradox of the set of definable decimals. This set is denumerable, and from an enumeration of it a Cantor-diagonal decimal can be formed. That decimal is not among the enumerated ones but is, it seems, definable and hence is in the set. Poincaré draws the vicious circle principle from Richard's own solution of the paradox (Richard 1905). The set E of definable decimals, Poincaré says, must be construed as composed just of the decimals that can be defined "without introducing the notion of the set E itself. Failing that, the definition of E would contain a vicious circle" (1906, 307). He then proposes as a general principle that only those definitions that do not contain a vicious circle should be taken to determine sets.

An analysis of the Richard paradox as resting on some type of circularity is extremely plausible, since the paradox exploits a notion of definability that allows definitions containing reference to the notion of definability itself. Most contemporary solutions of the paradox recognize this and proceed by stratifying notions of definability. On such views, the circle that would otherwise exist is peculiar to semantic or intensional notions like "definability." Poincaré's analysis, however, has a different cast. He focuses his challenge not on the notion of definability but on the set of decimals specified by invoking the notion.[7] This leads Poincaré to apply the analysis more widely; he descries the same circularity in the set-theoretic paradoxes as well.

Thus Poincaré immediately goes on to claim that the vicious circle principle solves the Burali-Forti paradox of the order type of the set of ordinals—a paradox that Poincaré, like Russell but unlike Burali-Forti himself, recognized from the first to be a true paradox. He says, "One introduces there the set E of all ordinal numbers; that means all ordinal

numbers that one can define without introducing the notion of the set
E itself" (1906, 307). Thus paradox is blocked. The extension of the vicious
circle analysis to this paradox is a bold step, for there is a real distinction
between this case and that of the Richard paradox. An item qualifies for
membership in the Richard set only if there is a sequence of words that
defines it. Since the truth or falsity of the latter condition may depend
on the composition of the Richard set, a circle is apparent. In contrast,
an item qualifies for membership in the Burali-Forti set if and only if it
is an ordinal, and this condition in no way depends on the Burali-Forti
set. Nonetheless, Poincaré alleges a vicious circle: one that arises if the
condition is applied universally. Therefore, he takes the vicious circle principle to restrict the range of candidates for membership in the set, namely, to items that can be given independently of the set.

Poincaré's treatment of the Burali-Forti paradox may have been influenced by a lesson Russell drew from it: "There are some properties
such that, given any class of terms all having such a property, we can
always define a new term also having the property in question. Hence we
can never collect all the terms having the said property into a whole"
(Russell 1905, 144). For Russell, this lesson does not constitute an analysis
or solution of the paradox; it is only an indication of what is to be analyzed.
Poincaré can be seen as taking it in a more definitive manner: a specification of a set must be interpreted as simply not applicable to any "new"
term producible from that set.

So far, then, Poincaré is using the vicious circle principle to bar from
membership in a set anything that in some sense presupposes that set. In
this form, the principle can also be used to block the Cantor and Russell
paradoxes. This is not yet, however, the full strength of the principle that
Poincaré urges. (Indeed, under a suitable interpretation of "presupposes,"
both the simple theory of types and the so-called iterative conception of
set wind up abiding by a restriction of this sort.)

The full force of Poincaré's principle first emerges in his use of it to
criticize the logicist definition of number. A finite number is defined as
a number that belongs to every inductive set, that is, every set that contains 0 and contains $n + 1$ whenever it contains n. Poincaré claims that
to avoid a vicious circle, the inductive sets invoked in the definition cannot include those specified by reference to the set of finite numbers. If
such a restriction is adopted, then mathematical induction cannot be deduced for such sets; many basic laws of arithmetic become unprovable,

and the logicist claim is defeated. Now, the vicious circle principle, so construed, restricts not just the candidates for membership in the set being defined, but also the range of the quantifiers in the definition. This yields the notion of predicativity in its modern sense: quantifiers occurring in specifications of sets must be construed as not including in their range of variation the set that is being defined, or anything defined by reference to that set.[8]

Poincaré makes one further application of the vicious circle principle in (1906). Earlier (1905, 25-29), he had pointed out that the then extant proof of the Schröder-Bernstein theorem invoked mathematical induction by utilizing an inductive construction of a sequence of sets. This belied the claim that cardinal number theory can be developed without special mention of the finite numbers. In 1906 Zermelo sends Poincaré a different proof: instead of an inductive construction, the intersection U of all sets with a certain property P is formed; the theorem is obtained by showing that a particular set defined from U has property P, and thus is among the sets intersected. Poincaré invokes the vicious circle principle to block the argument, for it rules out from the sets intersected any set in whose definition the intersection U itself figures. Here again, the principle is taken to restrict the range of quantifiers in the definition of a set.

The restriction Poincaré urges can be put, in the formal mode, thus: the quantifiers in a specification of a set cannot legitimately be instantiated by names that contain reference to that set. In (1906) he gives a general reason for his restriction, namely, that a purely logical proof must start from identities and definitions, and proceed in such a way that when definienda are replaced by definientia, one obtains an "immense tautology" (1906, 316). Impredicative definitions block such a replacement; hence they cannot be allowed in logic. In the polemical setting of 1906, this remark cuts no ice: logicists and set theorists did not think that their proofs must reduce to tautologies. On the technical level, however, it is a tremendously prescient remark. Although adherence to a predicativity constraint does not ensure reducibility to tautology in a literal sense, in many settings it does yield conservative extension results, which fail once impredicative definitions are allowed.

Poincaré's proposal met much criticism. In particular, Russell and Zermelo both found the same flaw in it: the very formulation of the vicious circle principle, they charge, violates the principle.

It is precisely the form of definition said to be predicative that contains something circular; for unless we already have the notion, we cannot know at all what objects might at some time be determined by it and would therefore have to be excluded. (Zermelo 1908, 191)

That is, if the specification "class of ϕs" is taken as meaning "class of things that have ϕ but do not presuppose the class of ϕs," then the specification involves the notion of the class of ϕs, and so a vicious circle has been introduced. Russell repeats the point often, and likens it to an attempt to avoid insulting a person with a long nose by remarking "When I speak of noses, I except such as are inordinately long" (Russell 1908, 155).

Now, as was pointed out above, for Poincaré the vicious circle principle amounts to a restriction on legitimate instantiations for the quantified variables in a specification of a set. In their criticism, Zermelo and Russell are assuming that such a restriction must arise from explicit limitations on the quantifiers, limitations that are part of the content of the specification. It is those explicit limitations that introduce the circle. Poincaré never answered this objection. For him, one presumes, a specification of a set stands as is, without explicit reference to the cases that are ruled out. The effect of the vicious circle principle comes in the *application* of the specification. The restriction on instantiations need not be grounded in an explicit anterior fixing of the quantifiers' ranges.

The difference between Poincaré and the logicians here is fundamental. For Russell and Zermelo, inference is a matter of universal and subject-neutral logical laws. An inference seemingly licensed by such a general law can be blocked only by so construing the content of the statement concerned as to make the law inapplicable. In this sense, logic is the arbiter of content. But, as we may surmise from section I, Poincaré does not share this conception of universal logical laws; hence he does not recognize any need for a notion of fixed content to ground restrictions on inference.

Such a denial is unintelligible from Russell's and Zermelo's point of view and is unsatisfactory from most contemporary ones. Contemporary predicative systems arrange their entities in a hierarchy; a definition with quantifiers ranging over entities of level at most n in the hierarchy automatically defines something at level $n + 1$. As a result, the defined entity need not be invoked to restrict the ranges of the quantifiers in its definition.

Zermelo presses another criticism as well, which has become a standard "classical" argument against the predicativity constraint.

> After all, an object is not created through. . .a "determination"; rather, every object can be determined in a wide variety of ways. . . . A definition may very well rely upon notions that are equivalent to [i.e., have the same extension as] the one to be defined. (1908, 191).

Zermelo thus totally rejects the vicious circle principle.

Russell, in contrast, adopts the vicious circle analysis and frames his logical system accordingly. The result is the ramified theory of types. Russell's disagreement with Zermelo can be focused on Zermelo's second objection. That objection requires a principle of extensionality about the entities under consideration, and here Russell parts company from the set theorists (and from Frege as well). Russell's logical theory concerns not extensional entities like sets, but intensional entities, namely, properties and propositions. Zermelo's point about equivalent notions is thereby rendered irrelevant; different definitions do determine different entities. Moreover, since the ranges of quantifiers in a definition are essential to the content of the defined proposition or property, support can be obtained for a hierarchy of propositions and properties that abides by the predicativity constraint. Vicious circles are thereby avoided, but not by ad hoc fiats; rather, the restrictions "result naturally and inevitably from our positive doctrines" (Russell 1908, 155). Thus, although Russell accepts Poincaré's proposal, philosophically his system is quite differently based.[9] Indeed, Poincaré is unsympathetic to Russell's theory (see 1909b, §§3-4); he continues to be hostile to formal systems, and to any call for more fundamental laws that would issue in the predicativity constraint.

Zermelo presses one further objection to Poincaré. He notes that many standard mathematical proofs involve impredicative reasoning, citing in particular the "fundamental theorem of algebra" (every algebraic equation has a root in the complex numbers), and, more generally, the panoply of theorems on maxima and minima (Zermelo 1908, 191). Thus the predicativity constraint would require the rejection of reasoning and results universally accepted by mathematicians. This criticism has particular force against Poincaré, who explicitly denies that foundational studies could affect mathematics: "[Mathematics] will pursue step by step its accustomed conquests, which are definitive and which it will never have to abandon" (1906, 307).

Poincaré is quick to reply (1909a, 199). He agrees that the standard proof of the fundamental theorem of algebra uses impredicative reasoning, and he sketches a maneuver that avoids it. (The maneuver relies on the replacement of a set of real numbers with a certain greatest lower bound by a sequence of rational numbers with the same bound.) Indeed, the same sort of maneuver is later used by Weyl (1918) to show that all the basic theorems of the theory of continuous functions, including those on maxima and minima, are predicatively provable. However, Poincaré ends his reply with a curious paragraph.

> More generally, if we envisage a set E of positive real numbers...one can prove that this set possesses a lower limit e; this lower limit is defined *after* the set E; and there is not petitio principii since e is not in general part of E. In certain particular cases, it can happen that e is part of E. In these cases, there is no more a petitio principii, since e is not part of E *in virtue of its definition*, but as a result of a proof posterior to both the definition of E and that of e. (1909a, 199)

The paragraph is odd in that it seems to have nothing to do with predicativity. In it, Poincaré notes the distinction between the greatest lower bound ("lower limit") of a set of reals and a minimum attained by a member of the set; and he warns that the assertion that a greatest lower bound is a minimum requires a further proof. But the impredicative reasoning to which Poincaré objects does not involve any claim that an object, defined by reference to a certain set, is a member of the set purely in virtue of its definition; a further proof is, in fact, supplied.

Zermelo is quick to exploit this oddity. In a reply, he shows that Poincaré's remark, altered by replacing the words "real numbers" and "lower limit" by "sets" and "intersection," would vindicate his logicist-style definition of finite number. He concludes, "I am able to evidence the legitimacy of my proof by thus appealing to the authority of M. Poincaré himself" (Zermelo 1909, 193).

The oddity of the passage can be dispelled, however, by reflection on what Poincaré means by "a proof posterior to both the definition of E and that e." The oddity arises if "posterior" is read as meaning, simply, "subsequent." Yet Poincaré probably intends "posterior" to be understood in light of the vicious circle principle, in which case a stronger stricture emerges. For a proof that is posterior in such a sense cannot contain certain instantiations; in particular, no quantifier in the specification of E can be instantiated by a term that makes reference to e. The original proof of the fundamental theorem does not abide by this stricture. Moreover, when so understood, the passage cannot be used to legitimate Zermelo's definition of finite number. In short, the passage should be read not as expounding the vicious circle principle, but as relying on it. As a result, Poincaré's charge that illicit conclusions are being drawn about an object "in virtue of its definition" is subtler than it appears. Indeed, in the passage Poincaré is using the vicious circle principle to push together the obvious fallacy of assuming a greatest lower bound to be a minimum and the controversial step of applying a definition impredicatively.[10] At bottom, he is urging, the same fallacy lurks in both.

This is further evidenced in Poincaré's characterization, in his last paper on foundations, of impredicative definitions as being composed of two *postulates*: "The object X to be defined has such-and-such relation to all individuals of genre G; X is of genre G" (1912, 154). Now if the genre G is taken to be a given set of reals and the relation is taken to be that of greatest lower bound, then the two postulates yield the conflation of greatest lower bound and minimum, an illegitimate conflation, to be sure, but not because of any impredicativity. However, if the genre represents not the set of reals but the range of quantifiers used in specifying the set, then the two postulates exhibit the form of an impredicative definition.

Among Poincaré's other late writings on foundations, (1909b) is particularly revealing. He starts with an extended discussion of the Richard paradox, taking it to show that whether a given sentence defines a real number, and if so which one, may depend on whether the Richard set has already been defined. Hence "definability" means something different before the Richard set is defined and after. As he puts it, the classification of real numbers into definable or not is "mutable." He then infers generally that impredicative definitions are fallacious, because after such a definition is enunciated, the entity taken to be defined by it may cause a change in meaning in that very definition. Impredicative definition thus engenders fallacies of equivocation. Thus, as before, Poincaré analyzes the Richard paradox as arising not from anything peculiar in the notion of definability, but from difficulties caused by an object introduced by a definition; the emphasis here, though, is on the general lesson learned from the paradox.

For Poincaré, the Richard paradox is not just one antinomy among many to be solved, as Russell viewed it. Rather, its solution exposes the mutability in mathematical definitions generally; hence the solution must apply to all of mathematics. Underlying this step is the idea that all entities that mathematics legitimately treats must be definable. Indeed, Poincaré urges, "the only objects about which it is permissible to reason are those which can be defined in a finite number of words (1909b, 464). Yet since, for Poincaré, definability has no formal analysis, no prior limit can be put on the objects that may, one day, be subject to our reasoning. Poincaré's rejection of the notion of a fixed range of quantifiers is explicit here. A universal theorem does not relate to all objects, imaginable or not, in such a range; rather, it asserts only that each particular case of the theorem—each case defined in a finite number of words that will be considered by the mathematicians or by succeeding generations of mathematicians—can be verified (1909b, 480).

This construal of generality, and the insistence on definability, reflect Poincaré's repeated denial of the existence of the actual infinite. "Every proposition about the infinite should be the translation, the abbreviated statement, of propositions about the finite" (1909b, 482). Poincaré is not here calling for some kind of finitist reduction, however. As a result of the lack of predetermined limits to "definability," the array of propositions about the finite that are abbreviated by a given proposition about the infinite is open-ended rather than fixed.

From all this, the view of mathematics that moves Poincaré may be surmised. Mathematics is a matter of our thinking and verbalizing mathematically. There is nothing beyond our words and thoughts that can anchor or ground the discipline. The idea that our classifications are what we reason about, is a direct rejection of the tenet, central to logicism (but not limited to it), that logic applies to a realm of fixed content. Once that tenet is rejected, logic cannot be thought of as the basic rationality-framing subject that Frege and Russell took it to be; indeed, logic loses all claim to priority. It is no wonder, then, that Poincaré is blind to the advances of the new logic and resists the purely logical sense of presupposition. Similarly, Poincaré's view of the vicious circle principle as governing applications of definitions, rather than as constraining the content of those definitions, becomes explicable. The immutability of our classifications is to be secured by our subsequent behavior: we must not do anything that would cause mutation. Immutability is *not* secured by reference to a logical structure present from the start.

The priority of the finite in Poincaré also stems from his psychologism. The finitude of our minds is, presumably, the basis for his insistence that everything about which we can reason must be definable in a finite number of words, where definability is left unanalyzed, its extent a matter of the conduct of future generations of mathematicians. Moreover, Poincaré pervasively assumes intuitive knowledge of the realm of the finite and the principles governing it. From this ground, the question of the justification of such principles applied to the infinite can first be launched. For logicists, in contrast, the distinction between finite and infinite is first made out after logic is in place; there might be questions of whether infinite structures exist, but there can be no question of whether logic is also "true of" the infinite. Poincaré does, in the end, recognize this contrast. Characteristically, he supports his own stance on priorities by reference to the order in which students can most effectively be instructed in mathematics. This is, as he puts it, "how the human spirit naturally proceeds" (1909b, 482). He concludes,

> M. Russell will doubtless tell me that these are not matters of psychology, but of logic and epistemology. I shall be driven to respond that there is not logic and epistemology independent of psychology. This profession of faith will probably close the discussion, since it will show an irremediable divergence of views.

Poincaré is here being far more perspicacious than many subsequent writers on the foundations of mathematics. The divergent views are not divergent answers to the same questions; they are deeply opposed construals of those questions. What counts as giving "foundations" for mathematics is entirely different for Poincaré than for Russell, Frege, or Zermelo. This difference is not readily adjudicable through the arguments found in the explicit controversies of the literature.

That is not to say that the differences are not susceptible to argument at all; the arguments, though, may have to operate at a far more general philosophical level. It is puzzling that Poincaré's apparently naturalistic view issues in prohibitions of forms of mathematical reasoning that are widely accepted and that have led, by themselves, to no mathematical crises. Thus, it can be suspected that Poincaré's strictures ultimately rest on more than straightforward psychological grounds. In that case, a foothold will be provided for Frege's powerful arguments against psychologism; if it is not to succumb to these arguments, Poincaré's view may well have to be elaborated into a full-blown metaphysical psychologism such as Brouwer's.

For all that, the notion of predicativity is undeniably illuminating and important. One can only admire Poincaré's genius in arriving at a notion that would prove to be technically so fecund. There remain, however, many questions as to its philosophical bases and the proper place of predicativity considerations in the philosophy of logic and mathematics; these questions require, and deserve, much further investigation.[11]

Notes

1. This marks a distinction between Poincaré's writings on geometry and those on foundations. (I use the latter term throughout this paper to mean foundations of analysis and number theory.)

2. Poincaré's psychological rendering of Kantian terms is explicit in an earlier article, where he says that mathematical induction is "an affirmation of a power of the mind" and that "the mind has a direct intuition of this power" (Poincaré 1894, 382).

3. For more on this theme, see Goldfarb (1982). In that article, I also note that the question of whether mathematics is analytic or synthetic in Kant's sense is not an issue for the logicists, since the new logic renders the distinction as Kant drew it philosophically insignificant. (Frege in [1884] redefines "analytic"; Russell in [1902] baldly asserts that logic is synthetic in Kant's sense.) Poincaré is unaware of this. He represents the logicists as claiming that mathematics is analytic in Kant's sense, and he often speaks of his own aim, against the logicists, as being that of showing this claim false.

4. In an article on the teaching of mathematics (1899), Poincaré stresses the importance of imparting intuition, in just this sense.

5. A similar point is made by Russell in explicit reply to Poincaré (Russell 1910, 252).

6. I have in mind here particularly the approach urged by Kitcher (1984). I do not mean to include Quine's very different sort of naturalism.

7. On this point Poincaré was quickly criticized by Giuseppe Peano (1906). The distinction Peano suggests, between semantic paradoxes and set-theoretic paradoxes, was given canonical form in Ramsey (1925).

8. A terminological note: Originally Russell used "predicative" for properties that define classes, i.e., those ϕ such that $[\chi \mid \phi \chi]$ exists. Poincaré proposes in (1906) that properties are predicative only if they contain no vicious circles. In later writings, Poincaré simply *defines* "predicative" to mean containing no vicious circles. Russell follows this later usage, which is the one that has come down to us.

9. The justifications for Russell's ramified theory of types are discussed in detail in Goldfarb (forthcoming).

10. The distinction between greatest lower bound and minimum became a well-known issue in late-nineteenth-century analysis, through critical scrutiny of Dirichlet's fallacious argument for the Dirichlet Principle, which relied on conflation of the two. The problem of finding a correct proof for the principle received much attention, and Poincaré himself contributed to its solution. Monna (1975) gives an excellent historical account of this issue.

11. I am grateful to Burton Dreben, Peter Hylton, and Thomas Ricketts for helpful suggestions and comments.

References

Burali-Forti, Cesare. 1897. Una questione sui numeri transfinit. *Rendiconti del Circolo Matematico di Palermo* 11: 154-64. Translation in (van Heijenoort 1967), pp. 105-11.

Couturat, Louis. 1904. Les principes des mathématiques. *Revue de Métaphysique et de Morale* 12: 19-50, 211-40, 341-83, 664-98, 810-44.

———1906. Pour la logistique (Réponse à M. Poincaré). *Revue de Métaphysique et de Morale* 14: 208-50.

Frege, Gottlob. 1884. *Die Grundlagen der Arithmetik*. Breslau: W. Koebner.

Goldfarb, Warren. 1982. Logicism and Logical Truth. *Journal of Philosophy* 79: 692-95.

——Russell's Reasons for Ramification. Forthcoming in *Bertrand Russell's Philosophy of Science*, ed. A. Anderson and W. Savage.

Hilbert, David. 1904. Über die Grundlagen der Logik und der Arithmetik. *Verhandlungen des Dritten Internationalen Mathematiker-Kongresses in Heidelberg vom 8 bis 13 August 1904* (Leipzig, 1905), pp. 174-85. Translation in (van Heijenoort 1967), pp. 129-38.

Kitcher, Philip. 1984. *The Nature of Mathematical Knowledge*, Oxford: Oxford University Press.

Monna, A. F. 1975. *Dirichlet's Principle*. Utrecht: Oosthoer, Scheltema, and Hoikema.

Parsons, Charles D. 1965. Frege's Theory of Number. In *Philosophy in America*, ed. Max Black. Ithaca, N.Y.: Cornell University Press, pp. 180-203.

Peano, Guiseppe. 1906. Super theorema de Cantor-Bernstein et additione. *Rivista di Matematica* 8: 136-57.

Poincaré, Henri. 1893. Le continue mathématique. *Revue de Métaphysique et de Morale* 1; 26-34.

——1894. Sur la nature du raisonnement mathématique. *Revue de Métaphysique et de Morale* 2: 371-84.

——1899. La logique et l'intuition dans la science mathématique et dans enseignement. *L'Enseignement Mathématique* 1: 157-62.

——1900. Dû role de l'intuition et de la logique en mathématiques. *Compte rendue du Deuxième congrès international des mathématiciens tenu à Paris due 6 au 12 août 1900* (Paris: Gauthier-Villars, 1902), pp. 115-30.

——1902. *La science et l'hypothese*. Paris: Flammarion.

——1905. Les mathématiques et la logique. *Revue de Métaphysique et de Morale* 13: 815-35; 14 (1906): 17-34.

——1906. Les mathématiques et la logique. *Revue de Métaphysique et de Morale* 14: 294-317.

——1908. *Science et méthode*. Paris: Flammarion.

——1909a. Réflexions sur les deux notes précedentes. *Acta Mathematica* 32: 195-200.

——1909b. La logique de l'infini. *Revue de Métaphysique et de Morale* 17: 461-82.

——1912. Les mathématiques et la logique. In *Dernières pensées*. Paris: Flammarion, 1913, pp. 143-62.

Ramsey, Frank Plumpton. 1925. The Foundations of Mathematics. *Proceedings of the London Mathematical Society* 25: 338-84.

Richard, Jules. 1905. Les principes des mathématiques et le problème des ensembles. *Revue Générale des Sciences Pures et Appliquées* 16: 541. Translation in (van Heijenoort 1967), pp. 142-44.

Russell, Bertrand. 1903. *The Principles of Mathematics*. Cambridge: Cambridge University Press.

——1905. On Some Difficulties in the Theory of Transfinite Numbers and Order Types. *Proceedings of the London Mathematical Society* 4: 29-53. Page citations are to reprint in (Russell 1973), pp. 135-64.

——1908. Mathematical Logic as Based on the Theory of Types. *American Journal of Mathematics* 30: 222-62. Page citations are to reprint in (van Heijenoort 1967), pp. 150-182.

——1910. The Theory of Logical Types. In (Russell 1973), pp. 215-52.

——1973. *Essays in Analysis*. Ed. D. Lackey. New York: Braziller.

van Heijenoort, Jean. 1967. *From Frege to Gödel: A Source Book in Mathematical Logic*. Cambridge, Mass.: Harvard University Press.

Weyl, Hermann. 1918. *Das Kontinuum*. Leipzig: Veit.

Whitehead, Alfred North. 1902. On Cardinal Numbers. *American Journal of Mathematics* 24: 367-94.

Zermelo, Ernst. 1908. Neuer Beweis für die Möglichkeit einer Wohlordnung. *Mathematische Annalen* 65: 107-28. Page citations are to translation in (van Heijenoort 1967), pp. 183-98.

——1909. Sur les ensembles finis et le principe de l'induction complète. *Acta Mathematica* 32: 183-93.

Logical Truth and Analyticity
in Carnap's
"Logical Syntax of Language"

Throughout his philosophical career, Carnap places the foundations of logic and mathematics at the center of his inquiries: he is concerned above all with the Kantian question "How is mathematics (both pure and applied) possible?"[1] Although he changes his mind about many particular issues, Carnap never gives up his belief in the importance and centrality of this question—nor does he ever waver in his conviction that he has the answer: the possibility of mathematics and logic is to be explained by a sharp distinction between formal and factual, analytic and synthetic truth. Thus, throughout his career Carnap calls for, and attempts to provide

> an explication for the distinction between logical and descriptive signs and that between logical and factual truth, because it seems to me that without these distinctions a satisfactory methodological analysis of science is not possible.[2]

For Carnap, it is this foundation for logic and mathematics that is distinctive of logical—as opposed to traditional—empiricism. As he puts it in his intellectual autobiography: "It became possible for the first time to combine the basic tenet of empiricism with a satisfactory explanation of the nature of logic and mathematics."[3] In particular, we can avoid the "non-empiricist" appeal to "pure intuition" or "pure reason" while, at the same time, avoiding the naive and excessively empiricist position of J. S. Mill.[4]

Indeed, from this point of view, Carnap's logicism and especially his debt to Frege become even more important than his empiricism and his connection with the Vienna Circle. The point has been put rather well, I think, by Beth in his insightful article in the Schilpp volume:

> His connection with the Vienna Circle is certainly characteristic of his way of thinking, but by no means did it determine his philosophy. It

seems to me that the influence of Frege's teachings and published work has been much deeper. In fact, this influence must have been decisive, and the development of Carnap's ideas may be considered as characteristic of Frege's philosophy as well.[5]

Carnap endorses this assessment in his reply to Beth,[6] and it is quite consistent with what he says about his debt to Frege elsewhere.[7]

Yet when one looks at *Logical Syntax*,[8] which is clearly Carnap's richest and most systematic discussion of these foundational questions, the idea that Carnap is continuing Frege's logicism appears to be quite problematic. Not only does Carnap put forward an extreme "formalistic" (purely syntactic) conception of the language of mathematics—a conception that, as he explicitly acknowledges, is derived from Hilbert and would be anathema to Frege (§84)—his actual construction of mathematical systems exhibits none of the characteristic features of logicism. No attempt is made to define the natural numbers: the numerals are simply introduced as primitive signs in both of Carnap's constructed languages. Similarly, no attempt is made to derive the principle of mathematical induction from underlying logical laws: in both systems it is introduced as a primitive axiom (in Language I it appears as a primitive [schematic] inference rule [R14 of §12]). In short, Carnap's construction of mathematics is thoroughly axiomatic and, as he explicitly acknowledges (§84), appears to be much closer to Hilbert's formalism than to Frege's logicism.

Carnap's official view of this question is that he is putting forward a reconciliation of logicism and formalism, a combination of Frege and Hilbert that somehow captures the best of both positions (§84).[9] In light of the above, however, it must strike the reader as doubtful that anything important in Frege's position has been retained. For that matter, although Carnap employs formalist rhetoric and an explicitly axiomatic formulation of mathematics, nothing essential to Hilbert's foundational program appears to be retained either. Thus, no attempt is made to give a finitary consistency or conservativeness proof for classical mathematics. Carnap takes Gödel's results to show that the possibility of such a proof is "at best very doubtful," and he puts forward a consistency proof in a metalanguage essentially richer than classical mathematics (containing, in effect, classical mathematics plus a truth-definition for classical mathematics), which, as Carnap again explicitly acknowledges (§§34h, 34i), is therefore of doubtful foundational significance. At this point, then, Carnap's claim to reconcile Frege and Hilbert appears hollow indeed. What he has actually done, it seems, is thrown away all that is most interesting and characteristic in both views.

Such an evaluation would be both premature and fundamentally unfair, however. To see why, we must look more closely at the centerpiece of Carnap's philosophy—his conception of *analytic truth*—and how that conception evolves from Frege's while incorporating post-Fregean advances in logic: in particular, advances due to Hilbert and Gödel.

The first point to bear in mind is the familiar one that Frege's construction of arithmetic is not simply the embedding of a special mathematical theory (arithmetic) in a more general one (set theory). Frege's *Begriffsschrift* is not intended to be a mathematical theory at all; rather, it is to function as the logical framework that governs all rational thinking (and therefore all particular theories) whatsoever. As such, it has no special subject matter (the universe of sets, for example) with which we are acquainted by "intuition" or any other special faculty. The principles and theorems of the *Begriffsschrift* are implicit in the requirements of any coherent thinking about anything at all, and this is how Frege's construction of arithmetic within the *Begriffsschrift* is to provide an answer to Kant: arithmetic is in no sense dependent on our spatiotemporal intuition but is built in to the most general conditions of thought itself. This, in the end, is the force of Frege's claim to have established the analyticity of arithmetic.

But why should we think that the principles of Frege's new logic delimit the most general conditions of all rational thinking? Wittgenstein's *Tractatus* attempts to provide an answer: this new logic is itself built in to any system of representation we are willing to call a language. For, from Wittgenstein's point of view, the *Begriffsschrift* rests on two basic ides: Frege's function/argument analysis of predication and quantification, and the iterative construction of complex expressions from simpler expressions via truth-functions. So any language in which we can discern both function/argument structure—in essence, where there are grammatical categories of intersubstitutable terms—and truth-functional iterative constructions will automatically contain all the logical forms and principles of the new logic as well. Since it is plausible to suppose that any system of representation lacking these two features cannot count as a language in any interesting sense, it makes perfectly good sense to view the new logic as delimiting the general conditions of any rational thinking whatsoever. For the new logic is now seen as embodying the most general conditions of meaningfulness (meaningful representation) as such.[10]

Carnap enthusiastically endorses this Wittgensteinian interpretation of Frege's conception of analyticity, and he is quite explicit about his debt to Wittgenstein throughout *Logical Syntax* (§§14, 34a, 52) and throughout his career.[11] Yet, at the same time, Carnap radically transforms the conception of the *Tractatus*, and he does this by emphasizing themes that are only implicit in Wittgenstein's thought. It is here, in fact, that Carnap brings to bear the work of Hilbert and Gödel in a most decisive fashion.

First of all, Carnap interprets Wittgenstein's elucidations of the notions of language, logical truth, logical form, and so on as definitions in *formal syntax*. They are themselves formulated in a metalanguage or "syntax-language," and they concern the syntactic structure either of some particular object-language or of languages in general:

> All questions of logic (taking this word in a very wide sense, but excluding all empirical and therewith all psychological reference) belong to syntax. *As soon as logic is formulated in an exact manner, it turns out to be nothing other than the syntax either of some particular language or of languages in general.* (§62)

This syntactic interpretation of logic is of course completely foreign to Wittgenstein himself. For Wittgenstein, there can be only one language— the single interconnected system of propositions within which everything that can be said must ultimately find a place; and there is no way to get "outside" this system so as to state or describe its logical structure: there can be no syntactic metalanguage. Hence logic and all "formal concepts" must remain ineffable in the *Tractatus*.[12] Yet Carnap takes the work of Hilbert and especially Gödel to have decisively refuted these Wittgensteinian ideas (see especially §73). Syntax (and therefore logic) can be exactly formulated; and, in particular, if our object-language contains primitive recursive arithmetic, the syntax of our language (and every other language) can be formulated within this language itself (§18).

Secondly, Carnap also clearly recognizes that the linguistic or "syntactic" conception of analyticity developed in the *Tractatus* is much too weak to embrace all of classical mathematics or all of Frege's *Begriffsschrift*. For the two devices of function/argument structure (substitution) and iterative truth-functional construction that were seen to underly Frege's distinctive analysis of predication and quantification do not lead us to the rich higher-order principles of classical analysis and set theory. As

Gödel's arithmetization of syntax again decisively shows, all that is forth-coming is primitive recursive arithmetic. Of course, the *Tractatus* is itself quite clear on the restricted scope of its conception of logic and mathematics in comparison with Frege's (and Russell's) conception. Witt-genstein's response to this difficulty is also all too clear: so much the worse for classical mathematics and set theory.[13]

Carnap's own response is quite different, however, for his aim throughout is not to replace or restrict classical mathematics but to pro-vide it with a philosophical foundation: to answer the question "How is classical mathematics possible?" And it is here that Carnap makes his most original and fundamental philosophical move: we are to give up the "ab-solutist" conception of logical truth and analyticity common to Frege and the *Tractatus*. For Carnap, there is no such thing as *the* logical framework governing all rational thought. Many such frameworks, many such systems of what Carnap calls *L-rules* are possible: and all have an equal claim to "correctness." Thus, we can imagine a linguistic framework whose L-rules are just those of primitive recursive arithmetic itself (such as Carnap's Language I); a second whose L-rules are given by set theory or some higher-order logic (such as Carnap's Language II); a third whose L-rules are given by intuitionistic logic; a fourth whose L-rules include part of what is in-tuitively physics (such as physical geometry: cf. §50); and so on. As long as the L-rules in question are clearly and precisely delimited within for-mal syntax, any such linguistic framework defines a perfectly legitimate language (*Principle of Tolerance*):

> *In logic there are no morals.* Everyone is at liberty to build up his own logic, i.e., his own form of language, as he wishes. All that is required of him is that, if he wishes to discuss it, he must state his methods clear-ly, and give syntactical rules instead of philosophical arguments. (§17)

Thus, Carnap's basic move is to relativize the "absolutist" and essential-ly Kantian program of Frege and the *Tractatus*.

Carnap's general strategy is then concretely executed as follows. First, within the class of all possible linguistic frameworks, one particular such framework stands out for special attention. A framework whose L-rules are just those of primitive recursive arithmetic has a relatively neutral and uncontroversial status—it is common to "Platonists," "intuitionists," and "constructivists" alike (§16); and, moreover, as Gödel's researches have shown, such a "minimal" framework is nonetheless adequate for for-mulating the logical syntax of any linguistic framework whatsoever—

including its own. So this linguistic framework, Carnap's Language I, can serve as an appropriate beginning and "fixed point" for all subsequent syntactic investigation—including the investigation of much richer and more controversial frameworks.

One such richer framework is Carnap's Language II: a higher-order system of types over the natural numbers including (higher-order) principles of induction, extensionality, and choice (§30). This framework will then be adequate for much of classical mathematics and mathematical physics. Nevertheless, despite the strength of this framework, we can exactly describe its logical structure within logical syntax; and, in particular, we can show that the mathematical principles in question are *analytic-in-Language-II*—in Carnap's technical terminology, they are included in the L-rules ("logical" rules), not the P-rules ("physical" rules) of Language II. So we can thereby explain the "mathematical knowledge" of anyone who adopts (who speaks, as it were) Language II. Such knowledge is implicit in the linguistic framework definitive of meaningfulness for such a person, and it is therefore formal, not factual.

We are now in a position to appreciate the extent to which Carnap has in fact combined the insights of Frege and Hilbert and has, in an important sense, attempted a genuine reconciliation of logicism and formalism. From Frege (and Wittgenstein), Carnap takes the idea that the possibility of mathematics is to be explained by showing how its principles are implicit in the general conditions definitive of meaningfulness and rationality. Mathematics is built in to the very structure of thought and language and is thereby forever distinguished from merely empirical truth. By relativizing the notion of logical truth, Carnap attempts to preserve this basic logicist insight in the face of all the well-known technical difficulties; and this is why questions of reducing mathematics to something else—to "logic" in some antecedently fixed sense—are no longer relevant. From Hilbert (and Gödel), Carnap takes the idea that primitive recursive arithmetic constitutes a privileged and relatively neutral "core" to mathematics and, moreover, that this neutral "core" can be used as a "metalogic" for investigating much richer and more controversial theories. The point, however, is not to provide consistency or conservativeness proofs for classical mathematics, but merely to delimit its logical structure: to show that the mathematical principles in question are analytic in a suitable language. Carnap hopes thereby to avoid the devastating impact of Gödel's Incompleteness Theorems.

Alas, however, it was not meant to be. For Gödel's results decisively undermine Carnap's program after all. To see this, we have to be a bit more explicit about the details of the program. For Carnap, a language or linguistic framework is syntactically specified by its formation and transformation rules, where these latter specify both axioms and rules of inference. The language in question is then characterized by its consequence-relation, which is defined in familiar ways from the underlying transformation rules. Now such a language or linguistic framework will contain both formal and empirical components, both "logical" and "physical' rules. Language II, for example, will not only contain classical mathematics but classical physics as well, including "physical" primitive terms (§40)—such as a functor representing the electromagnetic field— and "physical" primitive axioms (§82)—such as Maxwell's equations. The task of defining analytic-for-a-language, then, is to show how to distinguish these two components: in Carnap's technical terminology, to distinguish L-rules from P-rules, L-consequence from P-consequence (§§51, 52).

How is this distinction to be drawn? Carnap proceeds on the basis of a prior distinction between *logical* and *descriptive* expressions (§50). Intuitively, logical expressions include logical constants in the usual sense (connectives and quantifiers) plus primitive expressions of arithmetic (the numerals, successor, addition, multiplication, and so on). Given the distinction between logical and descriptive expressions, we then define the analytic (L-true) sentences of a language as those theorems (L- or P-consequences of the null set) that remain theorems under all possible substitutions of *descriptive* expressions (§51). In other worlds, what we might call "descriptive invariance" separates the L-consequences from the wider class of consequences *simpliciter*. But how is the distinction between logical and descriptive expressions itself to be drawn? Here Carnap appeals to the *determinacy* of logic and mathematics (§50): logical expressions are just those expressions such that every sentence built up from them alone is decided one way or another by the rules (L-rules or P-rules) of the language. That is, every sentence built up from logical expressions alone is provable or refutable on the basis of these rules. In the case of descriptive expressions, by contrast, although some sentences built up from them will no doubt be provable or refutable as well (in virtue of P-rules, for example), this will not be true for all such sentences—for sentences ascribing particular values of the electromagnetic field to particular space-time points, for example. In this way, Carnap intends to capture the idea that logic and mathematics are thoroughly a priori.

It is precisely here, of course, that Gödelian complications arise. For, if our consequence-relation is specified in terms of what Carnap calls *definite* syntactic concepts—that is, if this relation is recursively enumerable—then even the theorems of primitive recusive arithmetic (Language I) fail to be analytic; and the situation is even worse, of course, for full classical mathematics (Language II). Indeed, as we would now put it, the set of (Gödel numbers of) analytic sentences of classical first-order number theory is not even an arithmetical set, so it certainly cannot be specified by definite (recursive) means. Carnap himself is perfectly aware of these facts, and this is why he explicitly adds what he calls *indefinite* concepts to syntax (§45). In particular, he explicitly distinguishes (recursive and recursively enumerable) d-terms or rules of derivation from (in general nonarithmetical) c-terms or rules of consequence (§§47, 48).

Moreover, it is here that Carnap is compelled to supplement his "syntactic" methods with techniques we now associate with the name of Tarski: techniques we now call "semantic." In particular, the definition of *analytic-in-Language-II* is, in effect, a truth-definition for classical mathematics (§§34a-34d). Thus, if we think of Language II as containing all types up to ω (all finite types), say, our definition of *analytic-in-Language-II* will be formulated in a stronger metalanguage containing quantification over arbitrary sets of type ω as well. In general, then, Carnap's definition of analyticity for a language of any order will require quantification over sets of still higher order. The extension of analytic-in-L for any L will therefore depend on how quantifiers in a metalanguage essentially richer than L are interpreted; the interpretation of quantifiers in this metalanguage can only be fixed in a still stronger language; and so on (§34d).[14]

But why should this circumstance cause any problems for Carnap? After all, he himself is quite clear about the technical situation; yet he nevertheless sees no difficulty whatever for his logicist program. It is explicitly granted that Gödel's Theorem thereby undermines Hilbert's formalism; but why should it refute Frege's logicism as well? The logicist has no special commitment to the "constructive" or primitive recursive fragment of mathematics: he is quite happy to embrace all of classical mathematics. Indeed, Carnap, in his Principle of Tolerance, explicitly rejects all questions concerning the legitimacy or justification of classical mathematics.

What the logicist wishes to maintain is not a reduction or justification of classical mathematics via its "constructive" fragment (as Hilbert attempts in his finitary consistency and conservativeness proofs), but simply that classical mathematics is analytic: that it is true in virtue of language or meaning, not fact. So why should Gödel's Theorem undermine *this* conception?[15]

To appreciate the full impact of Gödel's results here, it is necessary to become clearer on the fundamental differences between Carnap's conception of analyticity or logical truth and that of his logicist predecessors. For precisely these differences are obscured by the notion of truth-in-virtue-of-meaning or truth-in-virtue-of-language—especially as this notion is wielded by Quine in his polemic against Carnap. Thus, the early pages of "Two Dogmas" distinguish two classes of logical truths.[16] A *general* logical truth—such as "No unmarried man is married"—is "a statement that is true and remains true under all reinterpretations of its components other than the logical particles." An *analytic* statement properly so-called—such as "No bachelor is unmarried"—arises from a general logical truth by substitution of synonyms for synonyms. This latter notion is then singled out for special criticism for relying on a problematic conception of meaning (synonymy); and this is the level on which Quine engages with Carnap.

The first point to notice is that *these* Quinean criticisms are indeed relevant to Carnap, but not at all to his logicist predecessors—for their analytic truths simply do not involve nonlogical constants in this sense. Thus, whereas Carnap's languages contain primitive arithmetical signs (the numerals, successor, addition, and so), Frege's *Begriffsschrift* and Russell's *Principia* do not. In these systems, the arithmetical signs are of course defined via the logical notions of truth-functions and quantifiers (including quantifiers over higher types). So Carnap needs to maintain that arithmetical truths are in some sense true in virtue of the meanings of 'plus' and 'times', say, whereas Frege and Russell do not—these latter signs simply do not occur in their systems.

But what about the logical notions Frege and Russell assume: the notions we now call logical constants and Quine calls logical particles? Does the same problem not arise for them? Do we not have to assume that the general logical truths of the *Begriffsschrift* and *Principia* are true in virtue of the meanings of 'and', 'or', 'not', 'all', and 'some'? According to Wittgenstein's position in the *Tractatus*, the so-called logical constants do not, properly speaking, have meaning at all. They are not words like others for which a "theory of meaning" is either possible or necessary. Indeed, for Wittgenstein, "there are no... 'logical constants' [in Frege's and Russell's sense]" (5.4). Rather, for any language, with any vocabulary of "constants" or primitive signs whatsoever, there are the purely combinatorial possibilities of building complex expressions from simpler expressions and of substituting one expression for another within such a complex expression. These abstract combinatorial possibilities are all that the so-called logical constants express: "Whenever there is compositeness, argument and function are present, and where these are present, we already have all the logical constants" (5.47). Thus, for Wittgenstein, logical truths are not true in virtue of the meanings of particular words—whether of 'and', 'or', 'not', or any others—but solely in virtue of "logical form" the general combinatorial possibilities common to all languages regardless of their particular vocabularies.

Now this conception—that logical truths are true in virtue of "logical form," and not in virtue of "meaning" in anything like Quine's sense— is essential to the *antipsychologism* of the *Tractatus*. For, if logic depends on the meanings of particular words—even "logical words" like 'and', 'not', and so on—then it rests, in the last analysis, on psychological facts about how these words are actually used. It then becomes possible to contest these alleged facts and to argue, for example, that a correct theory of meaning supports intuitionistic rather than classical logic, say. For Wittgenstein, this debate, in these terms, simply does not make sense. Logic rests on no facts whatsoever, and certainly not on facts about the meanings or usages of English (or German) words. Rather, logic rests on the abstract combinatorial possibilities common to all languages as such. In this sense, logic is absolutely presuppositionless and thus absolutely uncontentious.

The problem for this Tractarian conception has nothing at all to do with the Quinean problem of truth-in-virtue-of-meaning or truth-in-virtue-of-language. Rather, the problem is that the logic realizing this conception is much too weak to accomplish the original aim of logicism: explaining how mathematics—classical mathematics—is possible. Frege's *Begriffsschrift* cannot provide the required realization, because of the paradoxes; and neither can Russell's *Principia*, because of the need for axioms like infinity and reducibility. The *Tractatus* itself ends up with a conception of logic that falls somewhere between truth-functional logic and a ramified type-theory without infinity or reducibility; and it ends up with a conception of mathematics apparently limited to primitive recursive arithmetic.[17] So Wittgenstein may have indeed achieved a genuinely presuppositionless standpoint, but only by failing completely to engage the foundational question that originally motivated logicism.

At this point Carnap has an extremely ingenious idea. We retain Wittgenstein's purely combinatorial conception of logic, but it is implemented at the level of the *metalanguage* and given an explicit subject matter: namely, the syntactic structure of any language whatsoever. At the same time, precisely because logic in this sense is implemented at the level of the metalanguage not the object-language, it no longer has the impoverishing and stultifying effect evident in the *Tractatus*. For, although our purely syntactic metalanguage is to have a very weak, and therefore uncontroversial, underlying logic, we can nonetheless use it to describe—but not to justify or reduce—much stronger systems: in particular, classical mathematics. In this way Carnap hopes to engage, and in fact, to neutralize the basic foundational question. Logic, in the sense of logical syntax, can in no way adjudicate this question. Indeed, from Carnap's point of view, there is no substantive question to be adjudicated. Rather, logic in this sense constitutes a neutral metaperspective from which we can represent the consequences of adopting any and all of the standpoints in question: "Platonist," "constructivist," "intuitionist," and so on.

Corresponding to any one of these standpoints is a notion of logic (analyticity) in a second sense: a notion of analytic-in-L. Sentences analytic-in-L are not true in virtue of the abstract combinatorial possibilities definitive of languages in general, but in virtue of conventions governing this particular L—specifically, on those linguistic conventions that establish some words as "logical" and others as "descriptive." Hence it is at this point, and only at this point, that we arrive at the Quinean problem of truth-in-virtue-of-meaning. And it is at this point, then, that Carnap's logicism threatens to collapse into its dialectical opponent, namely psychologism.

Carnap hopes to avoid such a collapse by rigorously enforcing the distinction between *pure* and *applied* (descriptive) syntax. The latter, to be sure, is an empirical discipline resting ultimately on psychological facts: it aims to determine whether and to what extent the speech dispositions of a given speaker or community realize or exemplify the rules of a given abstractly characterized language L—including and especially those rules definitive of the notion of analytic-in-L. Pure syntax, on the other hand, is where we develop such abstract characterizations in the first place. We are concerned neither with the question of which linguistic framework is exemplified by a given community or speaker, nor with recommending one linguistic framework over others—classical over "constructive" mathematics, say. Rather, our aim is to step back from all such questions and simply articulate the consequences of adopting any and all such frameworks. The propositions of pure syntax are therefore logical or analytic propositions in the first sense: propositions of the abstract, purely combinatorial metadiscipline of logical syntax.

Here is where Gödel's Theorem strikes a fatal blow. For, as we have seen, Carnap's general notion of analytic-in-L is simply not definable in logical syntax so conceived, that is, conceived in the above "Wittgensteinian" fashion as concerned with the general combinatorial properties of any language whatsoever. *Analytic-in-L* fails to be captured in what Carnap calls the *"combinatorial analysis. . .* of finite, discrete serial structures" (§2): that is, primitive recursive arithmetic. Hence the very notion that supports, and is indeed essential to, Carnap's logicism simply does not occur in pure syntax as he understands it. If this notion is to have any place at all, then, it can only be within the explicitly empirical and psychological discipline of applied syntax; and the dialectic leading to Quine's challenge is now irresistible.[18] In this sense, Gödel's results knock away the last slender reed on which Carnap's logicism (and antipsychologism) rests.

In the end, what is perhaps most striking about *Logical Syntax* is the way it combines a grasp of the technical situation that is truly remarkable in 1934 with a seemingly unaccountable blindness to the full implications of that situation. Later, under Tarski's direct influence, Carnap of course came to see that his definition of analytic-in-L is not a properly "syntactic" definition at all; and in *Introduction to Semantics*[19] he officially renounces the definitions of *Logical Syntax* (§39) and admits that no satisfactory delimitation of L-truth in "general semantics" is yet known (§§13, 16). Instead, he offers two tentative suggestions: either we can suppose that our metalanguage contains a necessity operator, so that a sentence S is analytic-in -L just in case we have N(S is true) in M^L; or we can suppose that we have been already given a distinction between logical and factual truth in our metalanguage, so that a sentence S is analytic-in-L just in case 'S is true' is analytic-in-M^L (§116).

From our present, post-Quinean vantage point, the triviality and circularity of these suggestions is painfully obvious; but it was never so for Carnap. He never lost his conviction that the notion of analytic truth, together with a fundamentally logicist conception of mathematics, stands firm and unshakable. And what this shows, finally, is that the Fregean roots of Carnap's philosophizing run deep indeed. Unfortunately, however, they have yet to issue in their intended fruit.

Notes

1. I am indebted to helpful suggestions, advice, and criticism from Warren Goldfarb, Peter Hylton, Thomas Ricketts, and the late Alberto Coffa.

2. R. Carnap, "Reply to E. W. Beth on Constructed Language Systems," in *The Philosophy of Rudolf Carnap*, ed. P. Schilpp. (La Salle: Open Court, 1963), p. 932.

3. R. Carnap, "Intellectual Autobiography," in *Philosophy of Rudolf Carnap*, p. 47.

4. Ibid.

5. E. W. Beth, "Carnap's Views on the Advantages of Constructed Systems over Natural Languages in the Philosophy of Science," in *Philosophy of Rudolf Carnap*, pp. 470-71.

6. "Reply to E. W. Beth," pp. 927-28.

7. See, for example, Carnap's "Intellectual Autobiography," pp. 4-6; and the preface to the second edition of *The Logical Structure of the World*, trans. R. George (Berkeley: University of California Press, 1967), pp. v-vi.

8. R. Carnap, *The Logical Syntax of Language*, trans. A. Smeaton (London: Routledge, 1937). References are given parenthetically in the text via section numbers.

9. Cf. also Beth, "Carnap's Views on Constructed Systems," pp. 475-82; and Carnap, "Reply to E. W. Beth," p. 928.

10. For the relationship between Frege and the *Tractatus*, see especially T. Ricketts, "Frege, the *Tractatus*, and the Logocentric Predicament," *Nous* 19 (1985): 3-15.

11. See especially his "Intellectual Autobiography," p. 25.

12. *Tractatus Logico-Philosophicus*, 4.12-4.128; this is the basis for the "logocentric predicament" of note 10 above. Further references to Wittgenstein in the text are also to the *Tractatus*. See also W. Goldfarb, "Logic in the Twenties: the Nature of the Quantifier," *Journal of Symbolic Logic* 44 (1979): 351-68.

13. For the rejection of set theory, see 6.031. Frege's and Russell's impredicative definition of the ancestral is rejected at 4.1273; the axiom of reducibility is rejected at 6.1232-6.1233. Apparently, then, we are limited to (at most) predicative analysis—and even this may be too much because of the doubtful status of the axiom of infinity (5.535). The discussion of mathematics and logic at 6.2-6.241 strongly suggests a conception of mathematics limited to primitive recursive arithmetic.

14. Cf. Beth, "Carnap's Views on Constructed Systems," who takes this situation to undermine both Carnap's logicism and his formalism—and, in fact, to require an appeal to some kind of "intuition." Carnap's ingenuous response in his "Reply to E. W. Beth," pp. 928-30, is most instructive.

15. I am indebted to Thomas Ricketts and especially to Alberto Coffa for pressing me on this point.

16. W. V. Quine, "Two Dogmas of Empiricism," in *From a Logical Point of View* (New York: Harper, 1963), pp. 22-23.

17. See note 13 above.

18. See T. Ricketts, "Rationality, Translation, and Epistemology Naturalized," *Journal of Philosophy* 79 (1982): 117-37.

19. R. Carnap, *Introduction to Semantics* (Cambridge, Mass.: Harvard University Press, 1942); parenthetical references in the text are to section numbers.

——————— Gregory H. Moore ———————

The Emergence of First-Order Logic

1. Introduction

To most mathematical logicians working in the 1980s, first-order logic is the proper and natural framework for mathematics. Yet it was not always so. In 1923, when a young Norwegian mathematician named Thoralf Skolem argued that set theory should be based on first-order logic, it was a radical and unprecedented proposal.

The radical nature of what Skolem proposed resulted, above all, from its effect on the notion of categoricity. During the 1870s, as part of what became known as the arithmetization of analysis, Cantor and Dedekind characterized the set $\mathbb{R}$ of real numbers (up to isomorphism) and thereby found a categorical axiomatization for $\mathbb{R}$. Likewise, during the 1880s Dedekind and Peano categorically axiomatized the set $\mathbb{N}$ of natural numbers by means of the Peano Postulates.[1] Yet in 1923, when Skolem insisted that set theory be treated within first-order logic, he knew (by the recently discovered Löwenheim-Skolem Theorem) that in first-order logic neither set theory nor the real numbers could be given a categorical axiomatization, since each would have both a countable model and an uncountable model. A decade later, Skolem (1933, 1934) also succeeded in proving, by the construction of a countable nonstandard model, that the Peano Postulates do not uniquely characterize the natural numbers within first-order logic. The Upward Löwenheim-Skolem Theorem of Tarski, the first version of which was published as an appendix to (Skolem 1934), made it clear that no axiom system having an infinite model is categorical in first-order logic.

The aim of the present article is to describe how first-order logic gradually emerged from the rest of logic and then became accepted by mathematical logicians as the proper basis for mathematics—despite the opposition of Zermelo and others. Consequently, I have pointed out where a logician used first-order logic and where, as more frequently occurred, he employed some richer form of logic. I have distinguished between a logician's use of first-order logic (where quantifiers range only over individuals), second-order logic (where quantifiers can also range over sets or relations), ω-order logic (essentially the simple theory of types), and various infinitary logics (having formulas of infinite length or rules of inference with infinitely many premises).

It will be shown that several versions of second-order logic (sometimes including an infinitary logic) were common before *Principia Mathematica*. First-order logic—stripped of all infinitary operations—emerged only with Hilbert in (1917), where it remained a subsystem of logic, and with Skolem in (1923), who treated it as *all* of logic.

During the nineteenth and early twentieth centuries, there was no generally accepted classification of the different kinds of logic, much less an acceptance of one kind as the correct and proper one. (Infinitary logic, in particular, appeared in many guises but did not begin to develop as a distinct branch of logic until the mid-1950s.) Only very gradually did it become evident that there is a reasonable such classification, as opposed merely to the cornucopia of different logical systems introduced by various researchers. Likewise, it only became clear over an extended period that in logic it is important to distinguish between syntax (including such notions as formal language, formula, proof, and consistency) and semantics (including such notions as truth, model, and satisfiability). This distinction led, in time, to Gödel's Incompleteness Theorem (1931) and thus to understanding the limitations of even categorical axiom systems.

2. Boole: The Emergence of Mathematical Logic

Before the twentieth century, there was no reason to believe that the kind of logic that a mathematician used would affect the mathematics that he did. Indeed, during the first half of the nineteenth century the Aristotelian syllogism was still regarded as the ultimate form of all reasoning. When the continuous development of mathematical logic began in 1847, with the publication of George Boole's *The Mathematical Analysis of Logic*, Aristotelian logic was treated as one interpretation of a logical calculus. Like his predecessors, Boole understood logic as "the laws of thought," and so his work lay on the boundary of philosophy, psychology, and mathematics (1854, 1).

After setting up an uninterpreted calculus of symbols and operations, Boole then gave it various interpretations—one in terms of classes, one in terms of propositions, and one in terms of probabilities. This calculus was based on conventional algebra suitably modified for use in logic. He pursued the analogy with algebra by introducing formal symbols for the four arithmetic operations of addition, subtraction, multiplication, and division as well as by using functional expansions. Deductions took the form of equations and of the transformation of equations. The symbol *v* played the role of an indefinite class that served in place of an existential quantifier (1854, 124).

Eventually, a modified version of Boole's logic would become propositional logic, the lowest level of modern logic. When Boole wrote, however, his system functioned in effect as all of logic, since, within this system, Aristotelian syllogistic logic could be interpreted.

Boole influenced the development of logic by his algebraic approach, by giving a calculus for logic, and by supplying various interpretations for this calculus. His algebraic approach was a distinctly British blend, following in the footsteps of Peacock's "permanence of form" with its emphasis on the laws holding for various algebraic structures. But it would soon take root in the United States with Peirce and in Germany with Schröder.

3. Peirce and Frege: Separating the Notion of Quantifier

A kind of logic that was adequate for mathematical reasoning began to emerge late in the nineteenth century when two related developments occurred: first, relations and functions were introduced into symbolic logic; second, the notion of quantifier was disentangled from the notion of propositional connective and was given an appropriate symbolic representation. These two developments were both brought about independently by two mathematicians having a strong philosophical bent—Charles Sanders Peirce and Gottlob Frege.

Nevertheless, the notion of quantifier has an ancient origin. Aristotle, whose writings (above all, the *Prior Analytics*) marked the first appearance of formal logic, made the notions of "some" and "all" central to logic by formulating the assertoric syllogism. Despite the persistence for over two thousand years of the belief that all reasoning can be formulated in syllogisms, they remained in fact a very restrictive mode of deduction.

What happened circa 1880, in the work of Peirce and Frege, was not that the notion of quantifier was invented but rather that it was separated from the Boolean connectives on the one hand and from the notion of predicate on the other.

Peirce's contributions to logic fell squarely within the Boolean tradition. In (1865), Peirce modified Boole's system in several ways, reinterpreting Boole's + (logical addition) as union in the case of classes and as inclusive "or" in the case of propositions. (Boole had regarded $A + B$ as defined only when A and B are disjoint.)

Five years later, Peirce investigated the notion of relation that Augustus De Morgan had introduced into formal logic in (1859) and began to adapt this notion to Boole's system:

> Boole's logical algebra has such singular beauty, so far as it goes, that it is interesting to inquire whether it cannot be extended over the whole realm of formal logic, instead of being restricted to that simplest and least useful part of the subject, the logic of absolute terms, which, when he wrote [1854], was the only formal logic known. (Peirce 1870, 317)

Thus Peirce developed the laws of the relative product, the relative sum, and the converse of a relation. When he left for Europe in June 1870, he took a copy of this article with him and delivered it to De Morgan (Fisch 1984, xxxiii). Unfortunately, De Morgan was already in the decline that led to his death the following March. Peirce did not find a better reception when he gave a copy of the article to Stanley Jevons, who had elaborated Boole's system in England. In a letter of August 1870 to Jevons, whom Peirce described as "the only active worker now, I suppose, upon mathematical logic," it is clear that Jevons rejected Peirce's extension of Boole's system to relations (Peirce 1984, 445). Nevertheless, Peirce's work on relations eventually found wide currency in mathematical logic.

It was through applying class sums and products (i.e., unions and intersections) to relations that Peirce (1883) obtained the notion of quantifier as something distinct from the Boolean connectives. By way of example, he let l_{ij} denote the relation stating that i is a lover of j. "Any proposition whatever," he explained,

> is equivalent to saying that some complexus of aggregates and products of such numerical coefficients is greater than zero. Thus,
>
> $$\Sigma_i \, \Sigma_j l_{ij} > 0$$
>
> means that something is a lover of something; and

$$\Pi_i \Sigma_j l_{ij} > 0$$

means that everything is a lover of something. We shall, however, naturally omit, in writing the inequalities, the > 0 which terminates them all; and the above two propositions will appear as

$$\Sigma_i \Sigma_j l_{ij} \text{ and } \Pi_i \Sigma_j l_{ij}$$

$$(1883, 200\text{-}201)$$

When Peirce returned to the subject of quantifiers in (1885), he treated them in two ways that were to have a pronounced effect on the subsequent development of logic. First of all, he defined quantifiers (as part of what he called the "first-intentional logic of relatives [relations]") in a way that emphasized their analogy with arithmetic:

Here, in order to render the notation as inconical as possible, we may use Σ for *some*, suggesting a sum, and Π for *all*, suggesting a product. Thus $\Sigma_i x_i$ means that x is true of some one of the individuals denoted by i or

$$\Sigma_i x_i = x_i + x_j + x_k + \text{ etc.}$$

In the same way,

$$\Pi_i x_i = x_i x_j x_k \text{ etc.}$$

If x is a simple relation, $\Pi_i \Pi_j x_{ij}$ means that every i is in this relation to every j, $\Sigma_i \Pi_j x_{ij}$ that some one i is in this relation to every j. . . . It is to be remarked that $\Sigma_i x_i$ and $\Pi_i x_i$ are only *similar* to a sum and a product; they are not strictly of that nature, because the individuals of the universe may be innumerable. (1885, 194-95)

Thus in certain cases Peirce regarded a formula with an existential quantifier as an infinitely long propositional formula, for example the infinitary disjunction

$A(i)$ or $A(j)$ or $A(k)$ or . . . ,

where i, j, k, etc. were names for all the individuals in the universe of discourse. An analogous connection held between "for all i, $A(i)$" and the infinitary conjunction

$A(i)$ and $A(j)$ and $A(k)$ and . . . .

Here, and in the writings of those who later followed Peirce's approach (such as Schröder and Löwenheim), the syntax was not totally distinct from the semantics because a particular domain, to which the quantifiers were to apply, was given in advance. When this domain was infinite, it was natural to treat quantifiers in such an infinitary fashion, since, for a finite domain of elements i_1 to i_n, "for some i, $A(i)$" reduced to the finite disjunction

$A(i_1)$ or $A(i_2)$ or ... or $A(i_n)$,

and "for all i, $A(i)$" reduced to the finite conjunction

$A(i_1)$ and $A(i_2)$ and ... and $A(i_n)$.

Thus, unlike the logic of Peano for example, the logic that stemmed from Peirce was not restricted to formulas of finite length.

A second way in which Peirce's treatment of quantifiers was significant occurred in what he called "second-intensional logic." This kind of logic permitted quantification over predicates and so was one version of second-order logic.[2] Peirce used this logic to define identity (something that can be done in second-order logic but not, in general, in first-order logic):

> Let us now consider the logic of terms taken in collective senses [second-intensional logic]. Our notation ... does not show us even how to express that two indices, i and j, denote one and the same thing. We may adopt a special token of second intention, say 1, to express identity, and may write 1_{ij} .... And identity is defined thus:
>
> $$1_{ij} = \prod_k (q_{ki} q_{kj} + \bar{q}_{ki} \bar{q}_{kj}).$$
>
> That is, to say that things are identical is to say that every predicate is true of both or false of both.... If we please, we can dispense with the token q, by using the index of a token and by referring to this in the Quantifier just as subjacent indices are referred to. That is to say, we may write
>
> $$1_{ij} = \prod_k (x_{ki} x_{kj} + \bar{x}_{ki} \bar{x}_{kj}).$$
>
> (1885, 199)

In effect, Peirce used a form of Leibniz's principle of the identity of indiscernibles in order to give a second-order definition of identity.

Peirce rarely returned to his second-intensional logic. It formed chapter 14 of his unpublished book of 1893, *Grand Logic* (see his [1933, 56-58]). He also used it, in a letter of 1900 to Cantor, to quantify over relations while defining the less-than relation for cardinal numbers (Peirce 1976, 776). Otherwise, he does not seem to have quantified over relations. Moreover, what Peirce glimpsed of second-order logic was minimal. He appears never to have applied his logic in detail to mathematical problems, except in (1885) to the beginnings of cardinal arithmetic—an omission that contrasts sharply with Frege's work.

The logic proposed by Frege differed significantly both from Boole's system and from first-order logic. Frege's *Begriffsschrift* (1879), his first publication on logic, was influenced by two of Leibniz's ideas: a *calculus ratiocinator* (a formal calculus of reasoning) and a *lingua characteristica* (a universal language). As a step in this direction, Frege introduced a formal language on which to found arithmetic. Frege's formal language was two-dimensional, unlike the linear languages used earlier by Boole and later by Peano and Hilbert. From mathematics Frege borrowed the notions of function and argument to replace the traditional logical notions of predicate and subject, and then he employed the resulting logic as a basis for constructing arithmetic.

Frege introduced his universal quantifier in such a way that functions could be quantified as well as arguments. He made essential use of such quantifiers of functions when he treated the Principle of Mathematical Induction (1879, sections 11 and 26). To develop the general properties of infinite sequences, Frege both wanted and believed that he needed a logic at least as strong as what was later called second-order logic.

Frege developed these ideas further in his *Foundations of Arithmetic*, where he wrote of making "one concept fall under a higher concept, so to say, a concept of second order" (1884, section 53). A "second-order" concept was analogous to a function of a function of individuals. He quantified over a relation in the course of defining the notion of equipotence, or having the same cardinal number (1884, section 72)—thereby relying again on second-order logic. In his article "Function and Concept" (1891), he revised his terminology from function (or concept) of second order to function of "second level" (*zweiter Stufe*). Although he discussed this notion in more detail than he had in (1884), he did not explicitly quantify over functions of second level (1891, 26-27).

Frege's most elaborate treatment of such functions was in his *Fundamental Laws of Arithmetic* (1893, 1903). There quantification over second-level functions played a central role. Of the six axioms for his logic, two unequivocally belonged to second-order logic:

(1) If $a = b$, then for every property f, a has the property f if and only if b has the property f.

(2) If a property $F(f)$ of properties f holds for every property f, then $F(f)$ holds for any particular property f.

Frege would have had to recast his system in a radically different form if he had wanted to dispense with second-order logic. At no point did he

CARL A. RUDISILL LIBRARY
LENOIR RHYNE COLLEGE

give any indication of wishing to do so. In particular, there was no way in which he could have defined the general notion of cardinal number as he did, deriving it from logic, without the use of second-order logic.

In the *Fundamental Laws* (1893), Frege also introduced a hierarchy of levels of quantification. After discussing first-order and second-order propositional functions in detail, he briefly treated third-order propositional functions. Nevertheless, he stated his axioms as second-order (not third-order or ω-order) propositional functions. Frege developed a second-order logic, rather than a third-order or ω-order logic, because, in his system, second-level concepts could be represented by their extensions as sets and thereby appear in predicates as objects (1893, 42). Unfortunately, this approach, when combined with his Axiom V (which was a second-order version of the Principle of Comprehension), made his system contradictory— as Russell was to inform him in 1902.

Although Frege introduced a kind of second-order logic and used it to found arithmetic, he did not separate the first-order part of his logic from the rest. Nor could he have undertaken such a separation without doing violence to his principles and his goals.[3]

4. Schröder: Quantifiers in the Algebra of Logic

Ernst Schröder, who in (1877) began his research in logic within the Boolean tradition, was not acquainted at first with Peirce's contributions. On the other hand, Schröder soon learned of Frege's *Begriffsschrift* and gave it a lengthy review.[4] This review (1880) praised the *Begriffsschrift* and added that it promised to help advance Leibniz's goal of a universal language. Nevertheless, Schröder criticized Frege for failing to take account of Boole's contributions. What Frege did, Schröder argued, could be done more perspicuously by using Boole's notation; in particular, Frege's two-dimensional notation was extremely wasteful of space.

Three years later Frege replied to Schröder, emphasizing the differences between Boole's symbolic language and his own: "I did not wish to represent an abstract logic by formulas but to express a content [*Inhalt*] by written signs in a more exact and clear fashion than is possible by words" (1883, 1). As Frege's remark intimated, in his logic propositional functions carried an intended interpretation. In conclusion, he stressed that his notation allowed a universal quantifier to apply to just a *part* of a formula, whereas Boole's notation did not. This was what Schröder had overlooked in his review and what he would eventually borrow from Peirce: the separation of quantifiers from the Boolean connectives.

Schröder adopted this separation in the second volume (1891) of his *Lectures on the Algebra of Logic*, a three-volume study of logic (within the tradition of Boole and Peirce) that was rich in algebraic techniques applied to semantics. In the first volume (1890), he discussed the "identity calculus," which was essentially Boolean algebra, and three related subjects: the propositional calculus, the calculus of classes, and the calculus of domains. When in (1891) he introduced Peirce's notation for quantifiers, he used it to quantify over all subdomains of a given domain (or manifold) called 1: "In order to express that a proposition concerning a domain x holds. . . *for every domain x* (in our manifold 1), we shall place the sign Π_x before [the proposition]. . ." (1891, 26). Schröder insisted that there is no manifold 1 containing all objects, since otherwise a contradiction would result (1890, 246)—a premonition of the later set-theoretic paradoxes.

Unfortunately, Schröder conflated the relations of membership and inclusion, denoting them both with $\in$. (Frege's review (1895) criticized Schröder severely on this point.) This ambiguity in Schröder's notation might cast doubt on the assertion that he quantified over all subdomains of a given manifold and hence used a version of second-order logic. In one case, however, he was clearly proceeding in such a fashion. For he defined $x = 0$ to be $\Pi_a(x \in a)$, adding that this expressed "that a domain x is to be named 0 *if and only if x* is included in every domain a. . ." (1891, 29). In other cases, his quantifiers were taken, quite explicitly, over an infinite sequence of domains (1891, 430-31). Often his quantifiers were first-order and ranged over the individuals of a given manifold 1, a case that he treated as part of the calculus of classes (1891, 312).

Schröder's third volume (1895), devoted to the "algebra and logic of relations," contained several kinds of infinitary and second-order propositions. One kind, a_0, was introduced by Schröder in order to discuss Dedekind's notion of chain (a mapping of a set into itself). More precisely, a_0 was defined to be the infinite disjunction of all the finite iterations of the relative product of the domain a with itself (1895, 325). Here Schröder's aim (1895, 355) was to derive the Principle of Mathematical Induction in the form found in (Dedekind 1888).

In the same volume Schröder made his most elaborate use of second-order logic, treating it mainly as a tool in "elimination problems" where the goal was to solve a logical equation for a given variable. He stated a second-order proposition

$$\Pi_u(\bar{u}_{hk} + \bar{u}_{hl}) = 1'_{lk}$$

that (following Peirce) he could have taken to be the definition of identity, but did not (1895, 511). However, in what he described as "a procedure that possesses a certain boldness," he considered an infinitary proposition that had a universal quantifier for uncountably many (in fact, continuum many) variables (1895, 512). Finally, in order to move an existential quantifier to the left of a universal quantifier (as would later be done in first-order logic by Skolem functions), he introduced a universal quantifier subscripted with relation variables (and so ranging over them), and then expanded this quantifier into an infinite product of quantifiers, one for each individual in the given infinite domain (1895, 514). This general procedure would play a fundamental role in the proof that Löwenheim was to give in 1915 of Löwenheim's Theorem (see section 9 below).

5. Hilbert: Early Researches on Foundations

During the winter semester of 1898-99, David Hilbert lectured at Göttingen on Euclidean geometry, soon publishing a revised version as a book (1899). At the beginning of this book, which became the source of the modern axiomatic method, he briefly stated his purpose:

> The following investigation is a new attempt to establish for geometry a system of axioms that is *complete* and *as simple as possible*, and to deduce from these axioms the most important theorems of geometry in such a way that the significance of the different groups of axioms and the scope of the consequences to be drawn from the individual axioms are brought out as clearly as possible. (Hilbert 1899, 1)

Hilbert did not specify precisely what "complete" meant in this context until a year later, when he remarked that the axioms for geometry are complete if all the theorems of Euclidean geometry are deducible from the axioms (1900a, 181). Presumably his intention was that all *known* theorems be so deducible.

On 27 December 1899, Frege initiated a correspondence with Hilbert about the foundations of geometry. Frege had read Hilbert's book, but found its approach odd. In particular, Frege insisted on the traditional view of geometric axioms, whereby axioms were justified by geometric intuition. Replying on 29 December, Hilbert proposed a more arbitrary and modern view, whereby an axiom system only determines up to isomorphism the objects described by it:

You write: "I call axioms propositions that are true but are not proved because our knowledge of them flows from a source very different from the logical source, a source which might be called spatial intuition. From the truth of the axioms it follows that they do not contradict each other." I found it very interesting to read this sentence in your letter, for as long as I have been thinking, writing, and lecturing on these things, I have been saying the exact opposite: if the arbitrarily given axioms do not contradict each other with all their consequences, then they are true and the things defined by the axioms exist. For me this is the criterion of truth and existence. (Hilbert in [Frege 1980a], 39-40)

Hilbert returned to this theme repeatedly over the following decades.

Frege, in a letter of 6 January 1900, objected vigorously to Hilbert's claim that the consistency of an axiom system implies the existence of a model of the system. The only way to prove the consistency of an axiom system, Frege insisted, is to give a model. He argued further that the crux of Hilbert's "error" was in conflating first-level and second-level concepts.[5] For Frege, existence was a second-level concept, and it is precisely here that his system of logic, as found in the *Fundamental Laws*, differs from second-order logic as it is now understood.

What is particularly striking about Hilbert's axiomatization of geometry is an axiom missing from the first edition of his book. There his Axiom Group V consisted solely of the Archimedean Axiom. He used a certain quadratic field to establish the consistency of his system, stressing that this proof required only a denumerable set. When the French translation of his book appeared in 1902, he added a new axiom that differed fundamentally from all his other axioms and that soon led him to try to establish the consistency of a nondenumerable set, namely the real numbers:

> Let us note that to the five preceding groups of axioms we may still adjoin the following axiom which is not of a purely geometric nature and which, from a theoretical point of view, merits particular attention:

Axiom of Completeness

> To the system of points, lines, and planes it is impossible to adjoin other objects in such a way that the system thus generalized forms a new geometry satisfying all the axioms in groups I-V. (Hilbert 1902, 25)

Hilbert introduced his Axiom of Completeness, which is false in first-order logic and which belongs either to second-order logic or to the metamathematics of his axiom system, in order to ensure that every interval on a line contains a limit point. All the same, he had some initial reservations, since he added: "In the course of the present work we have not used this 'Axiom of Completeness' anywhere" (1902, 26). When the second German edition of the book appeared a year later (1903), his reservations had abated, and he designated the Axiom of Completeness as Axiom V2; so it remained in the many editions published during his lifetime.

Hilbert's initial version of the Axiom of Completeness, which referred to the real numbers rather than to geometry, stated that it is not possible to extend $\mathbb{R}$ to a larger Archimedean ordered field. This version formed part of his (1900a) axiomatization of the real number system. There he asserted that his Axiom of Completeness implies the Bolzano-Weierstrass Theorem and thus that his system characterizes the usual real numbers.

The fact that Hilbert formulated his Axiom of Completeness in his (1900a), completed in October 1899, lends credence to the suggestion that he may have done so as a response to J. Sommer's review, written at Göttingen in October 1899, of Hilbert's book (1899).[6] Sommer criticized Hilbert for introducing the Archimedean Axiom as an axiom of continuity—an assumption that was inadequate for such a purpose:

> Indeed, the axiom of Archimedes does not relieve us from the necessity of introducing explicitly an axiom of continuity, it merely makes the introduction of such an axiom possible. Thus, for the whole domain of geometry, Professor Hilbert's system of axioms is not sufficient. For instance,... it would be impossible to decide geometrically whether a straight line that has some of its points within and some outside a circle will meet the circle. (Sommer 1900, 291)

In effect, Hilbert met this objection with his new Axiom of Completeness. It is unclear why he formulated this axiom as an assertion about maximal models rather than in a more mathematically conventional way (such as the existence of a least upper bound for every bounded set).

That same year Hilbert gave his famous lecture, "Mathematical Problems," at the International Congress of Mathematicians held at Paris. As his second problem, he proposed that one prove the consistency of his axioms for the real numbers. At the same time he emphasized three assumptions underlying his foundational position: the utility of the axiomatic method, his belief that every well-formulated mathematical problem can be solved, and his conviction that the consistency of a set S of axioms implies the existence of a model for S (1900b, 264-66). When he gave this address, his view that consistency implies existence was only an article of faith—albeit one to which Poincaré subscribed as well (Poincaré 1905, 819). Yet in 1930 Gödel was to turn this article of faith into a theorem, indeed, into one version of his Completeness Theorem for first-order logic.

In 1904, when Hilbert addressed the International Congress of Mathematicians at Heidelberg, he was still trying to secure the foundations of the real number system. As a first step, he turned to providing a foundation for the positive integers. While discussing Frege's work, he considered the paradoxes of logic and set theory for the first time in print. To Hilbert these paradoxes showed that "the conceptions and research methods of logic, conceived in the traditional sense, do not measure up to the rigorous demands that set theory makes" (1905, 175). His remedy separated him sharply from Frege:

> Yet if we observe attentively, we realize that in the traditional treatment of the laws of logic certain fundamental notions from arithmetic are already used, such as the notion of set and, to some extent, that of number as well. Thus we find ourselves on the horns of a dilemma, and so, in order to avoid paradoxes, one must simultaneously develop both the laws of logic and of arithmetic to some extent. (1905, 176)

This absorption of part of arithmetic into logic remained in Hilbert's later work.

Hilbert excused himself from giving more than an indication of how such a simultaneous development would proceed, but for the first time he used a formal language. Within that language his quantifiers were in the Peirce-Schröder tradition, although he did not explicitly cite those authors. Indeed, he regarded "for some x, $A(x)$" merely as an abbreviation for the infinitary formula

$$A(1) \text{ o. } A(2) \text{ o. } A(3) \text{ o. } \ldots \, ,$$

where o. stood for "oder" (or), and analogously for the universal quantifier with respect to "und" (and) (1905, 178). Likewise, he followed Peirce and Schröder (as well as the geometric tradition) by letting his quantifiers range over a *fixed* domain. Hilbert's aim was to show the consistency of his axioms for the positive integers (the Peano Postulates without the Principle of Mathematical Induction). He did so by finding a combinatorial property that held for all theorems but did not hold for a contradiction. This marked the beginning of what, over a decade later, would become his proof theory.

Thus Hilbert's conception of mathematical logic, circa 1904, embodied certain elements of first-order logic but not others. Above all, his use of infinitary formulas and his restriction of quantifiers to a fixed domain differed fundamentally from first-order logic as it was eventually formulated. When in 1918 he began to publish again on logic, his basic perspective did not change but was supplemented by *Principia Mathematica*.

6. Huntington and Veblen: Categoricity

At the turn of the century the concept of the categoricity of an axiom system was made explicit by Edward Huntington and Oswald Veblen, both of whom belonged to the group of mathematicians sometimes called the American Postulate Theorists. Huntington, while stating an essentially second-order axiomatization for $\mathbb{R}$ by means of sequences, introduced the term "sufficient" to mean that "there is essentially *only one* such assemblage [set] possible" that satisfies a given set of axioms (1902, 264). As he made clear later in his article, his term meant that any two models are isomorphic (1902, 277).

In 1904 Veblen, while investigating the foundations of geometry, discussed Huntington's term. John Dewey had suggested to Veblen the use of the term "categorical" for an axiom system such that any two of its models are isomorphic. Veblen mentioned Hilbert's axiomatization of geometry (with the Axiom of Completeness) as being categorical, and added that, for such a categorical system, "the validity of any possible statement in these terms is therefore completely determined by the axioms; and so any further axiom would have to be considered redundant, even were it not deducible from the axioms by a finite number of syllogisms" (Veblen 1904, 346).

After adopting Veblen's term "categorical," Huntington made a further observation:

> In the case of any categorical set of postulates one is tempted to assert the theorem that if any proposition can be stated in terms of the fundamental concepts, either it is itself deducible from the postulates, or

else its contradictory is so deducible; it must be admitted, however, that our mastery of the processes of logical deduction is not yet, and possibly never can be, sufficiently complete to justify this assertion. (1905, 210)

Thus Huntington was convinced that any categorical axiom system is deductively complete: every sentence expressible in the system is either provable or disprovable. On the other hand, Veblen was aware, however fleetingly, of the possibility that in a categorical axiom system there might exist propositions true in the only model of the system but unprovable in the system itself.[7] In 1931 Gödel's Incompleteness Theorem would show that this possibility was realized, in second-order logic, for *every* categorical axiom system rich enough to include the arithmetic of the natural numbers.

7. Peano and Russell: Toward *Principia Mathematica*

In (1888) Guiseppe Peano began his work in logic by describing that of Boole (1854) and Schröder (1877). Frege, in a letter to Peano probably written in 1894, described Peano as a follower of Boole, but one what had gone further than Boole by adding a symbol for generalization.[8] Peano introduced this symbol in (1889), in the form $a \supset_{x,y,\ldots} b$ for "whatever $x, y, \ldots$ may be, b is deduced from a." Thus he introduced the notion of universal quantifier (independently of Frege and Peirce), although he did not separate it from his symbol $\supset$ for implication (1889, section II).

Peano's work gave no indication of levels of logic. In particular, he expressed "x is a positive integer" by $x \, \varepsilon \, N$ and so felt no need to quantify over predicates such as N. On the other hand, his Peano Postulates were essentially a second-order axiomatization for the positive integers, since these postulates included the Principle of Mathematical Induction. Beginning in 1900, Peano's formal language was adopted and extended by Bertrand Russell, and thereby achieved a longevity denied by Frege's.

When he became an advocate of Peano's logic, Russell entered an entirely new phase of his development. In contrast to his earlier and more traditional views, Russell now accepted Cantor's transfinite ordinal and cardinal numbers. Then in May 1901 he discovered what became known as Russell's Paradox and, after trying sporadically to solve it for a year, wrote to Frege about it on 16 June 1902. Frege was devastated. Although his original (1879) system of logic was not threatened, he realized that the system developed in his *Fundamental Laws* (1893, 1903a) was in grave danger. There he had permitted a set ("the extension of a concept," in his words) to be the argument of a first-level function; in this way Russell's Paradox arose in his system.

On 8 August, Russell sent Frege a letter containing the first known version of the theory of types—Russell's solution to the paradoxes of logic and set theory:

> The contradiction [Russell's Paradox] could be resolved with the help of the assumption that ranges [classes] of values are not objects of the ordinary kind: i.e., that $\phi(x)$ needs to be completed (except in special circumstances) either by an object or by a range of values of objects [class of objects] or a range of values of ranges of values [class of classes of objects], etc. This theory is analogous to your theory about functions of the first, second, etc. levels. (Russell in [Frege 1980a], 144)

This passage suggests that the seed of the theory of types grew directly from the soil of Frege's *Fundamental Laws*. Indeed, Philip Jourdain later asked Frege (in a letter of 15 January 1914) whether Frege's theory was not the same as Russell's theory of types. In a draft of his reply to Jourdain on 28 January, Frege answered a qualified yes:

> Unfortunately I do not understand the English language well enough to be able to say definitely that Russell's theory (*Principia Mathematica* I, 54ff) agrees with my theory of functions of the first, second, etc., levels. It does seem so.[9]

In 1903 Russell's Paradox appeared in print, both in the second volume of Frege's *Fundamental Laws* (1903a, 253) and in Russell's *Principles of Mathematics*. Frege dealt only with Russell's Paradox, whereas Russell also discussed in detail the paradox of the largest cardinal and the paradox of the largest ordinal. In an appendix to his book, Russell proposed a preliminary version of the theory of types as a way to resolve these paradoxes, but he remained uncertain, as he had when writing to Frege on 29 September 1902, whether this theory eliminated *all* paradoxes.[10]

When Russell completed the *Principles*, he praised Frege highly. Nevertheless, Russell's book, which had been five years in the writing, retained an earlier division of logic that was more in the tradition of Boole, Peirce, and Schröder than in Frege's. There Russell divided logic into three parts: the propositional calculus, the calculus of classes, and the calculus of relations.

In 1907, after several detours through other ways of avoiding the paradoxes,[11] Russell wrote an exposition (1908) of his mature theory of types, the basis for *Principia Mathematica*. A type was defined to be the range of significance of some propositional function. The first type consisted of the individuals and the second of what he called "first-order propositions": those propositions whose quantifiers ranged only over the first type. The third logical type consisted of "second-order prepositions," whose quantifiers ranged only over the first or second types (i.e., individuals or first-order propositions). In this manner, he defined a type for each finite index n. After introducing an analogous hierarchy of propositional functions, he avoided classes by using such propositional functions instead. Like Peirce, he defined the identity of individuals x and y by the condition that every first-order proposition holding for x also holds for y (1908, sections IV-VI).

Thus the theory of types, outlined in (Russell 1908) and developed in detail in *Principia Mathematica*, included a kind of first-order logic, second-order logic, and so on. But the first-order logic that it included differed from first-order logic as it is now understood, among other ways, in that a proposition about classes of classes could not be treated in his first-order logic. As we shall see, this privileged position of the membership relation was later attacked by Skolem.

One aspect of the logic found in *Principia Mathematica* requires further comment. For Russell and Whitehead, as for Frege, logic served as a foundation for all of mathematics. From their perspective it was impossible to stand outside of logic and thereby to study it as a system (in the way that one might, for example, study the real numbers). Given this state of affairs, it is not surprising that Russell and Whitehead lacked any conception of a metalanguage. They would surely have rejected such a conception if it had been proposed to them, for they explicitly denied the possibility of independence proofs for their axioms (Whitehead and Russell 1910, 95), and they believed it impossible to prove that substitution is generally applicable in the theory of types (1910, 120). Indeed, they insisted that the Principle of Mathematical Induction cannot be used to prove theorems *about* their system of logic (1910, 135). Metatheoretical research about the theory of types had to come from those schooled in a different tradition. When the consistency of the simple theory of types was eventually proved, without the Axiom of Infinity, in (1936), it was done by

Gerhard Gentzen, a member of Hilbert's school, and not by someone within the logicist tradition of Frege, Russell, and Whitehead.

Russell and Whitehead held that the theory of types can be viewed in two ways—as a deductive system (with theorems proved from the axioms) and as a formal calculus (1910, 91). Concerning the latter, they wrote:

> Considered as a formal calculus, mathematical logic has three analogous branches, namely (1) the calculus of propositions, (2) the calculus of classes, (3) the calculus of relations. (1910, 92)

Here they preserved the division of logic found in Russell's *Principles of Mathematics* and thereby continued in part the tradition of Boole, Peirce, and Schröder—a tradition that they were much less willing to acknowledge than those of Peano and Frege.

Russell and Whitehead lacked the notion of model or interpretation. Instead, they employed the genetic method of constructing, for instance, the natural numbers, rather than using the axiomatic approach of Peano or Hilbert. Finally, Russell and Whitehead shared with many other logicians of the time the tendency to conflate syntax and semantics, as when they stated their first axiom in the form that "anything implied by a true elementary proposition is true" (1910, 98).

Principia Mathematica provided the logic used by most mathematical logicians in the 1910s and 1920s. But even some of those who used its ideas were still influenced by the Peirce-Schröder tradition. This was the case for *A Survey of Symbolic Logic* by C. I. Lewis (1918). In that book, which analyzed the work of Boole, Jevons, Peirce, and Schröder (as well as that of Russell and Whitehead), the logical notation remained that of Peirce, as did the definition of the existential quantifier Σ_x and the universal quantifier Π_x in terms of logical expressions that could be infinitely long:

> We shall let $\Sigma_x \phi x$ represent $\phi x_1 + \phi x_2 + \phi x_3 + \ldots$ to as many terms as there are distinct values of x in ϕ. And $\Pi_x \phi x$ will represent $\phi x_1 \times \phi x_2 \times \phi x_3 \times \ldots$ to as many terms as there are distinct values of x in ϕx. . . . The fact that there might be an infinite set of values of x in ϕx does not affect the theoretical adequacy of our definitions. (Lewis 1918, 234-35)

Lewis sought to escape the difficulties that he found in such an infinitary logic by reducing it to the finite:

We can assume that any law of the algebra [of logic] which holds *whatever finite* number of elements be involved holds for any number of elements whatever.... This also resolves our difficulty concerning the possibility that the number of values of x in ϕx might not be even denumerable.... (1918, 236)

He made no attempt, however, to justify his assertion, which amounted to a kind of compactness theorem.

A second logician who introduced an infinitary logic in the context of *Principia Mathematica* was Frank Ramsey. In 1925 Ramsey presented a critique of *Principia*, arguing that the Axiom of Reducibility should be abandoned and proposing instead the simple theory of types. He entertained the possibility that a truth function may have infinitely many arguments (1925, 367), and he cited Wittgenstein as having recognized that such truth functions are legitimate (1925, 343). Further, Ramsey argued that "owing to our inability to write propositions of infinite length, which is logically a mere accident, $(\phi).\phi a$ cannot, like $p.q$, be elementarily expressed, but must be expressed as the logical product of a set of which it is also a member" (1925, 368-69). It appears that Ramsey was not concerned with infinitary formulas per se but only with using them in his heuristic argument for the simple theory of types. Yet his proposal was sufficiently serious that Gödel later cited it when arguing against such infinitary formulas (Gödel 1944, 144-46).

8. Hilbert: Later Foundational Research

Hilbert did not abandon foundational questions after his 1904 lecture at the Heidelberg congress, though he published nothing further on them for more than a decade. Rather, he gave lecture courses at Göttingen on such questions repeatedly—in 1905, 1908, 1910, and 1913. It was the course given in the winter semester of 1917-18, "Principles of Mathematics and Logic," that first exhibited his mature conception of logic.[12]

That course began shortly after Hilbert delivered a lecture, "Axiomatic Thinking," at Zurich on 11 September 1917. The Zurich lecture stressed the role of the axiomatic method in various branches of mathematics and physics. Returning to an earlier theme, he noted how the consistency of several axiomatic systems (such as that for geometry) had been reduced to a more specialized axiom system (such as that for $\mathbb{R}$, reduced in turn to the axioms for $\mathbb{N}$ and those for set theory). Hilbert concluded by stating that the "full-scale undertaking of Russell's to axiomatize logic can be seen as the crowning achievement of axiomatization" (1918, 412).

The 1917 course treated the axiomatic method as applied to two disciplines: geometry and mathematical logic. When considering logic, Hilbert emphasized not only questions of independence and consistency, as he had in the Zurich lecture, but also "completeness" (which, however, was handled as in his [1899]). Influenced by *Principia Mathematica*, he treated propositional logic as a distinct level of logic. But he deviated from Russell and Whitehead by offering a proof for the consistency of propositional logic. After discussing the calculus of monadic predicates and the corresponding calculus of classes, he turned to first-order logic, which he named the "functional calculus." Stating primitive symbols and axioms for it, he developed it at some length. In conclusion, he noted:

> With what we have considered thus far [first-order logic], foundational discussions about the calculus of logic come to an end—if we have no other goal than formalizing logical deduction. We, however, are not content with this application of symbolic logic. We wish not only to be in a position to develop individual theories from their principles purely formally but also to make the foundations of mathematical theories themselves an object of investigation—to examine in what relation they stand to logic and to what extent they can be obtained from purely logical operations and concepts. To this end the calculus of logic must serve as our tool.
>
> Now if we make use of the calculus of logic in this sense, then we will be compelled to extend in a certain direction the rules governing the formal operations. In particular, while we previously separated propositions and [propositional] functions completely from objects and, accordingly, distinguished the signs for indefinite propositions and functions rigorously from the variables, which take arguments, now we permit propositions and functions to be taken as logical variables in a way similar to that for proper objects, and we permit signs for indefinite propositions and functions to appear as arguments in symbolic expressions. (Hilbert 1917, 188)

Here Hilbert argued, in effect, for a logic at least as strong as second-order logic. But his views on this matter did not appear in print until his book *Principles of Mathematical Logic*, written jointly with Ackermann, was published in 1928. In that book, which consisted largely of a revision of Hilbert's 1917 course, he expressed himself even more strongly:

As soon as the object of investigation becomes the foundation of...mathematical theories, as soon as one wishes to determine in what relation the theory stands to logic and to what extent it can be obtained from purely logical operations and concepts, then the extended calculus [of logic] is essential. (Hilbert and Ackermann 1928, 86)

In the 1917 course (and again in the book), this extended functional calculus permitted quantification over propositions and included explicitly such expressions as

(X)(if X, then X or X).

Hilbert cited the Principle of Mathematical Induction as an axiom that, if fully expressed, requires a quantifier varying over propositions. Likewise, he defined the identity relation in his extended logic in a manner reminiscent of Peirce: two objects x and y are identical if, for every proposition P, P holds of x if and only if P holds of y (1917, 189-91). Finally, he treated set-theoretic notions (such as union and power set) by means of quantifiers over propositions.

What, in 1917, was Hilbert's extended calculus of logic? At first glance it might appear to be second-order logic. Yet he stated his preliminary version of the extended calculus in a way that permitted a function of propositional functions to occupy an argument place for individuals in a propositional function—an act that is illegitimate in second-order logic and that, as he knew, gave rise to Russell's Paradox. Hilbert used this paradox to motivate his adoption of the ramified theory of types as the definitive version of his extended calculus. After indicating how the real numbers can be constructed via Dedekind cuts by means of the Axiom of Reducibility, Hilbert concluded his course: "Thus we have shown that introducing the Axiom of Reducibility is the appropriate means to mold the calculus of levels [the theory of types] into a system in which the foundations of higher mathematics can be developed" (1917, 246).

In the 1917 course Hilbert explicitly treated first-order logic (without any infinitary trappings) as a subsystem of the ramified theory of types:

In this way is founded a new form of the calculus of logic, the "calculus of levels," which represents an extension of the original functional calculus [first-order logic], since this is contained in it as a theory of first order, but which implies an essential restriction as compared with our previous extension of the functional calculus. (1917, 222-23)

The same statement appeared verbatim in (Hilbert and Ackermann 1928,

101). Thus in 1917, and still in 1928, Hilbert treated first-order logic as a subsystem of all of logic (for him, the ramified theory of types), regarding set theory and the Principle of Mathematical Induction as incapable of adequate treatment in first-order logic (1917, 189, 200; Hilbert and Ackermann 1928, 83, 92).

On the occasion of his Zurich lecture, Hilbert invited Paul Bernays to give up his position as *Privatdozent* at the University of Zurich and come to Göttingen as Hilbert's assistant on the foundations of arithmetic. Bernays accepted. Thus he became Hilbert's principal collaborator in logic, serving first as official note-taker for Hilbert's 1917 course.

In 1918 Bernays wrote a *Habilitationsschrift* in which was proved the completeness of propositional logic. This work was influenced by (Schröder 1890), (Frege 1893), and *Principia Mathematica*. For the first time the completeness problem for a subsystem of logic was expressed precisely. Bernays stated and proved that "every provable formula is a valid formula, and conversely" (1918, 6). In addition he established the deductive completeness of propositional logic as well as what was later called Post completeness: no unprovable formula can be added to the axioms of propositional logic without giving rise to a contradiction (1918, 9).

Hilbert published nothing further on logic until 1922, when he reacted strongly against the claims of L. E. J. Brouwer and Hermann Weyl that the foundations of analysis were built on sand. As a countermeasure, Hilbert introduced his proof theory (*Beweistheorie*). This theory treated the axiom systems of mathematics as pure syntax, distinguishing them from what he called metamathematics, where meaning was permitted. His two chief aims were to show the consistency of both analysis and set theory and to establish the decidability of each mathematical question (*Entscheidungsproblem*)—aims already expressed in his (1900b).

Hilbert began by proving the consistency of a very weak subsystem of arithmetic, closely related to the one whose consistency he had established in 1904 (1922, 170). He claimed to have a proof for the consistency of arithmetic proper, including the Principle of Mathematical Induction. He considered this principle to be a second-order axiom, as he had previously and did subsequently.[13] Likewise, he introduced not only individual variables (*Grundvariable*) but variables for functions and even for functions of functions (1922, 166).

To defend mathematics against the attacks of Brouwer and Weyl, Hilbert intended to establish the consistency of mathematics from the bottom up. Having shown that a subsystem of arithmetic was consistent, he hoped soon to prove the consistency of the whole of arithmetic (including mathematical induction) and then of the theory of real numbers. Thus Hilbert was thinking in terms of a sequence of levels to be secured successively. In a handwritten appendix to the lecture course he gave during the winter semester of 1922-23,[14] eight of these levels were listed. Analysis occurred at level four, and higher-order logic was considered in part at level five.

Already in (1922, 157), Hilbert had spoken of the need to formulate Zermelo's Axiom of Choice in such a way that it becomes as evident as $2 + 2 = 4$. In (1923) Hilbert utilized a form of the Axiom of Choice as the cornerstone of his proof theory, which was to be "finitary." He did so as a way of eliminating the direct use of quantifiers, which he regarded as an essentially infinitary feature of logic: "Now where does there appear for the first time something going beyond the concretely intuitive and the finitary? Obviously already in the use of the concepts '*all*' and '*there exists*'" (1923, 154). For finite collections, he noted, the universal and existential quantifiers reduce to finite conjunctions and disjunctions, yielding the Principle of the Excluded Middle in the form that $\sim(\forall x)A(x)$ is equivalent to $(\exists x)\sim A(x)$ and that $\sim(\exists x)A(x)$ is equivalent to $(\forall x)\sim A(x)$. "These equivalences," he continued,

> are commonly assumed, without further ado, to be valid in mathematics for infinitely many individuals as well. In this way, however, we abandon the ground of the finitary and enter the domain of transfinite inferences. If we were always to use for infinite sets a procedure admissible for finite sets, we would open the gates to error.... In analysis... the theorems valid for finite sums and products can be translated into theorems valid for infinite sums and products only if the inference is secured, in the particular case, by convergence. Likewise, we must not treat the infinite logical sums and products
>
> $A(1)$ & $A(2)$ & $A(3)$ & ...
>
> and
>
> $A(1)$ v $A(2)$ v $A(3)$ v ...
>
> in the same way as finite ones. My proof theory ... provides such a treatment. (1923, 155)

Hilbert then introduced what he called the Transfinite Axiom, which he regarded as a form of the Axiom of Choice:

$A(\tau A) \rightarrow A(x)$.

The intended meaning was that if the proposition $A(x)$ holds when x is τA, then $A(x)$ holds for an arbitrary x, say a. He thus defined $(\forall x)A(x)$ as $A(\tau A)$ and similarly for $(\exists x)A(x)$ (1923, 157).

Soon Hilbert modified the Transfinite Axiom, changing it into the ε-axiom:

$A(x) \rightarrow A(\varepsilon_x (A(x)))$,

where ε was a universal choice function acting on properties. The first sign of this change occurred in handwritten notes that Hilbert prepared for his lecture course on logic during the winter semester of 1922-23;[15] the new ε-axiom appeared in print in (Ackermann 1925) and (Hilbert 1926). Hilbert had used the Transfinite Axiom, acting on number-theoretic functions, to quantify over number-theoretic functions and to obtain a proof of the Axiom of Choice for families of sets of real numbers (1923, 158, 164). Likewise, Ackermann used the ε-axiom, acting on number-theoretic functions, to quantify over such functions while trying to establish the consistency of number theory (1925, 32).

In 1923 Hilbert stressed not only that his proof theory could give mathematics a firm foundation by showing the consistency of analysis and set theory but also that it could settle such classical unsolved problems of set theory as Cantor's Continuum Problem (1923, 151). In his (1926), Hilbert attempted to sketch a proof for the Continuum Hypothesis (CH). Here one of his main tools was the hierarchy of functions that he called variable-types. The first level of this hierarchy consisted of the functions from $\mathbb{N}$ to $\mathbb{N}$—i.e., the number-theoretic functions. The second level contained functionals—those functions whose argument was a number-theoretic function and whose value was in $\mathbb{N}$. In general, the level $n+1$ consisted of functions whose arguments was a function of level n and whose value was in $\mathbb{N}$. He permitted quantification over functions of any level (1926, 183-84). In (1928, 75) he again discussed his argument for CH, which he explicitly treated as a second-order proposition. This argument met with little favor from other mathematicians.

From 1917 to about 1928, Hilbert worked in a variant of the ramified theory of types, one in which functions increasingly played the principal role. But by 1928 he had rejected Russell's Axiom of Reducibility as dubious—a significant change from his support for this axiom in (1917).

Instead, he asserted that the same purpose was served by his treatment of function variables (1928, 77). He still insisted, in (Hilbert and Ackermann 1928, 114-15), that the theory of types was the appropriate logical vehicle for studying the theory of real numbers. But, he added, logic could be founded so as to be free of the difficulties posed by the Axiom of Reducibility, as he had done in his various papers. Thus Hilbert opted for a version of the simple theory of types (in effect, ω-order logic).

Hilbert's co-workers in proof theory used essentially the same system of logic as he did. Von Neumann (1927) gave a proof of the consistency of a weak form of number theory, working mainly in a first-order subsystem. However, he discussed Ackermann's work involving both a second-order ε-axiom and quantification over number-theoretic functions (von Neumann 1927, 41-46). During the same period, Bernays discussed in some detail "the extended formalism of 'second order'," mentioning Löwenheim's (1915) decision procedure for monadic second-order logic and remarking how questions of first-order validity could be expressed by a second-order formula (Bernays and Schönfinkel 1928, 347-48).[16]

In (1929), Hilbert looked back with pride and forward with hope at what had been accomplished in proof theory. He thought that Ackermann and von Neumann had established the consistency of the ε-axiom restricted to natural numbers, not realizing that this would soon be an empty victory. Hilbert posed four problems as important to his program. The first of these was essentially second-order: to prove the consistency of the ε-axiom acting on number-theoretic functions (1929, 4). The second was the same problem for higher-order functions, whereas the third and fourth concerned completeness. The third, noting that the Peano Postulates are categorical, asked for a proof that if a number-theoretic sentence is shown consistent with number theory, then its negation cannot be shown consistent with number theory. The fourth was more complex, asking for a demonstration that, on the one hand, the axioms of number theory are deductively complete and that, on the other, first-order logic with identity is complete (1929, 8).

Similarly, in (Hilbert and Ackermann 1928, 69) there was posed the problem of establishing the completeness of first-order logic (without identity). The following year Gödel solved this problem for first-order logic, with and without identity (1929). His abstract (1930b) of this result spoke of the "restricted functional calculus" (first-order logic without identity) as a subsystem of logic, since no bound function variables were permitted.

Although the Completeness Theorem for first-order logic solved one of Hilbert's problems, Gödel soon published a second abstract (1930c) that threatened to demolish Hilbert's program. This abstract gave the First Incompleteness Theorem: in the theory of types, the Peano Postulates are not deductively complete. Furthermore, Gödel's Second Incompleteness Theorem stated that the consistency of the theory of types cannot be proved in the theory of types, provided that this theory is ω-consistent. This seemed to destroy any hope of proving the consistency of set theory and analysis with Hilbert's finitary methods.

Probably motivated by Gödel's incompleteness results, Hilbert (1931) introduced a version of the ω-rule, an infinitary rule of inference.[17] In his last statement on proof theory in 1934, Hilbert attempted to limit the damage done by those results. He wrote of

> the final goal of knowing that our customary methods in mathematics are utterly consistent.Concerning this goal, I would like to stress that the view temporarily widespread—that certain recent results of Gödel imply that my proof theory is not feasible—has turned out to be erroneous. In fact those results show only that, in order to obtain an adequate proof of consistency, one must use the finitary standpoint in a sharper way than is necessary in treating the elementary formalism. (Hilbert in [Hilbert and Bernays 1934], v)

9. Löwenheim: First-Order (Infinitary) Logic as a Subsystem

It was Leopold Löwenheim, aware of *Principia Mathematica* but firmly placed in the Peirce-Schröder tradition, who first established significant metamathematical results about the semantics of logic. His 1915 result, the earliest version of Löwenheim's Theorem (every satisfiable sentence has a countable model), is now considered to be about first-order logic. As will become evident, however, in one sense Löwenheim in 1915 was even further from first-order logic than Hilbert had been in 1904.

In an unpublished autobiographical note, Löwenheim remarked that shortly after he began teaching at a *Gymnasium* (about 1900) he "became acquainted with the calculus of logic through reviews and from Schröder's books" (Löwenheim in [Thiel 1977], 237). His first published paper (1908) was devoted to the solvability of certain symmetric equations in Schröder's calculus of classes, and he soon turned to related questions in Schröder's calculus of domains (1910, 1913a) and of relations (1913b).

In 1915 there appeared Löwenheim's most influential article, "On Possibilities in the Calculus of Relations." His later opinion of this article, as expressed in unpublished autobiographical notes, was ambivalent:

> I...have pointed out new paths for science in the field of the calculus of logic, which had been founded by Leibniz but which had come to a deadlock.... On an outing, a somewhat grotesque landscape stimulated my fantasy, and I had an insight that the thoughts I had already developed in the calculus of domains might lead me to make a breakthrough in the calculus of relations. Now I could find no rest until the idea was completely proved, and this gave me a host of troubles.... This breakthrough has scarcely been noticed, but some other breakthroughs which I made in my paper "On Possibilities in the Calculus of Relations" were noticed all the same. This became the foundation of the modern calculus of relations. But I did not take much pride in this paper, since the point had only been to ask the right questions while the proofs could be found easily without ingenuity or imagination. (Löwenheim in [Thiel 1977], 246-47)

Löwenheim's article (1915) continued Schröder's work (1895) on the logic of relations but made a number of distinctions that Schröder lacked, the most important being between a *Relativausdruck* (relational expression) and a *Zählausdruck* (individual expression). Löwenheim's definitions of individual expression and relational expression (1915, 447-48) differed from those for first-order and second-order formula in that, following Peirce and Schröder, such expressions were allowed to have a quantifier for each individual of the domain, or for each relation over the domain, respectively. In effect, Löwenheim's logic permitted infinitely long strings of quantifiers and Boolean connectives, but these infinitary formulas occurred only as the expansion of finitary formulas, to which they were equivalent in his system.

In (1895) Schröder had used a logic that was essentially the same as Löwenheim's, namely, a logic that was second-order but permitted quantifiers to be expanded as infinitary conjunctions (or disjunctions). After distinguishing carefully between the (infinitary) first-order part of his logic and the entire logic, Löwenheim asserted that all important problems of mathematics can be handled in this (infinitary) second-order logic.

Apparently, his discovery of Löwenheim's Theorem was motivated by the two ways in which Schröder's calculus of relations could express propositions: (1) by means of quantifiers and individual variables and (2) by means of relations with no individual variables (cf. the combinatory logic developed by Schönfinkel [1924] and later by Curry). Whereas Schröder regarded every proposition of form (1) as capable of being "condensed" (that is, written in form (2)) because he permitted individuals to be interpreted as a certain kind of relation, Löwenheim dispensed with this interpretation. He established that not all propositions can be condensed, by showing that condensable first-order propositions could not express that there are at least four elements.

After treating condensation, Löwenheim next gave an analogous result for denumerable domains: For an infinite domain M (possibly denumerable and possibly of higher cardinality), if a first-order proposition is valid in every finite domain but not in every domain, then the proposition is not valid in M. This was his original statement of Löwenheim's Theorem. His proof used a second-order formula stating that

$$(\forall x)(\exists y)\phi(x,y) \longleftrightarrow (\exists f)(\forall x)\phi(x, f(x)),$$

where f was a function variable, and he regarded this formula (for the given domain M) as expandable to an infinitary formula with a first-order existential quantifier for each individual of M (1915, section 2). His next theorem stated that, since Schröder's logic can express that a domain is finite or denumerable, then Löwenheim's Theorem cannot be extended to Schröder's (second-order) logic.

Some historians of mathematics have regarded Löwenheim's argument for his theorem as odd and unnatural.[18] But his argument appears so only because they have considered it within first-order logic. Although Löwenheim's Theorem holds for first-order logic (as Skolem was to show), this was not the logic in which Löwenheim worked.

10. Skolem: First-Order Logic as All of Logic

In 1913, after writing a thesis (1913a) on Schröder's algebra of logic, Skolem received his undergraduate degree in mathematics. He soon wrote several papers, beginning with (1913b), on Schröder's calculus of classes. During the winter of 1915-16, Skolem visited Göttingen, where he discussed set theory with Felix Bernstein. By that time Skolem was already acquainted with Löwenheim's Theorem and had seen how to extend it to a countable set of formulas. Furthermore, Skolem had realized that this extended version of the theorem could be applied to set theory. Thus he found what was later called Skolem's Paradox: Zermelo's system of set theory has a countable model (within first-order logic) even though this system implies the existence of uncountable sets (Skolem 1923, 232, 219).

Nevertheless, Skolem did not lecture on this result until the Fifth Scandanavian Congress of Mathematicians in July 1922. When it appeared in print the following year, he stated that he had not published it earlier because he had been occupied with other problems and because

> I believed that it was so clear that the axiomatization of set theory would not be satisfactory as an ultimate foundation for mathematics that, by and large, mathematicians would not bother themselves with it very much. To my astonishment I have seen recently that many mathematicians regard these axioms for set theory as the ideal foundation for mathematics. For this reason it seemed to me that the time had come to publish a critique. (Skolem 1923, 232)

It was precisely in order to establish the relativity of set-theoretic notions that Skolem proposed that set theory be formulated within first-order logic. At first glance, given the historical context, this was a strange suggestion. Set theory appeared to require quantifiers not only over individuals, as in first-order logic, but also quantifiers over sets of individuals, over sets of sets of individuals, and so on. Skolem's radical proposal was that the membership relation ε be treated not as a part of logic (as Peano and Russell had done) but like any other relation on a domain. Such relations could be given a variety of interpretations in various domains, and so should ε. In this way the membership relation began to lose its privileged position within logic.

Skolem made Löwenheim's article (1915) the starting point for several of his papers. This process began when Skolem used the terms *Zählausdruck* and *Zählgleichung* in an article (1919), completed in 1917, on Schröder's calculus of classes. But it was only in (1920) that Skolem began to discuss Löwenheim's Theorem, a subject to which he returned many times over the next forty years. Skolem first supplied a new proof for this theorem (relying on Skolem functions) in order to avoid the occurrence in the proof of second-order propositions (in the form of subsubscripts on Löwenheim's relational expressions [Skolem 1920, 1]). Extending this result, Skolem obtained what became known as the Löwenheim-Skolem Theorem (a countable and satisfiable set of first-order propositions has a countable model). Surprisingly, he did not even state the Löwenheim-Skolem Theorem for first-order logic as a separate theorem but immediately proved Löwenheim's Theorem for any countable conjunction of first-order propositions. His culminating generalization was to show Löwenheim's Theorem for a countable conjunction of countable disjunctions of first-order propositions. Thus, in effect, he established Löwenheim's Theorem (and hence the Löwenheim-Skolem Theorem) for what is now called $L_{\omega_1, \omega}$. [19]

Nevertheless, when in (1923) Skolem published his application of Löwenheim's Theorem to set theory, he stated the Löwenheim-Skolem Theory explicitly for first-order logic and did not mention any of his other generalizations of Löwenheim's Theorem for an infinitary logic. Never again did he return to the infinitary logic that he had adopted from Schröder and Löwenheim. Instead he argued in (1923), as he would for the rest of his life, that first-order logic is the proper basis for set theory and, indeed, for all of mathematics.

The reception of the Löwenheim-Skolem Theorem, insofar as it concerned set theory, was mixed. Abraham Fraenkel, when in (1927) he reviewed Skolem's paper of 1923 in the abstracting journal of the day, the *Jahrbuch über die Fortschritte der Mathematik*, did not mention first-order logic but instead held Skolem's result to be about Schröder's calculus of logic. Fraenkel concluded from Skolem's Paradox that the relativity of the notion of cardinal number is inherent in *any* axiomatic system. Thus Fraenkel lacked a clear understanding of the divergent effects of first-order and second-order logic on the notion of cardinal number.

In 1925, when John von Neumann published his axiomatization for set theory, he too was unclear about the difference between first-order and second-order logic. He specified his desire to axiomatize set theory by using only "a finite number of purely formal operations" (1925, section 2), but nowhere did he specify the logic in which his axiomatization was to be formulated. His concern, rather, was to give a finite characterization of Zermelo's notion of "definite property." At the end of his paper, von Neumann considered the question of categoricity in detail and noted various steps, such as the elimination of inaccessible cardinals, that had to be taken to arrive at a categorical axiomatization. After remarking that the axioms for Euclidean geometry were categorical, he observed that, because of the Löwenheim-Skolem Theorem,

> no categorical axiomatization of set theory seems to exist at all....
> And since there is no axiom system for mathematics, geometry, and so forth that does not presupppose set theory, there probably cannot

be any categorically axiomatized infinite systems at all. This circumstance seems to me to be an argument for intuitionism. (1925, section 5)

Yet von Neumann then claimed that the boundary between the finite and infinite was also blurred. He exhibited no awareness that these difficulties did not arise in second-order logic.

The reason for Fraenkel's and von Neumann's confusion was that, circa 1925, the distinction between first-order logic and second-order logic was still unclear, and it was equally unclear just how widely the Löwenheim-Skolem Theorem applied. Fraenkel had spoken of "the uncertainty of general logic" (1922, 101), and Zermelo added, concerning his 1908 axiomatization of set theory, that "a generally recognized 'mathematical logic', to which I could have referred, did not exist then—any more than it does today, when every foundational researcher has his own logistical system" (1929, 340)

What is surprising is that Gödel seemed unclear about the question of categoricity when he proved the Completeness Theorem for first-order logic in his doctoral dissertation (1929). Indeed, he made certain enigmatic comments that foreshadow his incompleteness results:

Brouwer, in particular, has emphatically stressed that from the consistency of an axiom system we cannot conclude without further ado that a model can be constructed. But one might perhaps think that the existence of the notions introduced through an axiom system is to be defined outright by the consistency of the axioms and that, therefore, a proof has to be rejected out of hand. This definition,...however, manifestly presupposes the axiom that every mathematical problem is solvable. Or, more precisely, it presupposes that we cannot prove the unsolvability of any problem. For, if the unsolvability of some problem (say, in the domain of real numbers) were proved, then, from the definition above, it would follow that there exist two non-isomorphic realizations of the axiom system for the real numbers, while on the other hand we can prove that any two realizations are isomorphic. (1929, section 1)

In his paper Gödel wrote of first-order logic, but then he introduced a notion, categoricity, that belongs essentially to second-order logic. Thus it appears that the distinction between first-order logic and second-order logic, insofar as it concerns the range of applicability of the Löwenheim-Skolem Theorem, remained unclear even to Gödel in 1929.

In (1929) Zermelo responded to the criticisms of his notion of "definite property" by Fraenkel, Skolem, and von Neumann. Zermelo chose to define "definite property," in effect, as any second-order propositional function built up from = and ε—although he did not specify the axioms or the semantics of his second-order logic. But in (1930) he was well aware that this logic permitted him to prove that any standard model of Zermelo-Fraenkel set theory without urelements consists of V_α for some level of Zermelo's cumulative type hierarchy if and only if α is a strongly inaccessible ordinal. This result holds in second-order logic, but it fails in first-order logic—as later shown in (Montague and Vaught 1959).

Zermelo was dismayed by the uncritical acceptance of "*Skolemism*, the doctrine that *every* mathematical theory, and set theory in particular, is satisfiable in a *countable model*" (1931, 85). He responded by proposing a powerful infinitary logic. It permitted infinitely long conjunctions and disjunctions (having any cardinal for their length) and no quantifiers—what is now called the propositional part of $L_{\infty,\omega}$. This was by far the strongest infinitary logic considered up to that time.

Zermelo was also dissatisfied with Gödel's Incompleteness Theorem, and wrote to him for clarification. There followed a spirited exchange of letters between the old set-theorist and the young logician. Zermelo, like Skolem and many others, failed to distinguish clearly between syntax and semantics. On the other hand, Zermelo insisted in these letters on extending the notion of proof sufficiently that every valid formula would be provable. In his last published paper (1935), he extended the notion of rule of inference to allow not only the ω-rule but well-founded, partially ordered strings of premises of any cardinality.

Zermelo's proposal about extending logic fell on deaf ears.[20] Gödel, though he adopted Zermelo's cumulative type hierarchy, formulated his own researches on set theory within first-order logic. Indeed, Gödel's constructible sets were the natural fusion of first-order logic with Zermelo's cumulative type hierarchy.

In December 1938, at a conference in Zurich on the foundations of mathematics, Skolem returned again to the existence of countable models for set theory and to Skolem's Paradox (1941, 37). On this occasion he emphasized the relativity, not only of set theory, but of mathematics as a whole. The discussion following Skolem's lecture revealed both interest in and ambivalence about the Löwenheim-Skolem Theorem, especially when Bernays commented:

The axiomatic restriction of the notion of set [to first-order logic] does not prevent one from obtaining all the usual theorems...of Cantorian set theory.... Nevertheless, one must observe that this way of making the notion of set (or that of predicate) precise has a consequence of another kind: the interpretation of the system is no longer necessarily unique.... It is to be observed that the impossibility of characterizing the finite with respect to the infinite comes from the restrictiveness of the [first-order] formalism. The impossibility of characterizing the denumerable with respect to the nondenumerable in a sense that is in some way unconditional—does this reveal, one might wonder, a certain inadequacy of the method under discussion here [first-order logic] for making axiomatizations precise? (Bernays in [Gonseth 1941], 49-50)

Skolem objected vigorously to Bernays's suggestion and insisted that a first-order axiomatization is surely the most appropriate.

In 1958, at a colloquium held in Paris, Skolem reiterated his views on the relativity of fundamental mathematical notions and criticized Tarski's contributions:[21]

It is self-evident that the dubious character of the notion of set renders other notions dubious as well. For example, the semantic definition of mathematical truth proposed by A. Tarski and other logicians presupposes the general notion of set. (Skolem 1958, 13)

In the discussion that followed Skolem's lecture, Tarski responded to this criticism:

[I] object to the desire shown by Mr. Skolem to reduce every theory to a denumerable model.... Because of a well-known generalization of the Löwenheim-Skolem Theorem, every formal system that has an infinite model has a model whose power is any transfinite cardinal given in advance. From this, one can just as well argue for excluding denumerable models from consideration in favor of nondenumerable models. (Tarski in [Skolem 1958], 17)

11. Conclusion

As we have seen, the logics considered from 1879 to 1923—such as those of Frege, Peirce, Schröder, Löwenheim, Skolem, Peano, and Russell— were generally richer than first-order logic. This richness took one of two forms: the use of infinitely long expressions (by Peirce, Schröder, Hilbert, Löwenheim, and Skolem) and the use of a logic at least as rich as second-order logic (by Frege, Peirce, Schröder, Löwenheim, Peano, Russell, and

Hilbert). The fact that no system of logic predominated—although the Peirce-Schröder tradition was strong until about 1920 and *Principia Mathematica* exerted a substantial influence during the 1920s and 1930s— encouraged both variety and richness in logic.

First-order logic emerged as a distinct subsystem of logic in Hilbert's lectures (1917) and, in print, in (Hilbert and Ackermann 1928). Nevertheless, Hilbert did not at any point regard first-order logic as the proper basis for mathematics. From 1917 on, he opted for the theory of types— at first the ramified theory with the Axiom of Reducibility and later a version of the simple theory of types (ω-order logic). Likewise, it is inaccurate to regard what Löwenheim did in (1915) as first-order logic. Not only did he consider second-order propositions, but even his first-order subsystem included infinitely long expressions.

It was in Skolem's work on set theory (1923) that first-order logic was first proposed as all of logic and that set theory was first formulated within first-order logic. (Beginning in [1928], Herbrand treated the theory of types as merely a mathematical system with an underlying first-order logic.) Over the next four decades Skolem attempted to convince the mathematical community that both of his proposals were correct. The first claim, that first-order logic is all of logic, was taken up (perhaps independently) by Quine, who argued that second-order logic is really set theory in disguise (1941, 144-45). This claim fared well for a while.[22] After the emergence of a distinct infinitary logic in the 1950s (thanks in good part to Tarski) and after the introduction of generalized quantifiers (thanks to Mostowski [1957]), first-order logic is clearly not all of logic.[23] Skolem's second claim, that set theory should be formulated in first-order logic, was much more successful, and today this is how almost all set theory is done.

When Gödel proved the completeness of first-order logic (1929, 1930a) and then the incompleteness of both second-order and ω-order logic (1931), he both stimulated first-order logic and inhibited the growth of second-order logic. On the other hand, his incompleteness results encouraged the search for an appropriate infinitary logic—by Carnap (1935) and Zermelo (1935). The acceptance of first-order logic as one basis on which to formulate all of mathematics came about gradually during the 1930s and 1940s, aided by Bernays's and Gödel's first-order formulations of set theory.

Yet Maltsev (1936), through the use of uncountable first-order languages, and Tarski, through the Upward Löwenheim-Skolem Theorem and the definition of truth, rejected the attempt by Skolem to restrict logic to countable first-order languages. In time, uncountable first-order languages and uncountable models became a standard part of the repertoire of first-order logic. Thus set theory entered logic through the back door, both syntactically and semantically, though it failed to enter through the front door of second-order logic.

Notes

1. Peano acknowledged (1891, 93) that his postulates for the natural numbers came from (Dedekind 1888).

2. Peirce's use of second-order logic was first pointed out by Martin (1965).

3. Van Heijenoort (1967, 3; 1986, 44) seems to imply that Frege did separate, or ought to have separated, first-order logic from the rest of logic. But for Frege to have done so would have been contrary to his entire approach to logic. Here van Heijenoort viewed the matter unhistorically, through the later perspectives of Skolem and Quine.

4. There is a widespread misconception, due largely to Russell (1919, 25n), that Frege's *Begriffsschrift* was unknown before Russell publicized it. In fact, Frege's book quickly received at least six reviews in major mathematical and philosophical journals by researchers such as Schröder in Germany and John Venn in England. These reviews were largely favorable, though they criticized various features of Frege's approach. The *Begriffsschrift* failed to persuade other logicians to adopt Frege's approach to logic because most of them (Schröder and Venn, for example) were already working in the Boolean tradition. (See [Bynum 1972, 209-35] for these reviews, and see [Nidditch 1963] on similar claims by Russell concerning Frege's work in general.)

5. Frege pointed this out to Hilbert in a letter of 6 January 1900 (Frege 1980a, 46, 91) and discussed the matter in print in (1903b, 370-71).

6. The suggestion that Sommer may have prompted Hilbert to introduce an axiom of continuity is due to Jongsma (1975, 5-6).

7. Forder, in a textbook on the foundations of geometry (1927, 6), defined the term "complete" to mean what Veblen called "categorical" and argued that a categorical set of axioms must be deductively complete. Here Forder presupposed that if a set of axioms is consistent, then it is satisfiable. This, as Gödel was to establish in (1930a) and (1931), is true for first-order logic but false for second-order logic. On categoricity, see (Corcoran 1980, 1981).

8. (Frege 1980a, 108). In an 1896 article Frege wrote: "I shall now inquire more closely into the essential nature of Peano's conceptual notation. It is presented as a descendant of Boole's logical calculus but, it may be said, as one different from the others.... By and large, I regard the divergences from Boole as improvements" (Frege 1896; translation in 1984, 242). In (1897), Peano introduced a separate symbol for the existential quantifier.

9. (Frege 1980a, 78). Nevertheless, Church (1976, 409) objected to the claim that Frege's system of 1893 is an anticipation of the simple theory of types. The basis of Church's objection is that for "Frege a function is not properly an (abstract) object at all, but is a sort of incompleted abstraction." The weaker claim made in the present paper is that Frege's system helped lead Russell to the theory of types when he dropped Frege's assumption that classes are objects of level 0 and allowed them to be objects of arbitrary finite level.

10. Russell in (Frege 1980a, 147; Russell 1903, 528). See (Bell 1984) for a detailed analysis of the Frege-Russell letters.

11. Three of these approaches are found in (Russell 1906): the zigzag theory, the theory of limitation of size, and the no-classes theory. In a note appended to this paper in February 1906, he opted for the no-classes theory. Three months later, in another paper read to the London Mathematical Society, the no-classes theory took a more concrete shape as the

substitutional theory. Yet in October 1906, when the Society accepted the paper for publication, he withdrew it. (It was eventually published as [Russell 1973].) The version of the theory of types given in the *Principles* was very close to his later no-classes theory. Indeed, he wrote that "*technically*, the theory of types suggested in Appendix B [1903] differs little from the no-classes theory. The only thing that induced me at that time to retain classes was the technical difficulty of stating the propositions of elementary arithmetic without them" (1973, 193).

12. Hilbert's courses were as follows: "Logische Prinzipien des mathematischen Denkens" (summer semester, 1905), "Prinzipien der Mathematik" (summer semester, 1908), "Elemente und Prinzipienfragen der Mathematik" (summer semester, 1910), "Einige Abschnitte aus der Vorlesung über die Grundlagen der Mathematik und Physik" (summer semester, 1913), and "Prinzipien der Mathematik und Logik" (winter semester, 1917). A copy of the 1913 lectures can be found in Hilbert's *Nachlass* in the Handschriftenabteilung of the Niedersächsische Staats- und Universitätsbibliothek in Göttingen: the others are kept in the "Giftschrank" at the Mathematische Institut in Göttingen. Likewise, all other lecture courses given by Hilbert and mentioned in this paper can be found in the "Giftschrank."

13. See (Hilbert 1917, 190) and (Hilbert and Ackermann 1928, 83). The editors of Hilbert's collected works were careful to distinguish the Principle of Mathematical Induction in (Hilbert 1922) from the first-order axiom schema of mathematical induction; see (Hilbert 1935, 176n). Herbrand also realized that in first-order logic this principle becomes an axiom schema (1929; 1930, chap. 4.8).

14. The course was entitled "Logische Grundlagen der Mathematik."

15. "Logische Grundlagen der Mathematik," a partial copy of which is kept in the university archives at Göttingen. On the history of the Axiom of Choice, see (Moore 1982).

16. Bernays also wrote about second-order logic briefly in his (1928).

17. On the early history of the ω-rule, see (Feferman 1986).

18. See, for example, (van Heijenoort 1967, 230; Vaught 1974, 156; Wang 1970, 27).

19. The recognition that Skolem in (1920) was primarily working in $L_{\omega_1, \omega}$ is due to Vaught (1974, 166).

20. For an analysis of Zermelo's views on logic, see (Moore 1980, 120-36).

21. During the same period Skolem (1961, 218) supported the interpretation of the theory of types as a many-sorted theory within first-order logic. Such an interpretation was given by Gilmore (1957), who showed that a many-sorted theory of types in first-order logic has the same valid sentences as the simple theory of types (whose semantics was to be based on Henkin's notion of general model rather than on the usual notion of higher-order model).

22. See (Quine 1970, 64-70). For a rebuttal of some of Quine's claims, see (Boolos 1975).

23. For the impressive body of recent research on stronger logics, see (Barwise and Feferman 1985).

References

Ackermann, W. 1925. Begründung des "Tertium non datur" mittels der Hilbertschen Theorie der Widerspruchsfreiheit. *Mathematische Annalen* 93: 1-36.

Barwise, J., and Feferman, S. eds. 1985. *Model-Theoretic Logics*. New York: Springer.

Bell, D. 1984. Russell's Correspondence with Frege. *Russell: The Journal of the Bertrand Russell Archives* (n.s.) 2: 159-70.

Bernays, P. 1918. *Beiträge zur axiomatischen Behandlung des Logik-Kalküls*. Habilitationsschrift, Göttingen.

———1928. Die Philosophie der Mathematik und die Hilbertsche Beweistheorie. *Blätter für Deutsche Philosophie* 4: 326-67. Reprinted in (Bernays 1976), pp. 17-61.

———1976. *Abhandlungen zur Philosophie der Mathematik*. Darmstadt: Wissenschaftliche Buchgesellschaft.

Bernays, P., and Schönfinkel, M. 1928. Zum Entscheidungsproblem der mathematischen Logik. *Mathematische Annalen* 99: 342-72.

Boole, G. 1847. *The Mathematical Analysis of Logic. Being an Essay toward a Calculus of Deductive Reasoning*. London.

——1854. *An Investigation of the Laws of Thought on Which Are Founded the Mathematical Theories of Logic and Probabilities.* London.

Boolos, G. 1975. On Second-Order Logic. *Journal of Philosophy* 72: 509-27.

Bynum, T. W. 1972. *Conceptual Notation and Related Articles.* Oxford: Clarendon Press.

Carnap, R. 1935. Ein Gültigkeitskriterium für die Sätze der klassischen Mathematik. *Monatshefte für Mathematik und Physik* 41: 263-84.

Church, A. 1976. Schröder's Anticipation of the Simple Theory of Types. *Erkenntnis* 10: 407-11.

Corcoran, J. 1980. Categoricity. *History and Philosophy of Logic* 1: 187-207.

——1981. From Categoricity to Completeness. *History and Philosophy of Logic* 2: 113-19.

Dedekind, R. 1888. *Was sind und was sollen die Zahlen?* Braunschweig.

De Morgan, A. 1859. On the Syllogism No. IV, and on the Logic of Relations. *Transactions of the Cambridge Philosophical Society* 10: 331-58.

Feferman, S. 1986. Introductory Note. In (Gödel 1986), pp. 208-13.

Fisch, M. H. 1984. The Decisive Year and Its Early Consequences. In (Peirce 1984), pp. xxi-xxxvi.

Forder, H. G. 1927. *The Foundations of Euclidean Geometry.* Cambridge: Cambridge University Press. Reprinted New York: Dover, 1958.

Fraenkel, A. A. 1922. Zu den Grundlagen der Mengenlehre. *Jahresbericht der Deutschen Mathematiker-Vereinigung* 31, Angelegenheiten, 101-2.

——1927. Review of (Skolem 1923). *Jahrbuch über die Fortschritte der Mathematik* 49 (vol. for 1923): 138-39.

Frege, G. 1879. *Begriffsschrift, eine der arithmetischen nachgebildete Formelsprache des reinen Denkens.* Halle: Louis Nebert. Translation in (Bynum 1972) and in (van Heijenoort 1967), pp. 1-82.

——1883. Über den Zweck der Begriffsschrift. *Sitzungsberichte der Jenaischen Gesellschaft für Medicin und Naturwissenschaft* 16: 1-10. Translation in (Bynum 1972), pp. 90-100.

——1884. *Grundlagen der Arithmetik: Ein logisch-mathematische Untersuchung über den Begriff der Zahl.* Breslau: Koebner.

——1891. *Function und Begriff. Vortrag gehalten in der Sitzung vom 9. Januar 1891 der Jenaischen Gesellschaft für Medecin und Naturwissenschaft.* Jena: Pohle. Translation in (Frege 1980b), pp. 21-41.

——1893. *Grundgesetze der Arithmetik, begriffsschriftlich abgeleitet.* Vol. 1. Jena: Pohle.

——1895. Review of (Schröder 1890). *Archiv für systematische Philosophie* 1: 433-56. Translation in (Frege 1980b, pp. 86-106.

——1896. Ueber die Begriffsschrift des Herrn Peano und meine eigene. *Verhandlungen der Königlich Sächsischen Gesellschaft der Wissenschaften zu Leipzig, Mathematisch-physicalische Klasse* 48: 362-68. Translation in (Frege 1984), pp. 234-48.

——1903a. Vol. 2 of (1893).

——1903b. Über die Grundlagen der Geometrie. *Jahresbericht der Deutschen Mathematiker-Vereinigung* 12: 319-24, 368-75.

——1980a. *Philosophical and Mathematical Correspondence.* Ed. G. Gabriel, H. Hermes, F. Kambartel, C. Thiel, and A. Veraart. Abr. B. McGuinness and trans. H. Kaal. Chicago: University of Chicago Press.

——1980b. *Translations from the Philosophical Writings of Gottlob Frege.* 3d ed. Ed. and trans. P. Geach and M. Black. Oxford: Blackwell.

——1984. *Collected Papers on Mathematics, Logic, and Philosophy.* Ed. B. McGuinness, and trans. M. Black et al. Oxford: Blackwell.

Gentzen, G. 1936. Die Widerspruchsfreiheit der Stufenlogik. *Mathematische Zeitschrift* 41: 357-66.

Gilmore, P. C. 1957. The Monadic Theory of Types in the Lower Predicate Calculus. *Summaries of Talks Presented at the Summer Institute of Symbolic Logic in 1957 at Cornell University* (Institute for Defense Analysis), pp. 309-12.

Gödel, K. 1929. Über die Vollständigkeit des Logikkalküls. Doctoral dissertation, University of Vienna. Printed, with translation, in (Gödel 1986), pp. 60-101.

——1930a. Die Vollständigkeit der Axiome des logischen Funktionenkalküls. *Monatshefte für Mathematik und Physik* 37: 349-60. Reprinted, with translation, in (Gödel 1986), pp. 102-23.

——1930b. Über die Vollständigkeit des Logikkalküls. *Die Naturwissenschaften* 18: 1068. Reprinted, with translation, in (Gödel 1986), pp. 124-25.

——1930c. Einige metamathematische Resultate über Entscheidungsdefinitheit und Widerspruchsfreiheit. *Anzeiger der Akademie der Wissenschaften in Wien* 67: 214-15. Reprinted, with translation, in (Gödel 1986), pp. 140-43.

——1931. Über formal unentscheidbare Sätze der Principia Mathematica und verwandter Systeme I. *Monatshefte für Mathematik und Physik* 38: 173-98. Reprinted, with translation, in (Gödel 1986), pp. 144-95.

——1944. Russell's Mathematical Logic. In *The Philosophy of Bertrand Russell*, ed. P. A. Schilpp. Evanston, Ill.: Northwestern Unviersity, pp. 123-53.

——1986. *Collected Works*. Ed. S. Feferman, J. W. Dawson, Jr., S. C. Kleene, G. H. Moore, R. M. Solovay, and J. van Heijenoort. Vol. I: *Publications 1929-1936*. New York: Oxford University Press.

Gonseth, F., ed. 1941. *Les entretiens de Zurich, 6-9 décembre 1938*. Zurich: Leeman.

Herbrand, J. 1928. Sur la théorie de la démonstration. *Comptes rendus hebdomadaires des séances de l'Académie des Sciences (Paris)* 186: 1274-76. Translation in (Herbrand 1971), pp. 29-34.

——1929. Sur quelques propriétés des propositions vrais et leurs applications. *Comptes rendus hebdomadaires des séances de l'Académie des Sciences (Paris)* 188: 1076-78. Translation in (Herbrand 1971), pp. 38-40.

——1930. *Recherche sur la théorie de la démonstration*. Doctoral dissertation, University of Paris. Translation in (Herbrand 1971), pp. 44-202.

——1971. *Logical Writings*. Ed. W. D. Goldfarb. Cambridge, Mass.: Harvard University Press.

Hilbert, D. 1899. *Grundlagen der Geometrie. Festschrift zur Feier der Enthüllung des Gauss-Weber Denkmals in Göttingen*. Leipzig: Teubner.

——1900a. Über den Zahlbegriff. *Jahresbericht der Deutschen Mathematiker-Vereinigung* 8: 180-94.

——1900b. Mathematische Probleme. Vortrag, gehalten auf dem internationalem Mathematiker-Kongress zu Paris, 1900. *Nachrichten von der Königlichen Gesellschaft der Wissenschaften zu Göttingen*, pp. 253-97.

——1902. *Les principes fondamentaux de la géometrie*. Paris: Gauthier-Villars. French translation of (Hilbert 1899) by L. Laugel.

——1903. Second German edition of (Hilbert 1899).

——1905. Über der Grundlagen der Logik und der Arithmetik. *Verhandlungen des dritten internationalen Mathematiker-Kongresses in Heidelberg vom 8. bis 13. August 1904*. Leipzig: Teubner. Translation in (van Heijenoort 1967), pp. 129-38.

——1917. *Prinzipien der Mathematik und Logik*. Unpublished lecture notes of a course given at Göttingen during the winter semester of 1917-18, (Math. Institut, Göttingen).

——1918. Axiomatisches Denken. *Mathematische Annalen* 78: 405-15. Reprinted in (Hilbert 1935), pp. 178-91.

——1922. Neubegründung der Mathematik (Erste Mitteilung). *Abhandlungen aus dem Mathematischen Seminar der Hamburgischen Universität* 1: 157-77. Reprinted in (Hilbert 1935), pp. 157-77.

——1923. Die logischen Grundlagen der Mathematik. *Mathematische Annalen* 88: 151-65. Reprinted in (Hilbert 1935), pp. 178-91.

——1926. Über das Unendliche. *Mathematische Annalen* 95: 161-90. Translation in (van Heijenoort 1967), pp. 367-92.

——1928. Die Grundlagen der Mathematik. *Abhandlungen aus dem Mathematischen Seminar der Hamburgischen Universität* 6: 65-85. Translation in (van Heijenoort 1967), pp. 464-79.

——1929. Probleme der Grundlegung der Mathematik. *Mathematische Annalen* 102: 1-9.

————1931. Die Grundlegung der elementaren Zahlenlehre. *Mathematische Annalen* 104: 485-94. Reprinted in part in (Hilbert 1935), pp. 192-95.

————1935. *Gesammelte Abhandlungen.* Vol. 3. Berlin: Springer.

Hilbert, D., and Ackermann, W. 1928. *Grundzüge der theoretischen Logik.* Berlin: Springer.

Hilbert, D., and Bernays, P. 1934. *Grundlagen der Mathematik.* Vol. 1. Berlin: Springer.

Huntington, E. V. 1902. A Complete Set of Postulates for the Theory of Absolute Continuous Magnitude. *Transactions of the American Mathematical Society* 3: 264-79.

————1905. A Set of Postulates for Ordinary Complex Algebra. *Transactions of the American Mathematical Society* 6: 209-29.

Jongsma, C. 1975. The Genesis of Some Completeness Notions in Mathematical Logic. Unpublished manuscript, Department of Philosophy, SUNY, Buffalo.

Lewis, C. I. 1918. *A Survey of Symbolic Logic.* Berkeley: University of California Press.

Löwenheim, L. 1908. Über das Auslösungsproblem im logischen Klassenkalkül. *Sitzungsberichte der Berliner Mathematischen Gesellschaft*, pp. 89-94.

————1910. Über die Auflösung von Gleichungen im logischen Gebietekalkül. *Mathematische Annalen* 68: 169-207.

————1913a. Über Transformationen im Gebietekalkül. *Mathematische Annalen* 73: 245-72.

————1913b. Potenzen im Relativkalkul und Potenzen allgemeiner endlicher Transformationen. *Sitzungsberichte der Berliner Mathematischen Gesellschaft* 12: 65-71.

————1915. Über Möglichkeiten im Relativkalkül. *Mathematische Annalen* 76: 447-70. Translation in (van Heijenoort 1967), pp. 228-51.

Maltsev, A. I. 1936. Untersuchungen aus dem Gebiete der mathematischen Logik. *Matematicheskii Sbornik* 1: 323-36.

Martin, R. M. 1965. On Peirce's Icons of Second Intention. *Transactions of the Charles S. Peirce Society* 1: 71-76.

Montague, R., and Vaught, R. L. 1959. Natural Models of Set Theories. *Fundamenta Mathematicae* 47: 219-42.

Moore, G. H. 1980. Beyond First-Order Logic: The Historical Interplay between Mathematical Logic and Axiomatic Set Theory. *History and Philosophy of Logic* 1: 95-137.

————1982. *Zermelo's Axiom of Choice: Its Origins, Development, and Influence.* Vol. 8: Studies in the History of Mathematics and Physical Sciences. New York: Springer.

Mostowski, A. 1957. On a Generalization of Quantifiers. *Fundamenta Mathematicae* 44: 12-36.

Nidditch, P. 1963. Peano and the Recognition of Frege. *Mind* (n.s.) 72: 103-10.

Peano, G. 1888. *Calcolo geometrico secondo l'Ausdehnungslehre di H. Grassmann, preceduto dalle Operazioni della logica deduttiva.* Turin: Bocca.

————1889. *Arithmetices principia, nova methodo exposita.* Turin: Bocca. Partial translation in (van Heijenoort 1967), pp. 83-97.

————1891. Sul concetto di numero. *Rivista di Matematica* 1: 87-102, 256-67.

————1897. Studii di logica matematica. *Atti della Reale Accademia delle Scienze di Torino* 32: 565-83.

Peirce, C. S. 1865. On an Improvement in Boole's Calculus of Logic. *Proceedings of the American Academy of Arts and Sciences* 7: 250-61.

————1870. Description of a Notation for the Logic of Relatives, Resulting from an Amplification of the Conceptions of Boole's Calculus of Logic. *Memoirs of the American Academy* 9: 317-78.

————1883. The Logic of Relatives. In *Studies in Logic*, ed. C. S. Peirce. Boston: Little, Brown, pp. 187-203.

————1885. On the Algebra of Logic: A Contribution to the Philosophy of Notation. *American Journal of Mathematics* 7: 180-202.

————1933. *Collected Papers of Charles Sanders Peirce.* Ed. C. Hartshorne and P. Weiss. Vol. 4: *The Simplest Mathematics.* Cambridge, Mass.: Harvard University Press.

————1976. *The New Elements of Mathematics.* Ed. Carolyn Eisele. Vol. 3: *Mathematical Miscellanea.* Paris: Mouton.

———1984. *Writings of Charles S. Peirce. A Chronological Edition*. Ed. E. C. Moore. Vol. 2: 1867-71. Bloomington: Indiana University Press.

Poincaré, H. 1905. Les mathématiques et la logique. *Revue de Métaphysique et de Morale* 13: 815-35.

Quine, W. V. 1941. Whitehead and the Rise of Modern Logic. In *The Philosophy of Alfred North Whitehead*, ed. P. A. Schilpp. Evanston, Ill.: Northwestern University, pp. 125-63.

———1970. *Philosophy of Logic*. Englewood Cliffs, N.J.: Prentice-Hall.

Ramsey, F. 1925. The Foundations of Mathematics. *Proceedings of the London Mathematical Society* (2) 25: 338-84.

Russell, B. 1903. *The Principles of Mathematics*. Cambridge: Cambridge University Press.

———1906. On Some Difficulties in the Theory of Transfinite Numbers and Order Types. *Proceedings of the London Mathematical Society* (2) 4: 29-53.

———1908. Mathematical Logic as Based on the Theory of Types. *American Journal of Mathematics* 30: 222-62.

———1919. *Introduction to Mathematical Philosophy*. London: Allen and Unwin.

———1973. On the Substitutional Theory of Classes and Relations. In *Essays in Analysis*, ed. D. Lackey. New York: Braziller, pp. 165-89.

Schönfinkel, M. 1924. Über die Bausteine der mathematischen Logik. *Mathematische Annalen* 92: 305-16. Translation in (van Heijenoort 1967), pp. 355-66.

Schröder, E. 1877. *Der Operationskreis des Logikkalkuls*. Leipzig: Teubner.

———1880. Review of (Frege 1879). *Zeitschrift für Mathematik und Physik* 25, Historisch-literarische Abteilung, pp. 81-94. Translation in (Bynum 1972), pp. 218-32.

———1890. *Vorlesungen über die Algebra der Logik (exacte Logik)*. Vol. 1. Leipzig. Reprinted in (Schröder 1966).

———1891. Vol. 2, part 1, of (Schröder 1890). Reprinted in (Schröder 1966).

———1895. Vol. 3 of (Schröder 1890). Reprinted in (Schröder 1966).

———1966. *Vorlesungen über die Algebra der Logik*. Reprint in 3 volumes of (Schröder 1890, 1891, and 1895). New York: Chelsea.

Skolem, T. 1913a. *Undersøkelser innenfor logikkens algebra* (Researches on the Algebra of Logic). Undergraduate thesis, University of Oslo.

———1913b. Om konstitusjonen av den identiske kalkyls grupper (On the Structure of Groups in the Identity Calculus). *Proceedings of the Third Scandinavian Mathematical Congress* (Kristiania), pp. 149-63. Translation in (Skolem 1970), pp. 53-65.

———1919. Untersuchungen über die Axiome des Klassenkalkuls und über Produktations- und Summationsproblem, welche gewisse Klassen von Aussagen betreffen. *Videnskapsselskapets skrifter, I. Matematisk-naturvidenskabelig klasse*, no. 3. Reprinted in (Skolem 1970), pp. 67-101.

———1920. Logisch-kombinatorische Untersuchungen über die Erfüllbarkeit oder Beweisbarkeit mathematischer Sätze nebst einem Theoreme über dichte Mengen. *Videnskapsselskapets skrifter, I. Matematisk-naturvidenskabelig klasse*, no. 4. Translation of §1 in (van Heijenoort 1967), pp. 252-63.

———1923. Einige Bemerkungen zur axiomatischen Begründung der Mengenlehre. *Videnskapsselskapets skrifter, I. Matematisk-naturvidenskabelig klasse*, no. 6. Translation in (van Heijenoort 1967), pp. 302-33.

———1933. Über die Möglichkeit einer vollständigen Charakterisierung der Zahlenreihe mittels eines endlichen Axiomensystems. *Norsk matematisk forenings skrifter*, ser. 2, no. 10, pp. 73-82. Reprinted in (Skolem 1970), pp. 345-54.

———1934. Über die Nicht-charakterisierbarkeit der Zahlenreihe mittels endlich oder abzählbar unendlich vieler Aussagen mit ausschliesslich Zahlenvariablen. *Fundamenta Mathematicae* 23: 150-61. Reprinted in (Skolem 1970), pp. 355-66.

———1941. Sur la portée du théorème de Löwenheim-Skolem. In (Gonseth 1941), pp. 25-52. Reprinted in (Skolem 1970), pp. 455-82.

———1958. Une relativisation des notions mathématiques fondamentales. *Colloques internationaux du Centre National de la Recherche Scientifique* (Paris), pp. 13-18. Reprinted in (Skolem 1970), pp. 633-38.

——1961. Interpretation of Mathematical Theories in the First Order Predicate Calculus. In *Essays on the Foundations of Mathematics, Dedicated to A. A. Fraenkel*, ed. Y. Bar-Hillel et al. Jerusalem: Magnes Press, pp. 218-25.

——1970. *Selected Works in Logic*. Ed. J. E. Fenstad. Oslo: Universitetsforlaget.

Sommer, J. 1900. Hilbert's Foundations of Geometry. *Bulletin of the American Mathematical Society* 6: 287-99.

Thiel, C. 1977. Leopold Löwenheim: Life, Work, and Early Influence. In *Logic Colloquim 76*, ed. R. Gandy and M. Hyland. Amsterdam: North-Holland, pp. 235-52.

van Heijenoort, J. 1967. *From Frege to Gödel: A Source Book in Mathematical Logic*. Cambridge, Mass.: Harvard University Press.

——1986. Introductory note. In (Gödel 1986), pp. 44-59.

Vaught, R. L. 1974. Model Theory before 1945. In *Proceedings of the Taski Symposium*, ed. L. Henkin. Vol. 25 of *Proceedings of Symposia in Pure Mathematics*. New York: American Mathematical Society, pp. 153-72.

Veblen, O. 1904. A System of Axioms for Geometry. *Transactions of the American Mathematical Society* 5: 343-84.

von Neumann, J. 1925. Eine Axiomatisierung der Mengenlehre. *Journal für die reine und angewandte Mathematik* 154: 219-40. Translation in (van Heijenoort 1967), pp. 393-413.

——1927. Zur Hilbertschen Beweistheorie. *Mathematische Zeitschrift* 26: 1-46.

Wang, H. 1970. Introduction to (Skolem 1970), pp. 17-52.

Whitehead, A. N., and Russell, B. 1910. *Principia Mathematica*. Vol. 1 Cambridge: Cambridge University Press.

Zermelo, E. 1929. Über den Begriff der Definitheit in der Axiomatik. *Fundamenta Mathematicae* 14: 339-44.

——1930. Über Grenzzahlen und Mengenbereiche: Neue Untersuchungen über die Grundlagen der Mengenlehre. *Fundamenta Mathematicae* 16: 29-47.

——1931. Über Stufen der Quantifikation und die Logik des Unendlichen. *Jahresbericht der Deutschen Mathematiker-Vereinigung* 41, Angelegenheiten, pp. 85-88.

——1935. Grundlagen einer allgemeinen Theorie der mathematischen Satzsysteme. *Fundamenta Mathematicae* 25: 136-46.

Added in Proof:

In a recent letter to me, Ulrich Majer has argued that Hermann Weyl was the first to formulate first-order logic, specifically in his book *Das Kontinuum* (1918). This is too strong a claim. I have already discussed Weyl's role in print (Moore 1980, 110-111; 1982, 260-61), but some further comments are called for here.

It is clear that Weyl (1918, 20-21) lets his quantifiers range only over objects (in his Fregean terminology) rather than concepts, and to this extent what he uses is first-order logic. But certain reservations must be made. For Weyl (1918, 19) takes the natural numbers as given, and has in mind something closer to ω-logic. Moreover, he rejects the unrestricted application of the Principle of the Excluded Middle in analysis (1918, 12), and hence he surely is not proposing *classical* first-order logic. Finally, one wonders about the interactions between Hilbert and Weyl during the crucial year 1917. What conversations about foundations took place between them in September 1917 when Hilbert lectured at Zurich and was preparing his 1917 course, while Weyl was finishing *Das Kontinuum* on a closely related subject?

II. REINTERPRETATIONS IN
THE HISTORY OF MATHEMATICS

Kronecker's Place in History

At a conference on the history and philosophy of mathematics, it seems especially appropriate to talk about Kronecker's place in history. My objective is to show that the prevalence of one philosophical viewpoint in contemporary mathematics has had a distorting influence on the way that that place—and with it several major issues in the history and philosophy of mathematics—have been viewed in our time.

Consider the story of Georg Cantor (1845-1918) and Leopold Kronecker (1823-91) as it is told by present-day writers:

Cantor was the founder of set theory—he even gave it is name. As such, he was the first to speak the language of modern mathematics and the first to deal rigorously with the transfinite entities that are the central notions of most of mathematics. His creation took great courage and had to overcome the prejudices of many generations of mathematicians who had regarded the notion of a completed infinite as a symptom of unrigorous thinking at best and, more likely, of complete nonsense. Chief among Cantor's opponents was Kronecker, "the uncrowned king of the German mathematical world" of those days. Kronecker was small in stature, but large in power. He was a wealthy man who could indulge in mathematics as a hobby, and he was the sort of difficult and prejudiced man who could make life unpleasant for those around him. He held extreme views on the foundations of mathematics, insisting that all of mathematics be based on the natural numbers, a limitation that would obviously nullify not only a large part of the mathematics of his own time but also almost all that has done since. His combativeness and his fanaticism poisoned his relations with his great contemporary in Berlin, Karl Weierstrass, and caused Weierstrass great unhappiness in his last years. But the primary victim of Kronecker's attacks was Cantor, a former student in Berlin, who, from his outpost in Halle, was trying to win that recognition for his theories. Cantor suffered a severe depression in 1884, verging on a nervous breakdown. One cannot determine a precise cause for such a breakdown, but he was under the dual strain of trying to solve the problem of the continuum hypothesis and of trying to withstand the opposition of Kronecker to his entire work. In the end, Cantor prevailed, although the effort nearly cost him his sanity. Kronecker died in 1891 and the torch of mathematics passed to a new generation, headed by David Hilbert, which recognized the true value of Cantor's theory and brought about the dawning of the modern age of mathematics, based on Cantor's firm foundation.

I'm sure you recognize the characters and the story. I would like you to consider, however, another interpretation of that controversy of a hundred years ago. Even Kronecker's enemies admit he was a superb mathematician, and he has many impressive friends. In the twentieth century, such outstanding mathematicians as Erich Hecke, Carl Ludwig Siegel, and Andre Weil have studied Kronecker's works intensively and have built on them in their own work. As far as Kronecker's personality is concerned, I have been able to find little evidence of his alleged hostility and aggressiveness. Although he was clearly a man who had strong opinions, there is every indication that his manners and his morals were thoroughly gentlemanly. I would rather accept the genial picture of him given by Frobenius in his *Gedaechtnisrede* than the ones given by E. T. Bell and Constance Reid. In the philosophy of mathematics, Kronecker's ideas have undeniably been part of a minority view, but it is a view that has continued to have its proponents, including Henri Poincaré, L. E. J. Brouwer, Hermann Weyl, and Errett Bishop. To have a name for the views that, roughly speaking, these men had in common I will use Brouwer's term "intuitionism," meaning the notion that mathematics must ultimately be based on the irreducible intuition of *counting*, of the natural numbers as a potentially infinite sequence.

As I understand it, the majority view rejects the intuitionistic views primarily on pragmatic grounds. It is thought that intuitionism would disallow too much of modern mathematics and would make that which remained too fussy and complicated. Cantor said, "The essence of mathematics lies in its freedom," and intuitionism is seen as a straitjacket.[1] But it is hard to imagine that anyone, even Cantor, believed that the essence of mathematics lay in its freedom. The essence of mathematics lies, rather, in its *truth*, by which I mean its power to convince us of its correctness.

If the use of transfinite logic diminishes its power to convince, and if, as the intuitionists maintain, virtually all of classical mathematics can be given an intuitionist foundation, then the use of transfinite logic in mathematics is both unwarranted and undesirable.

Kronecker believed that "someday people will succeed in 'arithmetizing' all of mathematics, that is, in founding it on the single foundation of the number-concept in its narrowest sense."[2] He said that this was his goal and that if he did not succeed then surely others who came after him would. This view inevitably put him in opposition not only to Cantor, whose "sets" could only appear to Kronecker as figments, but also to his colleague Weierstrass, who likewise had undertaken the "arithmetization of analysis." We have seen in our own time in the example of Errett Bishop how an outstanding mathematician can arrive at views of mathematics that place him, like it or not, in complete opposition to his colleagues, however much he might respect them. Quite simply, there is no way for a Bishop or a Kronecker to say politely, "I'm sorry, but your proofs do not prove what you think they do, and for you to convince me that your mathematics is worth my time you have to rethink in a fundamental way all you have done and show me that you have understood and can answer my objections. Meanwhile, I will pursue what seems to me a much more worthwhile project, that of establishing all of mathematics on a firm foundation, something I am confident I can do."

Kronecker, unlike Bishop, published nothing on his constructive program. He was engrossed in the last years of his life with his work on the theory of algebraic numbers and algebraic functions, and his intention to give an intuitionist development of the foundations of mathematics was never, as far as we know, seriously begun. Nonetheless, his thinking on foundational questions is evident in much of his work on mathematics proper, and, in my opinion, it is worthy of the highest respect. Therefore, I think it would give a truer picture of what happened between him and Cantor if the story were told in something like the following way.

The Weierstrass school, of which Cantor was a member, found that, in developing the theory of functions and in following up Cauchy's efforts to put the calculus on a firm foundation, they needed to use arguments dealing with infinity such as those that had been introduced by Bolzano earlier in the nineteenth century. Cantor went even further, completely disregarding the taboo against competed infinities and dealing with infinite collections that he called *Menge* or sets. Kronecker believed that none of these ways of dealing with infinity was acceptable or indeed necessary. He had a grand conception that all of mathematics would be based on the intuition of the natural numbers, but he never carried this conception to fruition. Naturally, Kronecker's denial of the validity of Weierstrass's arguments deeply wounded Weierstrass, and their relations went from great friendship in the early years to almost total alienation in Kronecker's last years, even though Kronecker insisted, perhaps insensitively, that a disagreement over mathematical questions should not affect their personal relations. Cantor's reaction to Kronecker's opposition was even stronger, in the first place because he was in a much weaker position vis-à-vis Kronecker because of his youth and his position at a provincial university, and in the second place because Cantor's personality contained a strain of paranoia that deeply disturbed at one time or another his relations with many contemporaries other than Kronecker, including such ostensible allies as Weierstrass, H. A. Schwarz, and G. Mittag-Leffler. Still, Cantor's ideas were taken up by Dedekind (who had in fact anticipated many of them) and Hilbert, and they became the basis of a new mode of mathematics that proved very fruitful and has dominated mathematics ever since, with only an occasional Brouwer or Weyl to oppose it.

I believe that the pendulum is beginning to swing back Kronecker's way, not least because of the appearance of computers on the scene, which has fostered a great upsurge of algorithmic thinking. Mathematicians are more and more interested in making computational sense of their abstractions. If I am right in thinking that Kronecker has been undervalued and caricatured by historians because they have been following the lead of the philosophers of mathematics and of mathematicians themselves, Kronecker's place in history may be about to improve considerably.

I will do what I can to bring this about. Surely one of the principal reasons Kronecker is so little studied today (except, it seems to me, by the best mathematicians) is that his works are so very difficult to read. Jordan very aptly called them in 1870 "l'envie et le désespoir des géometres."[3] This shows that the difficulty we experience today in reading Kronecker's works is not merely the difficulty of reading an old text written in an outdated terminology. Kronecker's difficult style was difficult for his contemporaries, too. I have been working at it for many years, especially his Kummer Festschrift entitled "Grundzuege einer arithmetischen Theorie der algebraischen Groessen."[4] I feel my efforts have been rewarded and that I will soon be able to publish papers that will convey something of what Kronecker was doing and that will, I hope, awaken the interest of others in studying Kronecker.

Let me conclude by giving a few indications of what it is that I find so valuable in Kronecker's works and how it relates to the history and philosophy of mathematics. It is said of Kronecker, as it is of Brouwer, that he succeeded in his mathematics by ignoring his intuitionist principles, but I think this allegation is completely untrue in both cases. Kronecker's principles permeate his mathematics—the problems he studies, the goals he sets for himself, the way he structures his theories. One of the parts of his Kummer Festschrift that I have studied most is his theory of what he calls "divisors," which is a first cousin of Dedekind's theory of "ideals." Two ways in which Kronecker's theory differs from Dedekind's show the difference in philosophy between the two men.

First, Dedekind regarded as the principal task of the theory the definition of "ideals" in such a way that the theorem on unique factorization into primes, which is false for algebraic numbers, becomes true for ideals. Kronecker took an altogether different view. He noted that the notion of "prime" was *relative to the field of numbers under consideration* and that if the field was extended things that had been prime might no longer be. For this reason, the theorem on unique factorization into primes properly belongs to a later part of the theory, after the basic concepts have been defined in a way that is independent of the field under consideration. Those of you who have studied algebraic number theory will undoubtedly have studied it using Dedekind's ideals and will remember that when you go from one field to another you have to manipulate the ideals— intersecting them with the lower field if the new field is smaller and taking the ideal in the larger field they generate if the new field is larger. In Kronecker's theory there is none of this. Divisors are defined and handled in such a way that nothing changes if the field is extended or if it is contracted.

Second, Kronecker defined his divisors, in essence, by telling how to *compute* with them. In Dedekind's terms, this amounts to giving an algorithm for determining, given a set of generators of an ideal, whether a given element of the field is or is not in the ideal. Nothing of the kind enters in Dedekind's theory because for Dedekind the definition was complete and satisfactory once the ideal was defined as a set, albeit an infinite set with no criterion for membership. For Kronecker, such a definition was by no means unthinkable, but it was certainly undesirable, and you can be sure he would settle for it only as a last resort. These are just two ways in which Kronecker's theory was affected—and improved—by his philosophy.

I have said above that, as far as we know, Kronecker never made a serious beginning in writing out his proposals for an intuitionist foundation of mathematics. This is not to say, however, that his hope that either he or a successor of his would do this was mere talk. I believe he had a unified view of all branches of mathematics and had, in many instances, fully thought-out ideas on how to base them on intuitionist principles. Had he devoted more time to this task and, I must add, had he been a better expositor, intuitionist ideas might not have had to endure a century of ostracism. It is my hope that this ostracism is now drawing to a close and that if not I then one of my successors will open the eyes of the mathematical world to the wealth of ideas hidden between the covers of Kronecker's collected works.

Notes

1. G. Cantor, *Gesammelte Abhandlugen* (Berlin: Lokay, 1932), p. 182.
2. L. Kronecker, *Werke*, vol. 3 (Leipzig: Teubner, 1899), p. 253.
3. C. Jordan, Introduction to "Traité des substitutions et des équations algébriques," Paris, 1870.
4. Kronecker's *Werke* was published in five volumes by Teubner, Leipzig, between 1895 and 1930. This paper is in vol. 2.

Felix Klein and
His "Erlanger Programm"

1. Introduction

Felix Klein's "Erlanger Programm" (E.P.), listed in our references as (Klein 1872), is generally accepted as a major landmark in the mathematics of the nineteenth century. In his obituary biography Courant (1925) termed it "perhaps the most influential and widely read paper in the second half of the nineteenth century." Coolidge (1940, 293) said that it "probably influenced geometrical thinking more than any other work since the time of Euclid, with the exception of Gauss and Riemann."

In a thoughtful recent article, Thomas Hawkins (1984) has challenged these assessments, pointing out that from 1872 to 1890 the E.P. had a very limited circulation; that it was "Lie, not Klein" who developed the theory of continuous groups; that "there is no evidence...that Poincaré ever studied the Programm;" that Killing's classification of Lie algebras (later "perfected by Cartan") bears little relation to the E.P.; and that Study, "the foremost contributor to...geometry in the sense of the Erlanger Programm,...had a strained and distant relationship with Klein."

Our paper should be viewed as a companion piece to the study by Hawkins. In our view, Klein's E.P. did have a major influence on later research, including especially his own. Moreover, Klein was the chief heir of five outstanding Germanic geometers, all of whom died within the decade preceding the E.P.: Möbius (1790-1868), Steiner (1796-1869), von Staudt (1798-1867), Plücker (1801-68), and Clebsch (1833-72).

Klein's close friendship with Lie at the time of the E.P. played an important role in the careers of both men. There is much truth in the frequently expressed idea that Klein's studies of 'discontinuous' groups were in some sense complementary to Lie's theory of 'continuous' groups (Coolidge 1940, 304). Less widely recognized is the fact that in the E.P. and related papers, Klein called attention to basic *global* aspects of geometry, whereas Lie's theorems were purely *local*. After reviewing these and other aspects of Klein's relationship to Lie, we will trace the influence of his ideas on Study, Hilbert, Killing, E. Cartan, and Hermann Weyl.

In discussing these developments, we have tried to follow a roughly chronological order. We hope that this will bring out the well-known evolution of Klein's scientific personality, from that of a brilliant, creative young geometer to that of a farsighted organizer and builder of institutions in his middle years, and finally to that of a mellow elder statesman or "doyen" and retrospective historian of ideas.[1]

2. Klein's Teachers

Klein's E.P. surely owes much to his two major teachers: Plücker and Clebsch. Already as a youth of 17, Klein became an assistant in Plücker's physics laboratory in Bonn. Though primarily a physicist, Plücker was also a very original geometer. Forty years earlier, he had written a brilliant monograph on analytic projective geometry (*Analytisch-geometrische Entwickelungen*, 1828, 1832), which established the use of homogeneous coordinates and the full meaning of duality.

Still more original (and more influential for Klein) was Plücker's "geometry of lines," first proposed in 1846. Plücker proposed taking the self-dual four-dimensional manifold of all *lines in* $\mathbf{R}^3$ as the set of basic "elements" of geometry. In it, the sets of all "points" and of all "planes" can be defined as what today would be called three-dimensional algebraic varieties. Klein's Ph.D. thesis (directed by Plücker) and several of his early papers dealt with this idea.

Shortly before Klein finished his thesis when still only 19, Plücker died. Clebsch, who had just gone to Göttingen from Giessen, invited Klein to join him there soon after. Barely 35 himself, Clebsch became Klein's second great teacher. After making significant contributions to the mechanics of continua (his book on elastic bodies was translated into French and edited by St. Venant), Clebsch had introduced together with his Giessen colleague Gordan (1831-1912) the *invariant theory* of the British mathematicians Boole (1815-64), Cayley (1821-95), and Sylvester (1814-97).

Clebsch assigned to Klein the task of completing and editing the second half of Plücker's work on the geometry of lines (*Neue Geometrie des Raumes gegrundet auf die Betrachtung der geraden Linie als Raumelemente*). As we shall see, Klein also absorbed from Clebsch both the concept of geometric invariant and Clebsch's interest in the "geometric function theory" of Riemann, whose intuitive approach contrasted sharply with the uncompromising rigor of Weierstrass.

Soon after coming to Göttingen from Giessen, Clebsch founded the *Mathematische Annalen* (M.A.) together with Carl Neumann in Leipzig. The first volume of this journal contained six papers by Clebsch and Gordan, along with others by Beltrami, Bessel, Brill, Cayley, Hankel, Jordan, Neumann, Sturm, Weber, and Zeuthen. Of these, Jordan's "Commentaire sur Galois" (151-60) and his paper on Abelian integrals (583-91) are especially relevant for the E.P., as forerunners of his 667-page treatise *Traité des substitutions*, which appeared a year later. In the preface of this great classic on the theory of substitution groups, Jordan thanked Clebsch for explaining how to "attack the geometric problems of Book III, Chap. III, the study of Steiner groups, and the trisection of hyperelliptic functions."

In 1869-70, Klein went to Berlin, against the advice of Clebsch. Although he found its intellectual atmosphere, dominated by Weierstrass and Kronecker, very uncongenial, it was there that Klein met two fellow students who deeply influenced his later development.

The first of these was Stolz. In his (EdM, vol. 1, 149), Klein states that he had been greatly impressed by Cayley's sententious dictum that "descriptive [i.e., projective] geometry is all geometry." Continuing, Klein states:

> In 1869, I had read Cayley's theory of metric projective geometry in Salmon's *Conics*, and heard for the first time about the work of Lobachewsky-Bolyai through Stolz in the winter of 1869-70. Although I understood very little of these indications, I immediately had the idea that there must be a connection between the two. (EdM, vol. 1, 151-52)

Klein then recalls a lecture he gave at Weierstrass's seminar on Cayley's *Massbestimmung* in projective geometry. He kept to himself his vaguely conceived thoughts about the connection of Cayley's work with non-Euclidean geometry, realizing how quickly the meticulous Weierstrass would have dismissed their vagueness, and how much he would have disliked their emphasis on the projective approach to geometry.[2]

Most important, it was during this visit to Berlin that Klein met Lie. In his own words (GMA, vol. 1, 50):

> The most important event of my stay in Berlin was certainly that, toward the end of October, at a meeting of the Berlin Mathematical Society,

I made the acquaintance of the Norwegian, Sophus Lie. Our work had led us from different points of view finally to the same questions, or at least to kindred ones. Thus it came about that we met every day and kept up an animated exchange of ideas. Our intimacy was all the closer because, at first, we found very little interest in our geometrical concerns in the immediate neighborhood.

3. Five Dazzling Years

The next five years were ones of incredible achievement for the young Klein. During this time he wrote 35 papers and supervised seven Ph.D. theses. It was while carrying on these activities that he wrote his E.P., and so it seems appropriate to recall some of his more important contemporary contacts.

About his continuing interactions with Stolz, Klein states:

In the summer of 1871, I was again with Stolz.... He familiarized me with the work of Lobachewsky and Bolyai, as well as with that of von Staudt. After endless debates with him, I finally overcame his resistance to my idea that non-Euclidean geometry was part of projective geometry, and published a note about it in...the Math. Annalen (1871). (EdM, vol. 1, 152)

After a digression on the significance of Gaussian curvature, Klein next recalls the background of his second paper on non-Euclidean geometry (Klein 1873). In it, Klein

investigated the foundations of von Staudt's [geometric] system, and had a first contact with modern axiomatics.... However, even this extended presentation did not lead to a general clarification.... Cayley himself mistrusted my reasoning, believing that a "vicious circle" was buried in it. (EdM, vol. 1, 153)

The connections of Clebsch with Jordan, whose monumental *Traité des substitutions* had just appeared, surely encouraged Klein to study groups and to go to Paris in 1870, where his new friend Lie rejoined him; the two even had adjacent rooms. In Paris, the two friends again talked daily, and also had frequent discussions with Gaston Darboux (1842-1917). There they extended an old theorem of Chasles, which states that orientation-preserving rigid transformations of space are, in general, 'screw motions' of translation along an axis and rotation about it. This theorem had been previously extended by Olinde Rodrigues (see Gray 1980) and Jordan (1869). Klein and Lie wrote two joint notes in Paris, the publication of which in the *Comptes Rendus* was sponsored by Chasles. They also published a third note in the *Berliner Monatshefte* of December 1870; by this time, however, the Franco-Prussian war had forced Klein to return to Germany.

The following year, Klein and Lie published two more joint papers in the *Mathematische Annalen* (GMA XXV and XXVI), of which the second (M.A. 4: 54-84) dealt with "systems of plane curves, transformed into each other by infinitely many permutable linear transformations." Thus it was around this time that Klein and Lie began to give a new direction to geometry, by emphasizing the importance for it of *continuous groups*, about which very little had been published previously (see §8).

It was also in 1871 that Lie's dissertation, *Over en Classe Geometriske Transformatione*, appeared. Originally published in Norwegian, a German translation (by Engel) is in (LGA, vol. 1, XI). It refers repeatedly to Plücker and cites related publications of Klein (LGA, vol. 1, 106, 127, 145, 149). Although seven years older than Klein, Lie had finished his dissertation two years later.

The following year (1872) was especially momentus for Klein. Partly through the influence of Clebsch, he obtained a full professorship at Erlangen when still only 23! It was for his inauguration to this chair that Klein wrote the E.P. Before it was finished, Clebsch died of diptheria at the age of 39, and most of his students moved from Göttingen to Erlangen to work with Klein. These students included Harnack, who later proved a famous theorem in potential theory, and Lindemann, who later published the first proof of the transcendence of π. Klein immediately gave Lindemann the task of writing Clebsch's unpublished lecture notes on geometry, a task that it would take Lindemann fifteen years to complete.

Not only Clebsch and Plücker, but also Möbius, Steiner, and von Staudt had died in the five-year period of 1867-72. Thus by accident, Klein had fallen heir to a great German tradition at age 23. The later sections of our paper will explain how he met this challenge successfully, in a most original and decisive way.

4. The "Erlanger Programm"

The E.P. was above all an affirmation of the key role played by groups in geometry. The most authentic description of its background is contained in Klein's prefatory notes to the relevant section of his *Collected Papers* (GMA, vol. 1, 411-16). There Klein writes:

In the following third part of the first volume of my *Collected Papers*...are collected those involving the concept of a *continuous transformation group.*

The...first two are joint publications with Lie in the summer of 1870 and the spring of 1871, on "W-curves."[3]

These were followed by my 1871 article "On Line and Metric Geometry".... As in Lie's great work "On Line and Sphere Complexes," new examples of continuous transformation groups were treated in this paper.

In this connection my two papers on non-Euclidean geometry should be cited.[4]

The E.P. itself...was composed in October, 1872. Two circumstances are relevant. First, that Lie visited me for two months beginning September 1. Lie, who on October 1 accompanied me to Erlangen... had daily discussions with me about his new theory of first-order partial differential equations (edited by me and published in the Gött. Nachr. of October 30). Second, Lie entered eagerly into my idea of classifying the different approaches to geometry on a group-theoretic basis.

The E.P. should not be judged as a research paper; it was a semitechnical presentation to the Erlangen philosophical faculty of ideas about geometry that Klein had discussed with Lie; it was primarily the exposition that was Klein's. The paper was published to fulfill an "obligation connected with Klein's appointment to the university" (Rowe 1983). As a new full professor, he was expected to present his colleagues with a printed exposition of some creative work. (In addition, he gave an oral *Antrittsrede* presenting his views on mathematical education; see [Rowe 1985].)

Klein's "Comparative Consideration of Recent Developments in Geometry," which was also intended to impress Clebsch, fulfilled this obligation in a most brilliant and original way. Its first section announced its main theme as follows (E.P., 67): "Given a manifold and a transformation group acting on it, to investigate those properties of figures [*Gebilde*] on that manifold which are invariant under [all] transformations of that group." In today's language, Klein proposed studying the concept of a *homogeneous manifold*: a structure [M,G] consisting of a manifold M and a *group G* acting transitively on M. This contrasts sharply with Riemann's concept of a structure [$M;d$] consisting of a manifold on

which a *metric d(p,q)* is defined by a local *distance* differential
$ds^2 = \Sigma g_{ij} dx_i dx_j$.

Two paragraphs later, Klein restated his proposal in a single terse sentence: "Given a manifold, and a transformation group acting on it, to study its *invariants.*" Thus Klein was also proposing to apply to geometry the concept of an 'invariant' that Clebsch, Jordan, and their predecessors had previously applied to algebra, and there only to the full linear group.

Klein began his essay by identifying each of the *continuous groups* of geometric transformations that was associated with some branch of geometry. Naturally, since the Euclidean group of rigid motions ("congruences") was familiar to his readers, he discussed it first, calling it the *"Hauptgruppe"* (chief group). Indeed, the importance of 'free mobility' had already been stressed by Helmholtz and Riemann. He then moved on to the larger projective group, the conformal group generated by inversions, the group of birational transformations leaving invariant the singularities of algebraic varieties, which contained all of the preceding groups, and the (still more general) group of all homeomorphisms leaving the topology of a manifold or space invariant. Curiously, the E.P. failed to mention the affine group and affine geometry, although Klein would later (1908) devote the first section of his chapter on geometric transformations in his *Elementarmathematik vom Höheren Standpunkt aus* to them. His failure to distinguish the group of homeomorphisms from the groups of bijections and of diffeomorphisms is less surprising, since Cantor's discovery that all $\mathbf{R}^n$ are bijective was still two years away.

Klein's audience surely did not notice these lapses. Indeed, his main new concepts, those of 'equivalence' and 'invariance' under a given group of transformations, are still hard to explain to novices today. One difficulty is that invariance and equivalence assume such a bewildering variety of forms depending on the group involved. (See H. Weyl, *The Classical Groups*, 23-26, and G. Birkhoff and S. Mac Lane, *Survey of Modern Algebra*, §9.5, for two more 'modern' attempts to explain these concepts.)

However, the main thrust of the E.P. was very clear. Jordan's *Traité des substitutions* had explained how the concept of a *group* of 'substitutions', properly applied, determines which polynomial equations can be solved by 'radicals'—i.e., by explicit formulas involving rational operations and taking of *n*-th roots. Klein showed similarly how the concept of a *continuous transformation group* gives an objective basis for classifying geometric theories and theorems. His few readers and listeners must have realized that they were being exposed to new and fundamental perspectives, even if they barely understood them.

5. Klein and Geometry

In 1872, the E.P. was 20 years ahead of its time; it would take at least that long for the new perspectives of Klein and Lie to gain general acceptance. (However, their ideas would take deep root; in fifty years it would become commonplace to refer also to "metric," "projective," "affine," and "conformal" *differential* geometry; see §13.)

In the meantime, Klein's rapid rise to leadership in Germanic geometry was based on his other writings. His preeminence rested on the impression he made on contemporaries, and not on what he might write for posterity. The value of his distinction between the familiar Riemann *sphere* (often associated with conformal geometry) and the *elliptic plane*, consisting of the sphere with opposite points identified, was immediately appreciated.

So was Klein's use of the name "elliptic geometry" for a manifold of constant positive curvature, as distinguished from a "hyperbolic geometry" for one having constant negative curvature, and "parabolic geometry" for Euclidean geometry and its cylindrical and toroidal "space forms" (Kline 1972, 913). (Not long after, du Bois-Reymond made an analogous classification of differential equations into those of "elliptic," "hyperbolic," and "parabolic" type.)

Klein's observations stimulated the British mathematician W. K. Clifford (1873) to call attention to the philosophical difference between local and global homogeneity, which had been overlooked by Riemann and Helmholtz. Clifford also called attention (as did Klein) to the connection between Plücker's line geometry and 'screws' (the Theorem of Chasles). However, he died before he could develop these ideas very far, and Klein would not return to them until 1890 (see §§7 and 8).

Klein also showed the logical incompleteness of von Staudt's path-breaking introduction of coordinates into axiomatically defined "projective geometries," an observation that stimulated Lüroth and others to clarify the assumptions underlying von Staudt's "algebra of throws."[5]

Although most mathematicians today would consider the E.P. as primarily a contribution to the *foundations* of geometry, Klein did not regard it as such, perhaps because in 1872 the concept of a continuous group was so novel, and even the theory of invariants still a research frontier. His prefatory remarks in his (GMA) about his contributions to the foundations of geometry concentrate on the topics that we have discussed above. By his own reckoning (GMA, vol. 1), Klein published ten papers on the foundations of geometry, fourteen on line geometry (an interest he had inherited from Plücker), and nine on the E.P. (of which the first was with Lie). He also lists (GMA, vol. 2) sixteen papers on intuitive (*Anschauliche*) geometry; in addition, he wrote several books on geometry. In particular, his *Einleitung in die Höhere Geometrie* of 1893 gave a general and quite comprehensive picture of the geometry of the day, and the second volume of his *Elementarmathematik vom Höheren Standpunkt aus* (1908) was also devoted to geometry. But of all his geometrical contributions over the years, Klein himself apparently regarded the E.P. as his "most notable achievement" (Young 1928, v).

6. Klein and Lie

The friendship formed by Klein and Lie in the unwelcoming atmosphere of Berlin in 1869, renewed in Paris in 1870 and again in Erlangen in 1872, proved invaluable for both men and for mathematics. We have already observed that much of the inspiration for the E.P. stemmed from their discussions during these years, and we shall now describe some of the less immediate and more worldly benefits derived by Lie from this friendship.[6]

Whereas Klein had eloquently expressed, within months of conceiving it, his idea that different branches of geometry were associated with invariance under different *groups of transformations* of underlying geometric manifolds, Lie's deeper ideas would mature much more slowly. He would spend the rest of his life in developing the intuitive concept of a 'continuous group of transformations' into a powerful general theory and in applying this concept to geometry and to partial differential equations.

Max Noether, in his obituary article (1900), described Lie's and Klein's 1869-70 sojourn in Berlin and the connections between the E.P. and Lie's later work:

> In the E.P. [we find expressed] for the first time the central role of the [appropriate] transformation group for all geometrical investigations...and that, with invariant properties, there is always associated such a group.... Lie, who had worked with the most varied groups, but to whom the meaning of classification had remained foreign, found the idea congenial from then on. (Noether 1900, 23-23)

Noether called attention in a footnote to alterations made by Lie in several of his 1872 articles, apparently as a result of Klein's new ideas.

During the decade 1872-82, Lie worked in isolation in Christiania (now Oslo), encouraged almost exclusively by Klein and Adolf Mayer (1839-1908). It was then that he published the striking fact that every finite continuous group ("Lie group" in today's terminology) acting on the line is locally equivalent (ähnlich) to either the translation group of all functions $x \mapsto x + b$, the affine group of all $x \mapsto ax + b$, or the projective group of all $x \mapsto (ax + b)/(cx + d)$, $ad \neq bc$ (LGA, vol. 5, 1-8; Gött. Nachr. 22 [1874]: 529-42). (For an annotated summary, see [Birkhoff 1973, 299-305]. Actually, the theorem is only true locally.)

Lie's thoughts at this time are revealed in a letter of 1873 to Mayer, which states: "I have obtained most interesting results and I expect very many more. They concern an idea whose origin may be found in my earlier works with Klein: namely, to apply the concepts of the theory of substitutions to differential equations" (LGA, vol. 5, 584). Four years later, Lie determined locally (almost) all finite continuous groups acting transitively on a two-dimensional manifold, the next step toward determining all the homogeneous manifolds (or "spaces") envisioned in the E.P.[7]

This was the third in a series of five definitive papers, published in Christiania in the years 1876-79, in which Lie laid the foundations of his theory of continuous groups. The introduction to the first of these states in part (LGA, vol. 5, 9):

> I plan to publish a series of articles, of which the present one is the first, on a new theory that I will call the *theory of transformation groups*. The investigations just mentioned have, as the reader will notice, many points of contact with several mathematical disciplines, especially with the theory of substitutions,[1] with geometry and modern manifold theory,[2] and finally also with the theory of differential equations.[3]

> These points of contact establish connections between these former separate fields.... I must prepare later articles to present the importance and scope of the new theory.

> [1]See Camille Jordan's *Traité des substitutions (Paris, 1870)*. Compare also Jordan's investigations of groups of motions.

> [2]See various geometric works by Klein and myself, especially Klein's [E.P.], which hitherto has perhaps not been studied sufficiently by mathematicians.

> [3]See my investigations on differential equations.

To reach a wider mathematical audience, Lie summarized his new theory of transformation groups in an 88 page paper (Lie 1880) published in German (LGA, vol. 6, I, III). However, even after this, it took another decade for Lie's ideas to be digested and their profound implications seen by most mathematicians. Lie became discouraged, and it seems clear that he owed much to Klein's continuing encouragement and insightful comments during these years (Rowe 1985).

As we shall see in §7, it was thanks largely to Klein's initiative that Lie obtained the cooperation of Friedrich Engel (1868-1941), the coauthor of his magnum opus (and of his posthumous collected works). Without Engel's expository cooperation, the dissemination of Lie's deep new ideas and methods would probably have been much slower.

Lie's relationship with Klein as well as the value for Lie of the latter's extensive correspondence with Mayer (much of it reproduced in [LGA]) has been described by Engel (1900). There Mayer is credited with persuading Clebsch of the value of Lie's new "integration method" (p. 37), and Lie's exchange with Mayer is presented as an example of two mathematicians "making the same discovery independently. . . and almost simultaneously."

Engel went on to describe Lie's "invariant theory of contact transformations," which appeared almost immediately after the E.P. as a completely new theory, entirely due to Lie. On the other hand, like Max Noether, he described Klein's idea that *many* domains of mathematics could be presented as invariant theories of appropriate groups as new and surprising to Lie.

Killing. Klein was also instrumental in establishing contact between Lie and Wilhelm Killing (1847-1923), who was a contemporary of Lie and Klein. As of 1884, Killing was groping toward concepts closely related to those of Lie. Hawkins (1982, §2) has described Klein's benign intervention as follows:

> Shortly after Killing posted a copy of his *Programmschrift* to Felix Klein in July of 1884, Klein informed him that its contents seemed to be closely related to the theory of transformation groups of his friend, Sophus Lie. . . .
>
> Upon learning about Lie, Killing sent him a copy of his *Programmschrift.* . . . Lie apparently did not respond to Killing's overture and definitely did not reciprocate and send Killing some copies of his own

work. . . . Lie quickly published a note in the *Archiv* [1884] in which he showed how a theorem stated without proof by Killing in [1884] could be derived from some of his previously published results.

Nevertheless, Killing was not discouraged, and with some help from Engel (see §7), he persevered in his effort to classify Lie algebras.

7. Leipzig and Lie Groups

Among the first to appreciate Lie's theory of contact transformations was Adolf Mayer at the University of Leipzig. Mayer had published papers developing Lie's ideas from 1872 on, first on contact transformations and then on Lie's methods for integrating differential equations invariant under a group.

In 1876, Mayer in Leipzig and Klein (then in Munich) had taken over from Carl Neumann the direction of the *Mathematische Annalen*, and in 1882 Klein joined Mayer in Leipzig. In the next year, Engel wrote there his inaugural dissertation *Zur Theorie der Berührungstransformationen* (Teubner, 1883); in the following year, Klein and Mayer sent Engel to Christiania to work with Lie. There Lie could inspire Engel, while Engel could help the lonely and somewhat disorganized Lie to organize and write up his profound discoveries in a systematic and readable form. Engel recalled later that in 1883, apart from his old friends F. Klein and A. Mayer, almost no one was interested in the group theory of which Lie was rightfully very proud. In 1883, only Picard recognized the significance of groups explicitly (*offentlich*) (Engel 1900, 42).[8]

Still later, Engel published (LGA, foreword to vol. 5) a vivid description of his year in Christiania with Lie, with whom he had "two conversations daily." Engel saw that Lie's seventeen published papers gave only a "very incomplete picture of the great buildings which he had in mind," and was inspired to "work with all his strengths" to bring them "as near to completion as possible." Future historians of mathematics should find Engel's careful analysis in this foreword of Lie's publications and Lie's opinions about them invaluable. A similar remark applies to the introduction to (Engel and Faber 1932).

Whereas Engel's inaugural dissertation had been merely competent, his *Habilitationsschrift* of 1885 (Teubner) showed genuine originality. In its introduction, Engel wrote:[9]

A long stay in Christiania gave me the opportunity, in personal exchange with Sophus Lie, to study in depth [*eingehend*] his theory of continuous transformation groups. Our common goal was to provide a coherent presentation of this theory, my share of the work being essentially only expository. In the process, however, I also occupied myself with some self-contained investigations in this area. My results will be explained in the course of this article. At present, our intended coherent presentation of the theory is not nearly finished... To understand what follows would, therefore, require knowing quite a few of Lie's papers on transformation groups. To minimize this difficulty, a brief summary will be given next to the principal concepts and theorems of Lie's theory, insofar as they will be needed below.

One year later, and to the intense displeasure of Weierstrass, Klein arranged to be replaced by Lie at the University of Leipzig upon his departure for Göttingen. Lie's move to Leipzig, where he was with both Mayer and Engel, proved to be most fruitful. It made it practical for Engel to serve as Lie's disciple, a role that lasted until he finished acting as coeditor with Paul Heegard of Lie's *Collected Works* (LGA) in 1924-34.

Indeed, the first volume of Lie-Engel was completed three years later, and in 1889 Fr. Schur (1856-1932) gave the first rigorous treatment of the *abstract* theory of (local) Lie groups. By 1893, all three volumes of Lie-Engel had appeared, and the (local) theory of Lie groups was firmly established.

In the meantime, Killing published his book *Zur Theorie der Lie'schen Transformationsgruppen* (1886), in which he determined almost all simple Lie algebras (see Hawkins 1982). Its preface states:

> Herr Klein kindly called my attention to the close connection between my investigations and [earlier papers of Lie on finite continuous transformation groups, their treatment by integration methods, and contact transformations].

> Now, at the time that my [1884] monograph appeared, I had begun another work, which occupied me longer than I had initially expected. After finishing this... I immediately began to study Lie's work.

> There can be no doubt of Lie's priority [as regards many parts of my earlier work]. I can only express my joy that the many-sided researches of Lie have so essentially advanced the general theory [of homogeneous spaces].

Unfortunately, Lie was not happy in Leipzig, and there he became excessively jealous of possible rivals. Thus the introduction of the third volume of Lie-Engel goes out of its way to assert that Lie was "not Klein's student," but that "rather, the contrary was the case."[10] Likewise, in a paper reproduced in (LGA, vol. 2, 472-79), he takes pains to identify gaps in the reasoning of de Tilly, Klein (M.A. 37: 364), Lindemann, Fr. Schur (who had, on the contrary, actually rigorized Lie's somewhat cavalier differentiability assumptions), Helmholtz, and Killing!

A turning point seems to have come when Lie was awarded the first Lobachewsky Prize in 1893 for solving the Riemann-Helmholtz problem (a celebrated problem, which, incidentally, Klein had suggested he work on). Namely, Lie had shown that any n-dimensional Riemannian manifold admitting an $n(n + 1)/2$-parameter group of rigid motions (the "free mobility" condition of Helmholtz) is locally isometric to either Euclidean n-space, the n-sphere, or the n-dimensional "hyperbolic" geometry of Lobachewsky-Bolyai. This classic *local* result of Lie stands in sharp contrast with Klein's continuing concern with the *global* Clifford-Klein problem, to which we will return in §8.

Appropriately, Klein was invited to write a suitable appreciation of Lie's solution for the occasion of the prize presentation, which he did with his usual imaginative, insightful style. His narrative contained, however, one complaint: the presence of an unmotivated, and to Klein unnatural, assumption of differentiability in the foundations of geometry.

Klein's complaint was given a positive interpretation by Hilbert. As the fifth in his famous list of unsolved problems proposed at the 1900 International Mathematical Congress, Hilbert proposed proving that any *continuous* (locally Euclidean) group was in fact an *analytical* group with respect to suitable parameters. Whether this problem should be attributed to Hilbert, Klein, or Lie, its successful solution took another 50 years, and Klein's E.P. was at least one of its indirect sources (see [Birkhoff and Bennett, forthcoming]).

8. Klein and Discontinuous Groups

Much as Lie found in invariance under continuous groups of transformations new ideas for integrating differential equations, Klein found in discontinuous groups new ideas not only for solving algebraic problems (as Jordan had before him), but also (later) for clarifying and extending the 'geometric function theory' originated by Riemann and advanced by

Clebsch. Indeed, already in 1875, Klein had supervised an Erlangen thesis by Harnack on elliptic functions and another by Wedekind in the next year entitled *On the Geometric Interpretation of Binary Forms*. A summary of Wedekind's thesis appeared (M.A. 9: 209-17),[11] immediately preceded by a paper (pp. 183-208) in which Klein used the Schwarz reflection principle to construct Riemann surfaces from regular polygons on the complex sphere. The special case of a rectangle, of course, leads to the elliptic function *sn z*, whose Riemann surface is a torus.

On page 193 of this paper, Klein first associated the regular octahedron with a biquadratic form, and then related the symmetric group of all permutations of five letters to the regular icosahedron as follows: "The 15 planes that pass through the center of the icosahedron and four pairwise antipodal vertices can be divided into 5 triples of orthogonal planes. The 15 lines in which these triples of planes intersect cut the sphere in 30 'doubly counted' points."[12]

Discussions of this paper with Gordan stimulated Klein to write a sequel entitled "On the Icosahedron" (M.A. 12 [1877]: 503-60), in which he "derived the theory of the quintic equation from geometric properties of the icosahedron." This turned out to have interesting connections with work of Jordan (see M.A. 11: 18) and Brioschi (M.A. 13: 109-60),[13] stemming from "Jacobi's berühmte Aufsätze in 3ten and 4ten Bände von Crelle's Journal" (ca. 1830).

Klein followed up these early efforts by a series of papers on elliptic modular functions and related topics during the years 1877-84 (GMA, vol. 3, 3-316). Klein refers to "the special lectures that, during the years 1877-80, I gave on number theory, elliptic functions, and algebraic equations, involving geometric group theory.... The audience included Gierster, Dyck, Bianchi, and Hurwitz" (p. 5). These lecture were given at the Technische Hochschule (now the Technische Universität) in Munich, for which Klein had left Erlangen in 1875. Thus in Munich Klein developed his ideas about discontinuous groups and geometric function theory "in a larger and more responsive circle of workers" (Courant 1925, 201).

It seems likely that Klein formed while in Munich the "genial idea" of becoming, in Courant's words:

> the pathfinder for the Mathematics and Mathematical Physics of the future.... In the depths of his soul,...he found the intuitive formulation of geometrical connections congenial. Klein was the most painstaking and effective apostle of Riemann's spirit... If mathematics can

build higher today on Riemann's foundations in tranquil clarity, this is thanks to Klein's special service. (Courant 1925, 202)

The connection of (periodic) trigonometric functions with the cylinder and of doubly periodic *elliptic functions* with the torus is obvious. It was apparently around 1880 that Klein first recognized analogous connections between other *automorphic functions* and the hyperbolic plane. We will return to this idea, which deserves a much more thorough historical study than we have had time to make, in the next section.

It was also at about this time that Klein, after taking over the editorship of the *Mathematische Annalen* with his friend A. Mayer, began to have assistants, Gierster and Dyck being the first two. Both wrote doctoral theses under Klein's guidance in the years 1879-81, as did Hurwitz. Dyck became coeditor of the M.A. in 1886, soon after becoming a professor in Munich, and remained in that position until 1919 when Einstein replaced him. (Hilbert had become coeditor in 1902). Dyck was also director of the Technische Hochschule during 1900-1906 and 1919-25, receiving the title of *von* Dyck (the German pre-1918 equivalent of being knighted) for his leadership.

Hurwitz went on to have an even more distinguished mathematical career; his collected works are available in two volumes. His early work shows clearly the influence of Klein's ideas, and his *Funktionentheorie* (1929), coauthored by Courant, also reflects Klein's ideas about geometric function theory, as seen in retrospect.

9. Klein and Poincaré

For a decade after writing the E.P., Klein's brilliance seemed unrivaled. By 1882 he had achieved leadership in geometry, with Gaston Darboux (1842-1917) as his closest rival. Lie's reputation was not yet comparable. Indeed, as we have seen, Klein was in 1882 in some sense Lie's patron, a status that was later to rankle Lie. Klein was also gaining in reputation as an analyst, through his contributions (and those of his students Hurwitz and Dyck) to Riemann's 'geometric function theory'. Indeed, in 1882, Teubner had already published Klein's first monograph on this subject.

Then suddenly, still in his early thirties, Klein became outshone by the incredibly original and versatile French mathematician Henri Poincaré (1854-1912). Although only five years younger than Klein, Poincaré was nearly ten years his mathematical junior. Poincaré had taken time off from his studies to help his physician-father during the Franco-Prussian war;

furthermore, it took longer to complete a doctorate in France than in Germany. As a result, Poincaré did not receive his Ph.D. from the University of Paris until 1879—ten years after Klein's degree was awarded.

Then, after two years in Caen, Poincaré returned to Paris. There he quickly published a series of notes on what are today called automorphic functions,[14] soon expanding on these notes in 1882-84 in a celebrated series of papers published in the first five volumes of *Acta Mathematica*. Morris Kline, in his informative discussion of automorphic functions (Kline 1972, 726-29), reports that Kronecker had tried to dissuade Mittag-Leffler from publishing Poincaré's first paper for fear that "this immature and obscure article would kill the journal" (p. 728).

In Poincaré's own words, the story of his first major breakthrough was as follows:

> For fifteen days I strove to prove that there could not be any functions like those I have since called Fuchsian functions. . . . One evening, contrary to my custom, I drank black coffee and could not sleep. Ideas rose in crowds; I felt them collide until pairs interlocked, so to speak, making a stable combination. By the next morning I had established the existence of a class of Fuchsian functions, those which come from the hypergeometric series: I had only to write out the results, which took but a few hours. (Poincaré 1907-8, 647-48)

Early on, Poincaré also published (in M.A. 19: 553-64) a synopsis of his ideas on the subject as of December 1881. This contains frequent references to "fuchsian" and "kleinian" functions, and it was followed by two short notes by Klein (M.A. 19: 565-68; 20: 49-51) on "functions that reproduce themselves under linear transformations," to which Klein would give their current name of "automorphic functions" in 1890. Of special historical interest are Klein's editorial comments (M.A. 19: 564) on Poincaré's synopsis. He questioned the relevance of the work of Fuchs; Poincaré later explained this relevance.

Curiously, Poincaré's "kleinian functions" refer to the parabolic case to which Klein had contributed little; this was a historical accident. Poincaré had been influenced by Fuchs's papers on solutions of (homogeneous) linear differential equations in the complex domain before he became aware of Klein's work, so he referred to "fuchsian functions" for the richer hyperbolic case to the half-plane. Klein complained of the inappropriateness, so Poincaré corrected his omission by later giving Klein's name to a much less novel class of functions.

It is clear that, at the time, Klein was having serious health problems and that he suffered a real breakdown. Young (1928, vii-viii) attributes the breakdown at least partially to "the antagonism he experienced at Leipzig" (at the hands of colleagues jealous of his rapid academic promotion). Young (1928, vii) and Row (1985, 288) both attribute the breakdown to overwork, aggravating an asthmatic condition. Young also reports that, while suffering from asthma, Klein proved the theorem that he himself "prized highest among his mathematical discoveries, known as the *'Grenzkreistheorem'* in the theory of automorphic functions."

Whatever the cause and nature of his problems, these years marked a turning point in Klein's career; he never regained the remarkable research activity of his earlier years. However, his GMA lists sixteen theses written at Leipzig under his direction, including those of Hurwitz (1881), Fine and Fiedler (1885), and his nephew Fricke (1886).

10. Years of Transition

Klein's first major act after recovering from his breakdown was to publish his famous book *Das Ikosaeder* (Klein 1884), which explained to mathematicians at large some of the connections that he had discovered between algebra, geometry (Euclidean and non-Euclidean), and analysis. Two years later, he left Leipzig to join H. A. Schwarz (1843-1921) in Göttingen.

In Göttingen, Klein continued to work on discontinuous groups. Especially, he constructed various new automorphic functions by multiple Schwarz reflections in the edges of regular circular polygons ("fundamental regions") and their images. In collaboration with his nephew Robert Fricke (1861-1930), he wrote up his ideas in classic treatises on the elliptic modular function and automorphic functions. The first of these is still studied today because of its applications to algebraic number theory.

By 1890, Lie's profound results on continuous groups and Klein's continuing applications of discontinuous groups had stimulated Fano to translate the E.P. into Italian. A French translation (by Padé) and an English translation by Klein's American student M. Haskell soon followed. Klein's foreword to the English translation states that his E.P.

> had but a limited circulation at first.... But...the general development of mathematics has taken, in the meanwhile, the direction corresponding precisely to these views, and particularly since *Lie* has begun the publication...of his *Theorie der Transformationsgruppen* (vol. I,

1888, vol. II, 1890) it seems proper to give a wider circulation to my Programme.

Klein's foreword to the German republication of the E.P. (M.A. 43 [1893]: 63-100) goes further, stating his desire "to include the collected applications of the theory of manifolds...not only to geometry, but also to mechanics and mathematical physics," and to work in "much material...which has been added in the intervening 20 years, naturally Lie's theory of continuous groups in particular, but also geometric connections that are implicit in the theory of automorphic functions." However, he concluded that this was simply too big a task.

American Influence. Klein's beneficial influence on American mathematicians seems to have begun in 1883-85, when F. N. Cole from Harvard and H. B. Fine from Princeton came to Leipzig (Archibald 1938, 100, 167). "There [Fine] attended lectures and seminars of Klein, Mayer, Fr. Schur, Carl Neumann, and W. Wundt (philosophy)." He also "wrote a thesis on a topic approved by Klein but suggested by Study, later one of Fine's closest friends."[15] After his return, Fine went on to lead the development of Princeton into one of the world's greatest mathematical centers. As regards Cole, Archibald states:

> After two years under Klein at Leipzig, Cole spent the next three years at Harvard, where his career as an undergraduate had been so brilliant. Aglow with enthusiasm, he gave courses in modern higher algebra, and in the theory of functions of a complex variable, geometrically treated, as in Klein's famous course of lectures at Leipzig in 1881-82. He was the first to open up modern mathematics to Prof. Osgood as a student, who characterized the lectures as "truly inspiring." Another student, M. Bôcher, as well as nearly all members of the Department, Profs. J. M. Peirce, B. O. Peirce and W. E. Byerly attended his lectures. He received the doctor's degree from Harvard on a topic suggested by Klein....

A professor at Columbia for 31 years, Cole served as secretary to the American Mathematical Society from 1896 to 1920.[16]

Before 1892 Klein had attracted a steady stream of American graduate students to Göttigen. These included M. W. Haskell, M. Bôcher, H. S. White, H. D. Thompson, and E. B. van Vleck, three of whom later became presidents of the American Mathematical Society. In 1893, Klein was chosen by the German government to head a delegation sent to the International Congress held in Chicago in conjunction with the World's Fair. One senses Klein's influence in the choice of those who presented papers. These included Weber, Hurwitz, Study, Meyer, Netto, Max Noether, Pringsheim, Fricke, Minkowski, and Hilbert (see *Bulletin of the American Mathematical Society* [1893]: 15-20).

Klein was made honorary president of the congress and gave a special series of lectures after it (Klein 1893b). These give a very readable account of Klein's views about many of the topics we have been discussing. To quote from W. F. Osgood's foreword to the 1911 edition: "His instinct for that which is vital in mathematics is sure, and the light with which his treatment illumines the problems here considered may well serve as a guide for the youth who is approaching the study of the problems of a later day."

11. Klein as a Leader

Meanwhile, in 1892, Schwartz left Göttingen to become the successor of Weierstrass in Berlin. From then on, Klein's gift for leadership increasingly dominated his activities. As Courant has written:

> When Schwarz went to Berlin in 1892, giving Klein a free hand in Göttingen, there began a new period of activity, in which his organizational involvement became more and more prominent. . . . The word *organize* meant for Klein not ruling by power: it was a symbol of deep insight and understanding. (Courant 1925, 207)

From that time until the outbreak of World War I, Klein was extraordinarily influential.

His influence on American mathematics continued. Two more Americans (F. S. Woods and V. Snyder) wrote Ph.D. theses under Klein's at least nominal direction. Fine, Bôcher, White, van Vleck, Woods, and Snyder were all active for many years in the American Mathematical Society, and Klein's influence on Haskell, Bôcher, and van Vleck in particular was considerable.

In addition, Klein was the key architect and organizer of: (i) a major expansion in the importance of Göttingen as a center of mathematical activity, (ii) the publication of the *Enzyklopädie der Mathematischen Wissenschaften* (EMW), and (iii) various 'reforms' in the style, standards, and substance of German mathematical education.

Klein and Göttingen. During these years, Klein was busy rebuilding Göttingen, the work place of Gauss and Riemann, into a preeminent world center of mathematics.[17] One of his first activities was to organize an international commission to fund a monument to Gauss and the physicist W. Weber, who had collaborated in constructing an early telegraph. (It was to honor the dedication of this monument that Hilbert, at Klein's invitation, wrote his famous *Grundlagen der Geometrie* [see §12].)

In 1895 Klein invited Hilbert to Göttingen, and Hilbert accepted with alacrity. Although Klein continued to give masterful advanced expository lectures, Hilbert was soon attracting the lion's share of doctoral candidates (see §12). Indeed, by 1900 Klein had become primarily a policy-maker and elder statesman, although barely 50. In this role, Klein obtained governmental and industrial support from an Institute of Applied Mathematics, with Prandtl and Runge as early faculty members. Sommerfeld, at one time his assistant and later coauthor with Klein of *Die Kreisel* ("The Top"), was another link of Klein with applied mathematics and physics.[18]

Klein and Education. Already in his *Antrittsrede* (see §4), Klein had expressed his concern about separation into humanistic and scientific education, stating that: "Mathematics and those fields connected with it are relegated to the natural sciences, and rightly so. . . . On the other hand, its conceptual content belongs to neither of the two categories" (Rowe 1985, 135). Klein's later involvement with German educational policy-making is described in (Pyenson 1983); in fact, Klein is the main subject of two of its chapters.

Klein was a universalist who believed strongly in integrating *pure* with *applied* mathematics, in the importance of both *logic* and *intuition* in geometry, and in the importance of having high-school teachers who understood and appreciated higher mathematics. Especially widely read by high-school teachers were his 1895 lectures on *Famous Problems in Elementary Geometry*, written for this purpose and translated into English, French, and Italian. (See R. C. Archibald, *American Mathematical Monthly* 21 [1914]: 247-59, where various slips were carefully corrected.)

In 1908, Klein became the president of the International Mathematical Teaching Commission. In this capacity he worked closely with the American David Eugene Smith and the Swiss Henri Fehr to improve mathematics education throughout the western world. As with other cooperative enterprises, this one was ended by World War I.

Klein and the Encyclopedia. One of Klein's most important legacies to the mathematical world was the EMW, to which we referred above. We have Young's first-hand description of Klein's rationale for his project (Young 1928, xiii):

> One day in the '90's the concept of the Enzyklopädie was formulated by Klein in the presence of the writer: the progress of mathematics, he said, using a favourite metaphor, was like the erection of a great tower; sometimes the growth in height is evident, sometimes it remains apparently stationary; those are the periods of general revision, when the advance, though invisible from the outside, is still real, consisting in underpinning and strengthening. And he suggested that such was the then period. What we want, he concluded, is a general view of the state of the edifice as it exists at present.

Klein himself edited the volume on mechanics.

Almost 35 years elapsed from the time Klein conceived his plan for an encyclopedia, to be published in French as well as German, to its completion in the late 1920s with articles surveying advances of the preceding two decades. From around 1905 to at least 1935, it truly lived up to its name. Not surprisingly, Fano wrote for it an article on "Continuous Groups and Geometry," which summarizes developments stemming from the E.P. Since the contents of this article cover somewhat the same topics as our §§1-7, we shall defer its discussion until we take up Élie Cartan's 1912 revision of it in §13 below.

12. Klein and Hilbert

The forty years that Klein spent at Göttingen transformed it into an almost legendary center of pure and applied mathematical research. One of the key figures in this transformation was David Hilbert (1862-1943), a mathematical genius who may have owed more to Klein than he cared to realize.[19] As we have stated, Klein brought Hilbert to Göttingen in 1895 to replace Heinrich Weber. Having studied with Lindemann and Hurwitz, Hilbert was in some sense Klein's "academic grandson"; moreover, he had gone to Paris on Klein's advice (in 1886), at that time arguably the world's greatest center of mathematical research.

Klein's example may also have stimulated Hilbert to broaden his research interests after coming to Göttingen, Hilbert having previously devoted his mathematical genius almost exclusively to invariant theory and algebraic number theory (Weyl 1944, 635). Soon after arriving in Göttingen, he showed in a letter to Klein that the Laguerre-Cayley-Klein projective metric defined by them in general ellipsoids had an analogue in arbitrary convex bodies. A few years later, he vindicated the "Dirichlet Principle" of Riemann, which had been discredited by Weierstrass.

By 1898, Hilbert had largely taken over the supervision of Ph.D. theses at Göttingen, of which no fewer than 60 were written under his direction between then and 1916. This was also the year in which Klein's plans for the Gauss-Weber Denkmal matured, and he invited Hilbert to be one of the two speakers to celebrate the great occasion. Hilbert chose to speak on the foundations of Euclidean geometry. His lecture notes of the previous winter on the subject, hastily polished, became the first edition of his famous *Grundlagen der Geometrie*. This book, now in its tenth edition, concluded with a study of 'constructability with ruler and compass' that is closely related in theme (though not in style) to Klein's beautiful exposition of the same subject in his "Ausgewählten Kapiteln..." of 1895.

Hilbert did not refer to this at all, an omission almost amounting to a discourtesy to a senior colleague. However, since a major stimulus for Klein's lectures had been the simplifications by Hilbert, Gordan, and Hurwitz of Lindemann's original (1882) proof of the transcendence of π, this was perhaps only fair, although a reference to Klein's brilliant booklet would have been gracious.

The purely *formal* approach of Hilbert's *Grundlagen* contrasts sharply with Klein's emphasis on the *intuitive* visualization of geometric ideas, and it is interesting to recall what Klein had to say about Hilbert's *Grundlagen* in his 1908 *Elementarmathematik vom Höhere Standpunkt aus*, vol. 2. After a brief review (pp. 130-59) of the E.P. and some of his later ideas (cf. §5), Klein discusses other approaches to geometry. Among these, his book takes up last the "modern theory of geometric axioms," observing that (p. 185):

> In it, we determine what parts of geometry can be set up without using certain axioms, and whether or not, by assuming the opposite of a given axiom, we can also secure a system free from contradiction, that is, a so-called 'pseudo-geometry.'

> As the most important work belonging here, I should mention Hilbert's *Grundlagen der Geometrie*. Its chief aim, as compared with earlier investigations, is to establish, in the manner indicated, the significance of the axioms of continuity.

It does seem curious that finally, near the end of his life, Hilbert should have apparently forsaken his extreme formalism and written (with S. Cohn-Vossen) a book entitled *Anschauliche Geometrie* ("Intuitive Geometry"). (For a fuller account of the *Grundlagen der Geometrie*, its background and influence, see [Birkhoff and Bennett, forthcoming].)

Hilbert's social and scientific personalities were very different from those of the dignified and highly intuitive Klein, and there is little doubt about Hilbert's restiveness as regards Klein's regal manner. Thus in a letter to his future wife, Courant wrote in 1907-8 that "Hilbert now rebels everywhere against Klein's assumed dictatorship" (Reid 1976, 19). Likewise, Ostrowski (coeditor of vol. 1 of Klein's GMA) wrote one of us in 1980: "As to Hilbert I do not think that you will find any reference to the Erlanger Programm. As a matter of fact, Hilbert did not think very much of it."

13. Study and Élie Cartan

Three more major mathematicians whose work reflects the influence of the E.P. are Eduard Study (1862-1930), Élie Cartan (1869-1951), and Hermann Weyl (1885-1955). Although they had related interests, they had very different backgrounds and tastes. We shall discuss next the influence of the E.P. on Study and Cartan, taking up its influence on Weyl in §14.

Eduard Study. As a geometer, Study was more influential than either Killing or Engel. Blaschke dedicated the first volume (1921) of his famous *Vorlesungen über Differentialgeometrie* to Study, and Study's Ph.D. students included not only Fine but also J. L. Coolidge, the author of several widely read books, whose *History of Geometrical Methods* (Coolidge 1940) is a standard reference. From the chapter "Higher Space Elements" in this treatise, we quote the following passage:

> The connecting thread in [this chapter] is the idea of treating directly as a space element some figure previously treated as a locus. The idea of doing this was dominant in geometrical circles, especially in the schools of Klein and Study, at the end of the nineteenth century. . . . It shades off imperceptibly into the theory of geometrical transformations. (Coolidge 1940, book 2, chap. 6)

This idea, obviously generalizing Plücker's "line geometry," can be used in the spirit of the E.P. to construct many "global" representations of continuous groups as transformation groups of manifolds.

Study began his career at the University of Munich, where he wrote his doctoral thesis *Ueber die Massbestimmungen Extensiver Grössen* in

1885. It was concerned with metric magnitudes in Grassmannian geometry, while his *Habilitationsschrift* dealt with the geometry of the conic section. Although Klein had been at Munich some years earlier, and Klein's student Dyck was to go there a year later, the only reference to Klein in either paper is the statement: "One sees that the geometric interest of this formulation of the problem has the closest connections with the researches of Riemann, Helmholtz, and Klein."

More relevant to the E.P. were Study's investigations on the three-dimensional Euclidean group and its subgroups (M.A. 39 [1891]: 444-566; 60 [1905]: 321-77). Its most general one-parameter subgroup is the group of helical or "screw" motions, previously studied by Klein and Lie. Stemming from these investigations, and hence indirectly from the Theorem of Chasles, was Study's major work *Geometrie der Dynamen* (1903). Concerned with the connected component of the Euclidean group and its one-parameter subgroups, this deals with the geometry of the "space" of force systems and rigid displacements, and its philosophy is akin to that of Plücker's "line geometry." It can be regarded as a sequel to Sir Robert Ball's *Geometry of Screws*, whose 1871 edition had excited Klein and Clifford, and whose third edition would appear in 1911, but Study's book says little about any earlier work.[20]

Study's beautiful researches on the Problem of Apollonius (to construct the circles tangent to three given circles) were also, according to his first paper on the subject (M.A. 49 [1897]: 497-542), inspired by Klein's use of inversions in circles to generate symmetrical patterns. Also related to the E.P. was Study's original but obscure and rambling *Methoden zur Theorie der Ternaerien Formen* (Teubner 1889), dedicated to "my dear friend Friedrich Engel." In this book, Study applied Lie's concept of an *infinitesimal* transformation to invariants and covariants, mentioning (p. 143) the problem of determining "all types of r-parameter subgroups" of the full linear group. He also distinguished "integral" invariants, and "algebraic" (as well as "integral algebraic") invariants from general invariants.

Élie Cartan. Perhaps the greatest geometer of the twentieth century, Élie Cartan's thesis (1894) was purely algebraic. In it, he determined all simple complex Lie algebras, thus completing and making precise the earlier results of Killing. Presumably inspired to undertake this task by Poincaré (cf. §8), Cartan describes his advances over Killing as follows:

Unfortunately Killing's research lacks rigor, particularly concerning groups which are not simple; it makes constant use of a theorem which is not proved in its generality; I show in this work an example where the theorem is not true, and when the occasion arises, a number of other errors of lesser importance. (Cartan 1952, partie I, 139)

Cartan's brilliant thesis was only the beginning of an outstanding career, which reached its climax when he was in his fifties and sixties, between World War I and World War II. His first major geometric effort was his translation and extension of Fano's article "Continuous Groups and Geometry" (ESM, vol. 3, 5; Cartan 1952, vol. 3, 2), which was concerned with developments stemming from the E.P. Written in 1912, Cartan's article notably amplified Fano's original by including the extension of the E.P. to space-time suggested by Einstein's then new theory of special relativity, and by summarizing Study's important but involved contributions.

Lorentz Group. Much as the E.P. notably extended the invariant theory of Cayley and Sylvester by pointing out "the possibility of constructing other than projective invariants," so a celebrated paper by Minkowski (written at Göttingen) identified the main contribution of Einstein's theory of special relativity as the replacement of the Galileo-Newton group by the "Lorentz group," actually first identified as a group by Poincaré. As Klein immediately recognized, this suggested that the E.P. might be applicable not only to geometry but to physics as well. Following a 1910 paper by Bateman, Cartan would later go further and describe the role of the conformal group, extended to space-time, in electromagnetic theory and special relativity.

Eight years later, in the preface to the second volume of his classic *Projective Geometry* (Veblen and Young 1917), Veblen would state:

We have in mind two principles for the classification of any theorem of geometry: (a) the axiomatic basis...from which it can be derived...; and (b) the group to which it belongs in a given space.

The two principles of classification, (a) and (b), give rise to a double sequence of geometries, most of which are of consequence in present-day mathematics.[21]

Chapter 3 of that volume was devoted to applying Klein's classification scheme (b).

General Relativity. Unfortunately for the E.P., Einstein's "general relativity" theory fits much less neatly into Klein's classification scheme than his "special theory." See (Bell 1940, chap. 20, esp. 443-49), where a quotation from Veblen's chapter 3 is contrasted with one made by Veblen ten years later, and one made in 1939 by J. H. C. Whitehead.[22]

However, the properties of Einstein's curved space-time are expressed by local *differential invariants*, which can be classified as 'conformal', 'metric', 'affine', 'projective', etc., very much in the spirit of the E.P.

J. A. Schouten (1926) spelled out this connection. Cartan's invited address on "Lie Groups and Geometry" at the Oslo congress (1936) and J. H. C. Whitehead's biography of Cartan ("Obituary Notices of the Fellows of the Royal Society" 8, 1952) give perhaps the most authoritative opinions on the subject.

In the 1920s and 1930s, Cartan also showed his consummate and creative command of old and new mathematics by his major contributions to the "globalization" of Lie's purely local theory of Lie groups. The resulting global theory has made both Klein's and Lie's expositions technically obsolete.

Between 1927 and 1935, Cartan published what Chern and Chevalley call "his most important work in Riemannian geometry...the theory of symmetric spaces" (*Bulletin of the American Mathematical Society* 58 [1952]: 244). These are Riemannian manifolds in which ds^2 is invariant under reflection in any point. In retrospect, Cartan's tortuous path (via the parallelism of Levi-Civita) to the recognition of this simple idea seems amazing.[23] Even more amazing is the failure of Klein, after applying reflections in lines in the hyperbolic plane in so many ways, to identify "symmetric Riemann spaces" at all!

14. Klein and Hermann Weyl

Hermann Weyl began his career in the Göttingen that Klein had built up, and Hilbert was his thesis advisor. Hence he was in some sense an academic great-grandson of Klein, whom he must have known. Weyl seems never to have been overawed by Hilbert; thus his 1908 thesis stated unequivocally (M.A. 66 [1909]: 273): "As will be shown below, the applicability of Hilbert's method is by no means limited to the continuous kernels treated by Hilbert...but also leads to interesting consequences in certain more general cases." Beginning in 1917, Weyl outdistanced Hilbert (again) in mathematicizing Einstein's then new general theory of relativity. In papers and in his famous book *Raum. Zeit. Materie* (later translated into

English as *Space, Time, Matter*), he introduced *local* generalizations of affine, projective, and conformal geometry that are related to their global counterparts as Riemannian geometry is related to Euclidean, as Schouten (1926) was later to explain. Still under 40, he then attacked Hilbert's "formalist" logic of mathematics in the early 1920s supporting the conflicting "intuitionist" logic of L. E. J. Brouwer.

Weyl showed more respect for Klein, to whom he dedicated the first (1912) edition of his *Die Idee der Reimannschen Fläche*.[24] This was because, as he stated in the preface to its 1955 edition, "Klein had been the first to develop the freer conception of a Riemann surface, . . . thereby he endowed Riemann's basic ideas with their full power."

Then, in the middle 1920s, Weyl was the spark plug of the famous Peter-Weyl theory of group representations. He used a very concrete theory of *group-invariant* measure on *compact Lie groups*, the existence of which permits one to extend to compact Lie groups the result of E. H. Moore 'and Maschke: that every group of linear transformations having a finite group-invariant measure is equivalent to a group of orthogonal transformations.[25] In his *Gruppentheorie und Quantenmechanik* (translated into English by H. P. Robertson), Weyl later applied the analogous result for the orthogonal group to the then new quantum mechanics.[26]

By an irony of fate, very little of Weyl's deep and influential research work was done at Göttingen, the source of much of his inspiration. It was only three years after he finally accepted a professorship there that Hitler seized power in Germany. In the next year, Weyl emigrated to the new Institute of Advanced Study in Princeton, where he spent his last twenty years, creative and versatile to the end. His later books, *The Classical Groups* (1939) and *Symmetry* (1952), show his spiritual affinity with Klein, and it seems fitting to conclude our review with two quotations from the former. First, Weyl remarks:

> This is not the place for repeating the string of elementary definitions and propositions concerning groups which fill the first pages of every treatise on group theory. Following Klein's "Erlanger Program" (1872), we prefer to describe in general terms the significance of groups for the idea of *relativity*, in particular in geometry. (1939, 14)

Later, he makes a more specific evaluation:

> The dictatorial regime of the projective idea in geometry was first broken by the German astronomer and geometer Möbius, but the

classical document of the democratic platform in geometry, establishing the group of transformations as the ruling principle in any kind of geometry, and yielding equal rights of independent consideration to each and every such group, is F. Klein's Erlanger Programm. (1939, 28)

We hope that the great influence of this classical document, through its extensions and interpretations by Klein himself, by Lie, by Weyl, and by many other mathematicians, has been clarified by our review.

Notes

1. In discussing this last phase of Klein's career, Rowe (1985, 278) has referred to him as the "doyen of German mathematics for nearly three decades."

2. Weierstrass himself always viewed geometry *metrically*.

3. For the significance of W-curves, Klein refers his reader to ##13-20 and #34 of the article of Scheffers (EMW, vol. 3, D4).

4. In the second of these (Klein 1873), submitted in June 1872, Klein describes briefly how "the different methods of geometry can be characterized by an associated transformation group." This paper and the E.P. were the first publications by Klein or Lie that used the phrase *transformation group*. The word *invariant* is conspicuous by its absence, although invariant theory had been studied in Germany (without explicit mention of the word *group*) for at least a decade.

5. See (Birkhoff and Bennett, forthcoming) for references to this work of Klein and Lüroth, as well as its influence on later theories.

6. A glance at the index of names in the relevant volumes of Lie's *Collected Papers* (LGA) makes Lie's scientific indebtedness to Klein obvious.

7. Lie's *local* theory (LGA, vol. 5) was finally extended into a rigorous *global* theory by G. D. Mostow (*Annals of Mathematics* 52 [1950]: 606-36). There Mostow showed that the two-dimensional manifolds are the plane, cylinder, torus, sphere, projective homogeneous plane, Möbius strip, and Klein bottle.

8. Picard had used the simplicity of the (Lie) projective group on two variables to prove the impossibility of solving $u'' + p(x)u' + q(x)u = 0$ by quadratures—a theorem that Lie wished he had discovered himself.

9. See also (Hawkins 1982, §3).

10. For a very human account of Lie's jealousy of Klein and their final reconciliation, which reproduces a moving letter from Frau Klein, see (Young 1928, xviii-xix).

11. Wedekind's summary mentions his use of a result in the E.P.; Klein entitled his paper "Binary Forms Invariant under Linear Transformations."

12. By this, Klein meant that the subgroup of the group of the icosahedron leaving each point invariant is of order 2, as contrasted with 5 for the vertices of the icosahedron and 3 for the dodecahedron.

13. As late as 1886, Brioschi (M.A. 26 [1886]: 108) would refer to the "icosahedral" hyper geometric equation:

$$x(x - 1)v'' + (7x - 4v')/6 + 11v/3^2 4^2 5^2 = 0.$$

14. See, for example, L. R. Ford, *Automorphic Functions* (New York: McGraw-Hill, 1929).

15. Another close friend of Fine was Woodrow Wilson, president of Princeton University, and later of the United States. Further information on H. B. Fine can be found in William Aspray's paper in this volume.

16. "For a quarter of a century no one could think of the American Mathematical Society apart from the personality of Professor Cole" (Archibald 1938, 101).

17. Many references to Klein's role in Göttingen may be found in (Reid 1970, 1976); see also chapter 11 of (Klein 1893b).

18. Klein had shown his belief in the importance of applied mathematics in 1875, by leaving a full professorship at Erlagen for a position at the Technische Hochschule in Munich and referring to it as a great advance ("einem grossen Sprung") (Young 1928, vii). There he held a seminar on pure and applied mathematics with von Linde (Pyenson 1979, 55-61).

19. For Klein's role in suggesting Hilbert's Fifth Problem, see (Birkhoff and Bennett, forthcoming).

20. See (Ziegler 1985) and (EMW, vol. 3, parts 1, 2).

21. Principle (a) clearly refers to the method used by Hilbert in his *Grundlagen der Geometrie*.

22. Actually, Whitehead did not lose interest in the E.P. as quickly as Bell suggests; see (*Annals of Mathematics* 33 [1932]: 681-87).

23. For Cartan's mature exposition of the theory of symmetric spaces, see Proceedings, International Congress of Mathematicians, Zurich, 1932, 152-61.

24. A few years earlier, Paul Koebe (also in Göttingen) had finally solved rigorously the "uniformization problem" that had eluded both Klein and Poincaré. Koebe's "Primenden," one of his major technical tools, seem related to the ideas of Klein's "Grenzkreistheorem" (see §9).

25. The Peter-Weyl theory surely helped to inspire Haar's 1933 theory of invariant *Lebesgue* measure on compact topological groups. It was von Neumann's subsequent application of Haar measure to solve the Klein-Hilbert Fifth Problem for compact groups, and Pontrjagin's parallel solution for Abelian groups, that paved the way for its complete solution.

26. One of us adapted Weyl's title to the chapter on group theory and fluid mechanics in his *Hydrodynamics* (Birkhoff, Princeton University Press, 1950). In this the notion of 'self-similar solution' (exploited earlier by Sedov and others) was generalized to arbitrary groups. The ultimate inspiration for this chapter was Klein's group-theoretic interpretation of special relativity in his (EdM), as an extension of the E.P.

References

[EdM] Klein, Felix. 1926-27. *Entwicklung der Mathematik in 19ten Jahrhundert*. Berlin: Springer. Reprinted New York: Chelsea, 1950.

[EMW] *Enzyklopädie der Mathematischen Wissenschaften*, esp. vol. 3.1.

[ESM] *Encyclopédie des Sciences Mathématiques*, esp. vol. 3.

[GMA] Klein, Felix. 1921-23. *Gesammelte Mathematische Abhandlungen*. 3 vols. Berlin: Springer.

[LGA] Lie, Sophus. 1922-37. *Gesammelte Abhandlungen*. 6 vols. Ed. F. Engel and P. Heegard. Berlin: Springer.

Archibald, R. C. 1938. *A semi-centennial History of the American Mathematical Society; 1888-1938*. Providence, R.I.: American Mathematical Society, vol. 1.

Bell, E. T. 1940. *The Development of Mathematics*. New York: McGraw-Hill.

Birkhoff, G. 1973. *Source Book in Classical Analysis*. Cambridge, Mass.: Harvard University Press.

Birkhoff, G., and Bennett, M. K. Hilbert's "Grundlagen der Geometrie." *Rendiconti del Circolo Mathematico di Palermo*. Forthcoming.

Cartan, E. 1952. *Oeuvres complètes*. 12 vols. Paris: Gauthier-Villars.

Clebsch, A., and Lindemann, F. 1876-91. *Vorlesungen uber Geometrie*, bearb. und hrsg. von F. Lindemann. Leipzig: Teubner.

Clifford, W. K. 1873. Preliminary Sketch of Biquaternions. *Proceedings of the London Mathematical Society* 4: 381-95.

Coolidge, J. L. 1940. *A History of Geometrical Methods*. Oxford: Clarendon Press.

Courant, R. 1925. Felix Klein. *Jahresbericht der Deutschen Mathematiker Vereiningung* 34: 197-213.

Engel, F. 1900. Sophus Lie. *Jahresbericht der Deutschen Mathematiker Vereiningung* 8: 30-46.

Engel, F. and Faber, K. 1932. *Die Liesche Theorie der Partiellen Differentialgleichungen Erster Ordnung*. Leipzig: Teubner.

Gray, Jeremy J. 1980. Olinde Rodrigues' Paper of 1840 on Transformation Groups. *Archive for History of Exact Sciences* 21: 375-85.

Hawkins, Thomas. 1981. Non-Euclidean Geometry and Weierstrassian Mathematics.... *Historia Mathematica* 7: 289-342.

———. 1981. Wilhelm Killing and the Structure of Lie Algebras. *Archive for History of Exact Sciences* 26: 127-92.

———. 1984. The *Erlanger Programm* of Felix Klein.... *Historia Mathematica* 11: 442-70.

Hilbert, David. 1899. *Grundlagen der Geometrie*. 1st ed. Liepzig: Teubner.

Hopf, Heinz. 1926. Zum Clifford-Kleinschen Raumproblem. *Mathematische Annalen* 95: 313-39.

Jordan, Camille. 1869. Mémoire sur les groupes de mouvements. *Annali di Matematica* (2) 2: 167-213 and 322-451.

———. 1870. *Traité des substitutions*. Paris: Gauthier-Villars.

Killing, Wilhelm. 1886. *Zur Theorie der Lie'schen Transformationsgruppen*. Braunschweig.

———. 1891. Uber die Clifford-Kleinsche Raumformen. *Mathematische Annalen* 39: 257-78.

Klein, Felix. 1872. *Vergleichende Bertrachtungen über neuere geometrische Forschungen Erlangen*. Reprinted with additional footnotes in *Mathematische Annalen* 43 (1893): 63-100 and in GMA, vol. 1, 460-97.

———. 1873. Uber die sogenannte nicht-Euklidische Geometrie. *Mathematische Annalen* 6: 112-45.

———. 1884. *Vorlesungen über das Ikosaeder*.... Leipzig: Teubner. (English translation by G. G. Morrice, 1888; cf. also F. N. Cole's review in *American Journal of Mathematics* 9 [1886]: 45-61.

———. 1892. *Einleitung in die geometrischen Funktionentheorie*. Lectures given in 1880-81, Leipzig, and reproduced in Göttingen a decade later.

———. 1893a. *Einleitung in die Höhere Geometrie*. 2 vols. Lithographed lecture notes by Fr. Schilling.

———. 1893b. The Evanston Colloquim "Lectures on Mathematics." Republished by the American Mathematical Society in 1911.

———. 1895. *Vorträge über ausgewählte Fragen der Elementargeometrie*. Göttingen. Translated as *Famous Problems of Elementary Geometry* by W. W. Beman and D. E. Smith in 1900; 2d ed. reprinted by Stechert, 1936.

———. 1908. *Elementarmathematik vom Höheren Standpunkt aus*. Vol. 2. Leipzig: Teubner.

Kline, Morris. 1972. *Mathematical Thought from Ancient to Modern Times*. New York: Oxford University Press.

Lie, Sophus. 1880. Theorie der Transformationsgruppen I. *Mathematische Annalen* 16: 441-528.

Lie, S., and Engel, F., 1888-93. *Theorie der Transformationsgruppen*. 3 vols. Leipzig: Teubner. Reprinted in 1930-36.

Lie, S., and Scheffers, G. 1896. *Vorlesungen über continuierliche Gruppen*. Leipzig; Teubner.

Noether, Max. 1900. Sophus Lie. *Mathematische Annalen* 53: 1-41.

Plücker, Julius. 1868-69. *Neue Geometrie des Raumes*..., hrsg. von Felix Klein. Leipzig: Teubner.

Poincaré, Henri. 1907-8. *Science et Méthode*. English translation published by the Science Press in 1921 with the title *Foundations of Science*.

———. 1916. *Oeuvres complètes*. 12 vols.; vol. 11 most relevant. Paris: Gauthier-Villars.

Pyenson, L. 1979. Mathematics, Education, and the Göttingen Approach to Physical Reality, 1890-1914. *Europa* 2(2): 91-127.

———. 1983. *Neohumanism and the Persistence of Pure Mathematics in Wilhelmian Germany*. Philadelphia: American Philosophical Society.

Reid, Constance. 1970. *Hilbert*. Berlin: Springer.

———. 1976. *Courant in Göttingen and New York*. Berlin: Springer.

Rowe, D. E. 1983. A Forgotten Chapter in the History of Felix Klein's *Erlanger Programm. Historia Mathematica* 10: 448-57.

———. 1984. Felix Klein's *Erlanger Antrittsrede. . . . Historia Mathematica* 12: 123-41.

———. 1985. Essay Review. *Historia Mathematica* 12: 278-91.

Schouten, J. A. 1926. Erlanger Programm and Uebertragungslehre. *Rendiconti del Circolo Mathematico di Palermo* 50: 142-69.

Study, E. 1903. *Geometrie der Dynamen*. Leipzig: Teubner.

Veblen, O., and Young, J. W. 1917. *Projective Geometry*. Vol. 2. (Actually written by Veblen.) Boston: Ginn.

Weyl, Hermann. 1912. *Die Idee der Riemannschen Fläche*. Leipzig: Teubner. (English translation of the 2d ed. by Gerald Mac Lane, Addison-Wesley, 1955.)

———. 1939. *The Classical Groups*. Princeton, N.J.: Princeton University Press.

———. 1944. David Hilbert and His Mathematical Work. *Bulletin of the American Mathematical Society* 50: 612-44.

Young, W. H. 1928. Christian Felix Klein: 1849-1925. *Proceedings of the Royal Society of London* A121: i-xix.

Ziegler, R. 1985. *Die Geschichte der Geomestrischen Mechanik in 19. Jahrhundert*. Stuttgart: Fr. Steiner Verlag Wiesbaden GMBH.

—Joseph W. Dauben —

Abraham Robinson and Nonstandard Analysis: History, Philosophy, and Foundations of Mathematics

> Mathematics is the subject in which we don't know* what we are talking about.
> —Bertrand Russell
>
> * Don't care would be more to the point.
> —Martin Davis
>
> I never understood why logic should be reliable everywhere else, but not in mathematics.
> —A. Heyting

1. Infinitesimals and the History of Mathematics

Historically, the dual concepts of infinitesimals and infinities have always been at the center of crises and foundations in mathematics, from the first "foundational crisis" that some, at least, have associated with discovery of irrational numbers (more properly speaking, incommensurable magnitudes) by the pre-socartic Pythagoreans[1], to the debates that are currently waged between intuitionists and formalist—between the descendants of Kronecker and Brouwer on the one hand, and of Cantor and Hilbert on the other. Recently, a new "crisis" has been identified by the constructivist Erret Bishop:

There is a crisis in contemporary mathematics, and anybody who has

This paper was first presented as the second of two Harvard Lectures on Robinson and his work delivered at Yale University on 7 May 1982. In revised versions, it has been presented to colleagues at the Boston Colloquim for the Philosophy of Science (27 April 1982), the American Mathematical Society meeting in Chicago (23 March 1985), the Conference on History and Philosophy of Modern Mathematics held at the University of Minnesota (17-19 May 1985), and, most recently, at the Centre National de Recherche Scientifique in Paris (4 June 1985) and the Department of Mathematics at the University of Strasbourg (7 June 1985). I am grateful for energetic and constructive discussions with many colleagues whose comments and suggestions have served to develop and sharpen the arguments presented here.

not noticed it is being willfully blind. *The crisis is due to our neglect of philosophical issues....*[2]

Bishop, too, relates his crisis in part to the subject of the infinite and infinitesimals. Arguing that formalists mistakenly concentrate on the "truth" rather than the "meaning" of a mathematical statement, he criticizes Abraham Robinson's nonstandard analysis as "formal finesse," adding that "it is difficult to believe that debasement of meaning could be carried so far."[3] Not all mathematicians, however, are prepared to agree that there is a crisis in modern mathematics, or that Robinson's work constitutes any debasement of meaning at all.

Kurt Gödel, for example, believed that Robinson more than anyone else had succeeded in bringing mathematics and logic together, and he praised Robinson's creation of nonstandard analysis for enlisting the techniques of modern logic to provide rigorous foundations for the calculus using actual infinitesimals. The new theory was first given wide publicity in 1961 when Robinson outlined the basic idea of his "nonstandard" analysis in a paper presented at a joint meeting of the American Mathematical Society and the Mathematical Association of America.[4] Subsequently, impressive applications of Robinson's approach to infinitesimals have confirmed his hopes that nonstandard analysis could enrich "standard" mathematics in important ways.

As for his success in defining infinitesimals in a rigorously mathematical way, Robinson saw his work not only in the tradition of others like Leibniz and Cauchy before him, but even as vindicating and justifying their views. The relation of their work, however, to Robinson's own research is equally significant, as Robinson himself realized, and this for reasons that are of particular interest to the historian of mathematics. Before returning to the question of a "new" crisis in mathematics due to Robinson's work, it is important to say something, briefly, about the history of infinitesimals, a history that Robinson took with the utmost seriousness.

This is not the place to rehearse the long history of infinitesimals in mathematics. There is one historical figure, however, who especially interested Robinson—namely, Cauchy—and in what follows Cauchy provides a focus for considering the historiographic significance of Robinson's own work. In fact, following Robinson's lead, others like J. P. Cleave, Charles Edwards, Detlef Laugwitz, and W. A. J. Luxemburg have used nonstandard analysis to rehabilitate or "vindicate" earlier infinitesimalists.[5] Leibniz, Euler, and Cauchy are among the more prominent mathematicians who have been rationally reconstructed—even to the point of having had, in the views of some commentators, "Robinsonian" nonstandard infinitesimals in mind from the beginning. The most detailed and methodically sophisticated of such treatments to date is that provided by Imre Lakatos; in what follows, it is his analysis of Cauchy that is emphasized.

2. Lakatos, Robinson, and Nonstandard Interpretations of Cauchy's Infinitesimal Calculus

In 1966, Lakatos read a paper that provoked considerable discussion at the International Logic Colloquium meeting that year in Hannover. The primary aim of Lakatos's paper was made clear in its title: "Cauchy and the Continuum: The Significance of Non-standard Analysis for the History and Philosophy of Mathematics."[6] Lakatos acknowledged his exchanges with Robinson on the subject of nonstandard analysis as he continued to revise the working draft of his paper. Although Lakatos never published the article, it enjoyed a rather wide private circulation and eventually appeared after Lakatos's death in volume 2 of his *Mathematics, Science and Epistemology*.

Lakatos realized that two important things had happened with the appearance of Robinson's new theory, indebted as it was to the results and techniques of modern mathematical logic. He took it above all as a sign that metamathematics was turning away from its original philosophical beginnings and was growing into an important branch of mathematics.[7] This view, now more than twenty years later, seems fully justified.

The second claim that Lakatos made, however, is that nonstandard analysis revolutionizes the historian's picture of the history of the calculus. The grounds for this assertion are less clear—and in fact, are subject to question. Lakatos explained his interpretation of Robinson's achievement as follows at the beginning of his paper:

> Robinson's work . . . offers a rational reconstruction of the discredited infinitesimal theory which satisfies modern requirements of rigour and which is no weaker than Weierstrass's theory. This reconstruction makes infinitesimal theory an almost respectable ancestor of a fully-fledged, powerful modern theory, lifts it from the status of pre-scientific gibberish and renews interest in its partly forgotten, partly falsified history.[8]

But consider the word *almost*. Robinson, says Lakatos, only makes the achievements of earlier infinitesimalists *almost* respectable. In fact,

Robinson's work in the twentieth century cannot vindicate Leibniz's work in the seventeenth century, Euler's in the eighteenth century, or Cauchy's in the nineteenth century. There is nothing in the language or thought of Leibniz, Euler, or Cauchy (to whom Lakatos devotes most of his attention) that would make them early Robinsonians. The difficulties of Lakatos's rational reconstruction, however, are clearer in some of the details he offers.

For example, consider Lakatos's interpretation of the famous theorem from Cauchy's *Cours d'analyse* of 1821, which purports to prove that the limit of a sequence of continuous functions $s_n(x)$ is continuous. This is what Lakatos, in the spirit of Robinson's own reading of Cauchy, has to say:

> In fact *Cauchy's theorem was true and his proof as correct as an informal proof can be.* Following Robinson . . . Cauchy's argument, if not interpreted as a proto-Weierstrassian argument but as a genuine Leibniz-Cauchy one, runs as follows: . . .
>
> $s_n(x)$ should be defined and continuous and converge not only at standard Weierstrassian points but at *every* point of the "denser" Cauchy continuum, and . . . the sequence $s_n(x)$ should be defined for infinitely large indices *n* and represent continuous functions at such indices.[9]

In one last sentence, this is all summarized in startling terms as follows:

> Cauchy made absolutely no mistake, he only proved a completely different theorem, about transfinite sequences of functions which Cauchy-converge on the Leibniz continuum.[10]

But upon reading Cauchy's *Cours d'analyse*—or either of his later presentations of the theorem in his *Résumés analytiques* of 1833 or in the *Comptes Rendues* for 1853—one finds no hint of *transfinite* indices, sequences, or Leibnizian continua made "denser" than standard intervals by the addition of infinitesimals. Cauchy, when referring to infinitely large numbers $n' > n$, has "very large"—but finite—numbers in mind, not *actually infinite* Cantorian-type transfinite numbers.[11]

This is unmistakably clear from another work Cauchy published in 1833—*Sept leçons de physique générale*—given at Turin in the same year he again published the continuous sum theorem. In the *Sept leçons*, however, Cauchy explicitly denies the existence of infinitely large numbers for their allegedly contradictory properties.[12]

Moreover, if Lakatos was mistaken about Cauchy's position concerning the actually infinite, he was also wrong about Cauchy's continuum being one of Leibnizian infinitesimals. If, by virtue of such infinitesimals, Cauchy's original proof had been correct all along, why would he then have issued a *revised* version in 1853, explicitly to improve upon the earlier proofs? Instead, were Lakatos and Robinson correct in their rational reconstructions, all Cauchy would need to have done was point out the nonstandard meaning of his infinitesimals—explaining how infinitely large and infinitely small numbers had given him a correct theorem, as well as a proof, all along.

Lakatos also draws some rather remarkable conclusions about *why* the Leibnizian version of nonstandard analysis failed:

> The downfall of Leibnizian theory was then not due to the fact that it was inconsistent, but that it was capable only of limited growth. It was the heuristic potential of growth—and explanatory power—of Weierstrass's theory that brought about the downfall of infinitesimals.[13]

This rational reconstruction may complement the overall view Lakatos takes of the importance of research programs in the history of science, but it does no justice to Leibniz or to the subsequent history of the calculus in the eighteenth and early nineteenth centuries, which (contrary to Lakatos) demonstrates that (i) in the eighteenth century the (basically Leibnizian) calculus constituted a theory of considerable power in the hands of the Bernoullis, Euler, and many others; and (ii) the real stumbling block to infinitesimals *was* their acknowledged inconsistency.

The first point is easily established by virtue of the remarkable achievements of eighteenth-century mathematicians who used the calculus because it was powerful—it produced striking results and was indispensable in applications.[14] But it was also suspect from the beginning, and precisely because of the question of the contradictory nature of infinitesimals.

This brings us to the second point: despite Lakatos's dismissal of their inconsistency, infinitesimals were perceived even by Newton and Leibniz, and certainly by their successors in the eighteenth century, as problematic precisely because of their contradictory qualities. Newton was specifically concerned with the fact that infinitesimals did not obey the Archimedean axiom and therefore could not be accepted as part of rigorous mathematics.[15] Leibniz was similarly concerned about the logical acceptability of infinitesimals. The first public presentation of his differential calculus in 1684 was severely determined by his attempt to *avoid* the logical difficulties connected with the infinitely small. His article in the *Acta Eruditorum* on maxima and minima, for example, presented the differential as a finite line segment rather than the infinitely small quantity that was used in practice.[16]

This confusion between theoretical considerations and practical applications carried over to Leibniz's metaphysics of the infinite, for he was never committed to any one view but made conflicting pronouncements. Philosophically, as Robinson himself has argued, Leibniz had to assume the reality of the infinite—the infinity of his monads, for example—or the reality of infinitesimals not as mathematical points but as substance- or force-points—namely, Leibniz's "monads" themselves.[17]

That the eighteenth century was concerned not with doubts about the potential of infinitesimals but primarily with fears about their logical consistency is clear from the proposal Lagrange drew up for a prize to be awarded by the Berlin Academy for a rigorous theory of infinitesimals. As the prize proposal put it:

> It is well known that higher mathematics continually uses infinitely large and infinitely small quantities. Nevertheless, geometers, and even the ancient analysts, have carefully avoided everything which approaches the infinite; and some great modern analysts hold that the terms of the expression *infinite magnitude* contradict one another.
>
> The Academy hopes, therefore, that it can be explained how so many true theorems have been deduced from a contradictory supposition, and that a principle can be delineated which is sure, clear—in a word, truly mathematical—which can appropriately be substituted for *the infinite*.[18]

Lakatos seems to appreciate all this—and even contradicts himself on the subject of Leibniz's theory and the significance of its perceived inconsistency. Recalling his earlier assertion that Leibniz's theory was not overthrown because of its inconsistency, consider the following line, just a few pages later, where Lakatos asserts that nonstandard analysis raises the problem of "how to appraise *inconsistent* theories like Leibniz's calculus, Frege's logic, and Dirac's delta function."[19]

Lakatos apparently had not made up his mind as to the significance of the inconsistency of Leibniz's theory, which raises questions about the historical value and appropriateness of the extreme sort of rational reconstruction that he has proposed to "vindicate" the work of earlier generations. In fact, neither Leibniz nor Euler nor Cauchy succeeded in giving a satisfactory foundation for an infinitesimal calculus that also demonstrated its logical consistency. Basically, Cauchy's "epsilontics" were a means of avoiding infinites and infinitesimals. Nowhere do Robinsonian infinitesimals or justifications appear in Cauchy's explanations of the rigorous acceptability of his work.[20]

Wholly apart from what Lakatos and others like Robinson have attempted in reinterpreting earlier results in terms of nonstandard analysis, it is still important to understand Robinson's own reasons for developing his historical knowledge in as much detail—and with as much scholarship—as he did. For Robinson, the history of infinitesimals was more than an antiquarian interest; it was not one that developed with advancing age or retirement, but was a *simultaneous* development that began with his discovery of nonstandard analysis in the early 1960s. Moreover, there seem to have been serious reasons for Robinson's keen attention to the history of mathematics as part of his own "research program" concerned with the future of nonstandard analysis.

3. Nonstandard Analysis and the History of Mathematics

In 1965, in a paper titled "On the Theory of Normal Families," Robinson began with a short look at the history of mathematics.[21] He noted that for about one hundred and fifty years after its inception in the seventeenth century, mathematical analysis developed vigorously on inadequate foundations. Despite this inadequacy, the precise, quantitative results produced by the leading mathematicians of that period have stood the test of time.

In the first half of the nineteenth century, however, the concept of the limit, advocated previously by Newton and d'Alembert, gained ascendancy. Cauchy, whose influence was instrumental in bringing about the change, still based his arguments on the intuitive concept of an infinitely small number as a variable tending to zero. At the same time, however, he set the stage for the formally more satisfactory theory of Weierstrass, and today deltas and epsilons are the everyday language of the calculus, at least for most mathematicians. It was this precise approach that paved the way for the formulation of more general and more abstract concepts. Robinson used this history to explain the importance of compactness as applied to functions of a complex variable, which had led to the theory of normal families developed largely by Paul Montel. There followed the qualitative development of complex variable theory, such as Picard theory, and, finally, against this background, more quantitative theories like those developed by Rolf Nevanlinna—to whom Robinson's paper was dedicated as part of a *Festschrift*.

The historical notes to be found at the beginning of Robinson's paper were echoed again at the end, when he turned to ask whether the results he had achieved using nonstandard analysis couldn't be achieved just as well by standard methods. Although he admitted that because of the transfer principle (developed in his paper of 1961, "Non-Standard Analysis") this was indeed possible, he added that such translations into standard terms usually complicated matters considerably. As for nonstandard analysis and the use of infinitesimals it permitted, his conclusion was emphatic:

> Nevertheless, we venture to suggest that our approach has a certain natural appeal, as shown by the fact that it was preceded in history by a long line of attempts to introduce infinitely small and infinitely large numbers into Analysis.[22]

And so the reason for the historical digression was its usefulness in serving a much broader purpose than merely introducing some rather remote historical connections between Newton, Leibniz, Paul Montel, and Rolf Nevanlinna. History could serve the mathematician as propaganda. Robinson was apparently concerned that many mathematicians were prepared to adopt a "so what" attitude toward nonstandard analysis because of the more familiar reduction that was always possible to classical foundations. There were several ways to outflank those who chose to minimize nonstandard analysis because, theoretically, it could do nothing that wasn't equally possible in standard analysis. Above all, nonstandard analysis was often simpler and more intuitive in a very direct, immediate way than standard approaches. But, as Robinson also began to argue with increasing frequency and in greater detail, *historically* the concept of infinitesimals had always seemed natural and intuitively preferable to more convoluted and less intuitive sorts of rigor. Now that nonstandard analysis showed *why* infinitesimals were safe for consumption in mathematics, there was no reason not to exploit their natural advantages. The paper for Rolf Nevanlinna was meant to exhibit both the technical applications and, at least in part through its appeal to history, the naturalness of nonstandard analysis in developing the theory of normal families.

4. Foundations and Philosophy of Mathematics

If Robinson regarded the history of infinitesimals as an aid to the justification in a very general way of nonstandard analysis, what contribution did it make, along with his results in model theory, to the foundations and philosophy of mathematics? Stephan Körner, who taught a philosophy of mathematics course with Robinson at Yale in the fall of 1973, shortly before Robinson's death early the following year, was doubtless closest to Robinson's maturest views on the subject.[23] Basically, Körner sees Robinson as a follower of—or at least working in the same spirit as—Leibniz and Hilbert. Like Leibniz (and Kant after him), Robinson rejected any empirical basis for knowledge about the infinite—whether in the form of infinitely large or infinitely small quantities, sets, whatever. Leibniz is famous for his view that infinitesimals are useful fictions—a position deplored by such critics as Nieuwentijt or the more flamboyant and popular Bishop Berkeley, whose condemnation of the Newtonian calculus might equally well have applied to Leibniz.[24] Leibniz adopted both the infinitely large and the infinitely small in mathematics for pragmatic reasons, as permitting an economy of expression and an intuitive, suggestive, heuristic picture. Ultimately, there was nothing to worry about since the mathematician could eliminate them from his final result after having infinitesimals and infinities to provide the machinery and do the work of a proof.

Leibniz and Robinson shared a similar view of the ontological status of infinities and infinitesimals. They are not just fictions, but well-found ones—"fictiones bene fondatae," in the sense that their applications prove useful in penetrating the complexity of natural phenomena and help to reveal relationships in nature that purely empirical investigations would never produce.

As Emil Borel once said of Georg Cantor's transfinite set theory (to paraphrase not too grossly): although he objected to transfinite numbers or inductions in the formal presentation of finished results, it was certainly permissible to use them to discover theorems and create proofs—again, whatever works.[25] It was only necessary to be sure that in the final version they were eliminated, thus making no official appearance. Robinson, however, was interested in more, especially in the reasons *why* the mathematics worked as it did, and in particular why infinities and infinitesimals were now admissible as rigorous entities despite centuries of doubts and attempts to eradicate them entirely.

Here Robinson succeeded where Leibniz and his successors failed. Leibniz, for example, never demonstrated the consistent foundations of his calculus, for which his work was sharply criticized by Nieuwentijt, among others. Throughout the eighteenth century, the troubling foundations (really, lack of foundations) of the Leibnizian infinitesmal calculus continued to bother mathematicians, until the epsilon-delta methods of Cauchy and the "arithmetic" rigor of Weierstrass reestablished analysis on acceptably finite terms. Because, as Körner remarks, "Leibniz's approach was considered irremediably inconsistent, hardly any efforts were made to improve this delimination."[26]

Robinson was clearly not convinced of the inconsistency of infinitesimals, and in developing the methods of Skolem (who had advanced the idea of nonstandard arithmetic) he was led to consider the possibility of nonstandard analysis. At the same time, his work in model theory and mathematical logic contributed not only to his creation of nonstandard analysis, but to his views on the foundations of mathematics as well.

5. Robinson and "Formalism 64"

In the 1950s, working under the influence of his teacher Abraham Fraenkel, Robinson seems to have been satisfied with a fairly straightforward philosophy of Platonic realism. But by 1964, Robinson's philosophical views had undergone considerable change. In a paper titled simply "Formalism 64," Robinson emphasized two factors in rejecting his earlier Platonism in favor of a formalist position:

(i) Infinite totalities do not exist in any sense of the word (i.e., either really or ideally). More precisely, any mention, or purported mention, of infinite totalities is, literally, *meaningless*.

(ii) Nevertheless, we should continue the business of Mathematics "as usual," i.e. we should act *as if* infinite totalities really existed.[27]

Georg Kreisel once commented that, as he read Robinson's "Formalism 64," it was not clear to him whether Robinson meant *1864* or *1964*! Robinson, however, was clearly responding in his views on formalism to research that had made a startling impression upon mathematicians only in the previous year—namely, Paul Cohen's important work in 1963 on forcing and the independence of the continuum hypothesis.

As long as it appeared that the accepted axiomatic systems of set theory (the Zermelo-Fraenkel axiomatization, for example) were able to cope with all set theoretical problems that were of interest to the working mathematician, belief in the existence of a unique "universe of sets" was almost unanimous. However, this simple view of the situation was severely shaken in the 1950s and early 1960s by two distinct developments. One of these was Cohen's proof of the independence of the continuum hypothesis, which revealed a great disparity between the scale of transfinite ordinals and the scale of cardinals—or power sets. As Robinson himself noted in an article in *Dialectica*, the relation "is so flexible that it seems to be quite beyond control, at least for now."[28]

The second development of concern to Robinson was the emergence of new and varied axioms of infinity. Although the orthodox Platonist believes that in the real world such axioms must either be true or false, Robinson found himself persuaded otherwise. Despite his new approach to foundations in "Formalism 64," he was not dogmatic, but remained flexible:

> The development of "meaningless" infinitistic theories may at some future date become so unsatisfactory to me that I shall be willing to acknowledge the greater intellectual seriousness of some form of constructivism. But I cannot imagine that I shall ever return to the creed of the true platonist, who sees the world of the actual infinite spread out before him and believes that he can comprehend the incomprehensible.[29]

6. Erret Bishop: Meaning, Truth, and Nonstandard Analysis

Incomprehensible, however, is what some of Robinson's critics have said, almost literally, of nonstandard analysis itself. Of all Robinson's opponents, at least in public, none has been more vocal—or more vehement—than Erret Bishop.

In the summer of 1974, it was hoped that Robinson and Bishop would actually have a chance to discuss their views in a forum of mathematicians and historians and philosophers of mathematics who were invited to a special Workshop on the Evolution of Modern Mathematics held at the American Academy of Arts and Sciences in Boston. Garrett Birkhoff, one of the workshop's organizers, had intended to feature Robinson as the keynote speaker for the section of the Academy's program devoted to foundations of mathematics, but Robinson's unexpected death in April of 1974 made this impossible. Instead, Erret Bishop presented the featured paper for the section on foundations. Birkhoff compared Robinson's ideas with those of Bishop in the following terms:

> During the past twenty years, significant contributions to the foundations of mathematics have been made by two opposing schools. One, led by Abraham Robinson, claims Leibnizian antecedents for a "nonstandard analysis" stemming from the "model theory" of Tarski. The other (smaller) school, led by Errett Bishop, attempts to reinterpret

Brouwer's "intuitionism" in terms of concepts of "constructive analysis."[30]

Birkhoff went on to describe briefly (in a written report of the session) the spirited discussions following Bishop's talk, marked, as he noted, "by the absence of positive reactions to Bishop's view."[31] Even so, Bishop's paper raised a fundamental question about the philosophy of mathematics, which he put simply as follows: "As pure mathematicians, we must decide whether we are playing a game, or whether our theorems describe an external reality."[32] If these are the only choices, then one's response is obviously limited. For Robinson, the excluded middle would have to come into play here—for he viewed mathematics, in particular the striking results he had achieved in model theory and nonstandard analysis, as constituting much more than a meaningless game, although he eventually came to believe that mathematics did *not* necessarily describe any external reality. But more of Robinson's own metaphysics in a moment.

Bishop made his concerns over the crisis he saw in contemporary mathematics quite clear in a dramatic characterization of what he took to be the pernicious efforts of historians and philosophers alike. Not only is there a crisis at the foundations of mathematics, according to Bishop, but a very real *danger* (as he put it) in the role that historians seemed to be playing, along with nonstandard analysis itself, in fueling the crisis:

> I think that it should be a fundamental concern to the historians that what they are doing is potentially dangerous. The superficial danger is that it will be and in fact has been systematically distorted in order to support the status quo. And there is a deeper danger: it is so easy to accept the problems that have historically been regarded as significant as actually being significant.[33]

Interestingly, in his own historical writing, Robinson sometimes made the same point concerning the triumph, as many historians (and mathematicians as well) have come to see it, of the success of Cauchy-Weierstrassian epsilontics over infinitesimals in making the calculus "rigorous" in the course of the nineteenth century. In fact, one of the most important achievements of Robinson's work in nonstandard analysis has been his conclusive demonstration of the poverty of this kind of historicism— of the mathematically Whiggish interpretation of increasing rigor over the mathematically unjustifiable "cholera baccillus" of infinitesimals, to use Georg Cantor's colorful description.[34]

As for nonstandard analysis, Bishop had this to say at the Boston meeting:

> A more recent attempt at mathematics by formal finesse is nonstandard analysis. I gather that it has met with some degree of success, whether at the expense of giving significantly less meaningful proofs I do not know. My interest in nonstandard analysis is that attempts are being made to introduce it into calculus courses. It is difficult to believe that debasement of meaning could be carried so far.[35]

Two things deserve comment here. The first is that Bishop (surprisingly, in light of some of his later comments about nonstandard analysis) does not dismiss it as *completely* meaningless, but only asks whether its proofs are "significantly less meaningful" than constructivist proofs. Leaving open for the moment what Bishop has in mind here for "meaningless" in terms of proofs, is seems clear that by one useful indicator to which Bishop refers, nonstandard analysis is year-by-year showing itself to be increasingly "meaningful."[36]

Consider, for example, the pragmatic value of nonstandard analysis in terms of its application in teaching the calculus. Here it is necessary to consider the success of Jerome Keisler's textbook *Elementary Calculus: An Approach Using Infinitesimals*, which uses nonstandard analysis to explain in an introductory course the basic ideas of calculus. The issue of its pedagogic value will also serve to reintroduce, in a moment, the question of meaning in a very direct way.

Bishop claims that the use of nonstandard analysis to teach the calculus is wholly pernicious. He says this explicitly:

> The technical complications introduced by Keisler's approach are of minor importance. The real damage lies in his obfuscation and devitalization of those wonderful ideas. No invocation of Newton and Leibniz is going to justify developing calculus using [nonstandard analysis] on the grounds that the usual definition of a limit is too complicated! . . .
>
> Although it seems to be futile, I always tell my calculus students that mathematics is not esoteric: it is commonsense. (Even the notorious e, d definition of limit is commonsense, and moreover is central to the important practical problems of approximation and estimation.) They do not believe me.[37]

One reason Bishop's students may not believe him is that what he claims, in fact, does not seem to be true. There is another side to this as well, for one may also ask whether there is any truth to the assertions made by Robinson (and emphatically by Keisler) that "the whole point of our infinitesimal approach to calculus is that it is easier to define and explain *limits* using infinitesimals."[38] Of course, this claim also deserves examination, in part because Bishop's own attempt to dismiss Keisler's methods as being equivalent to the axiom "0 = 1" is simply nonsense.[39] In fact, there are concrete indications that despite the allegations made by Bishop about obfuscation and the nonintuitiveness of basic ideas in nonstandard terms, exactly the *opposite* is true.

Not long ago a study was undertaken to assess the validity of the claim that "from this nonstandard approach, the definitions of the basic concepts [of the calculus] become simpler and the arguments more intuitive."[40] Kathleen Sullivan reported the results of her dissertation, written at the University of Wisconsin and designed to determine the pedagogical usefulness of nonstandard analysis in teaching calculus, in the *American Mathematical Monthly* in 1976. This study, therefore, was presumably available to Bishop when his review of Keisler's book appeared in 1977, in which he attacked the pedagogical validity of nonstandard analysis. What did Sullivan's study reveal? Basically, she set out to answer the following questions:

> Will the students acquire the basic calculus skills? Will they really understand the fundamental concepts any differently? How difficult will it be for them to make the transition into standard analysis courses if they want to study more mathematics? Is the nonstandard approach only suitable for gifted mathematics students?[41]

To answer these questions, Sullivan studied classes at five schools in the Chicago-Milwaukee area during the years 1973-74. Four of them were small private colleges, the fifth a public high school in a suburb of Milwaukee. The same instructors who had taught the course previously agreed to teach one introductory course using Keisler's book (the 1971 edition) as well as another introductory course using a standard approach (thus serving as a control group) to the calculus. Comparison of SAT scores showed that both the experimental (nonstandard) group and the standard (control) group were comparable in ability before the courses began. At the end of the course, a calculus test was given to both groups. Instructors teaching the courses were interviewed, and a questionnaire was filled out by everyone who has used Keisler's book within the last five years.

The single question that brought out the greatest difference between the two groups was question 3:

Define $f(x)$ by the rule $\begin{cases} f(x) = x^2 \text{ for } x \neq 2; \\ f(x) = 0 \text{ for } x = 2. \end{cases}$

Prove using the definition of limit that $\lim_{x \to 2} f(x) = 4$.

	Control Group (68 students)	Experimental Group (68 students)
Did not attempt	22	4
Standard arguments:		
Satisfactory proof	2	14
Correct statement, faulty proof	15	14
Incorrect arguments	29	23
Nonstandard arguments:		
Satisfactory proof		25
Incorrect arguments		2

The results, as shown in the accompanying tabulation, seem to be striking; but, as Sullivan cautions:

> Seeking to determine whether or not students really do perceive the basic concepts any differently is not simply a matter of tabulating how many students can formulate proper mathematical definitions. Most teachers would probably agree that this would be a very imperfect instrument for measuring understanding in a college freshman. But further light on this and other questions can be sought in the comments of the instructors.[42]

Here, too, the results are remarkable in their support of the heuristic value of using nonstandard analysis in the classroom. It would seem that, contrary to Bishop's views, the traditional approach to the calculus may be the more pernicious. Instead, the new nonstandard approach was praised in strong terms by those who actually used it:

> The group as a whole responded in a way favorable to the experimental method on every item: the students learned the basic concepts of the calculus more easily, proofs were easier to explain and closer to intuition, and most felt that the students end up with a better understanding of the basic concepts of the calculus.[43]

As to Bishop's claim that the *d,e* method is "commonsense,"[44] this too is open to question. As one teacher having successfully used Keisler's book remarked, "When my most recent classes were presented with the epsilon-delta definition of limit, they were outraged by its obscurity compared to what they had learned [via nonstandard analysis]."[45]

But as G. R. Blackley warned Keisler's publishers (Prindle, Weber and Schmidt) in a letter when he was asked to review the new textbook prior to its publication:

> Such problems as might arise with the book will be *political*. It is revolutionary. Revolutions are seldom welcomed by the established party, although revolutionaries often are.[46]

The point to all of this is simply that, if one take *meaning* as the standard, as Bishop urges, rather than *truth*, then it seems clear that by its own success nonstandard analysis has indeed proven itself *meaningful* at the most elementary level at which it could be introduced—namely, that at which calculus is taught for the first time. But there is also a *deeper* level of meaning at which nonstandard analysis operates—one that also touches on some of Bishop's criticisms. Here again Bishop's views can also be questioned and shown to be as unfounded as his objections to nonstandard analysis pedagogically.

Recall that Bishop began his remarks in Boston at the American Academy of Arts and Sciences workshop in 1974 by stressing the crisis in contemporary mathematics that stemmed from what he perceived as a misplaced emphasis upon formal systems and a lack of distinction between the ideas of "truth" and "meaning." The choice Bishop gave in Boston was between mathematics as a meaningless game or as a discipline describing some objective reality. Leaving aside the question of whether mathematics *actually* describes reality, in some objective sense, consider Robinson's own hopes for nonstandard analysis, those beyond the purely technical results he expected the theory to produce. In the preface to his book on the subject, he hoped that "some branches of modern Theoretical Physics might benefit directly from the application of nonstandard analysis."[47]

In fact, the practical advantages of using nonstandard analysis as a branch of applied mathematics have been considerable. Although this is not the place to go into detail about the increasing number of results arising from nonstandard analysis in diverse contexts, it suffices here to mention impressive research using nonstandard analysis in physics, especially quantum theory and thermodynamics, and in economics, where study of exchange economies has been particularly amenable to nonstandard interpretation.[48]

7. Conclusion

There is another purely theoretical context in which Robinson considered the importance of the history of mathematics that also warrants consideration. In 1973, Robinson wrote an expository article that drew its title from a famous monograph written in the nineteenth century by Richard Dedekind: *Was sind und was sollen die Zahlen?* This title was roughly translated—or transformed in Robinson's version—as "Numbers—What Are They and What Are They Good For?" As Robinson put it: "Number systems, like hair styles, go in and out of fashion—it's what's underneath that counts."[49]

This might well be taken as the leitmotiv of much of Robinson's mathematical career, for his surpassing interest since the days of his dissertation written at the University of London in the late 1940s was model theory, and especially the ways in which mathematical logic could not only illuminate mathematics, but have very real and useful applications within virtually all of its branches. In discussing number systems, he wanted to demonstrate, as he put it, that

> the collection of all number systems is not a finished totality whose discovery was complete around 1600, or 1700, or 1800, but that it has been and still is a growing and changing area, sometimes absorbing new systems and sometimes discarding old ones, or relegating them to the attic.[50]

Robinson, of course, was leading up in his paper to the way in which nonstandard analysis had again broken the bounds of the traditional Cantor-Dedekind understanding of the real numbers, especially as they had been augmented by Cantorian transfinite ordinals and cardinals.

To make his point, Robinson turned momentarily to the nineteenth century and noted that Hamilton had been the first to demonstrate that there was a larger arithmetical system than that of the complex numbers—namely, that represented by his quaternions. These were soon supplanted by the system of vectors developed by Josiah Willard Gibbs of Yale and eventually transformed into a vector calculus. This was a more useful system, one more advantageous in the sorts of applications for which quaternions had been invented.

Somewhat later, another approach to the concept of number was taken by Georg Cantor, who used the idea of equinumerosity in terms of one-to-one correspondences to define numbers. In fact, for Cantor a cardinal number was a symbol assigned to a set, and the same symbol represented all sets equivalent to the base set. The advantage of this view of the nature of numbers, of course, was that it could be applied to infinite sets, producing transfinite numbers and eventually leading to an entire system of transfinite arithmetic. Its major disadvantage, however, was that it led Cantor to reject adamantly any mathematical concept of infinitesimal.[51]

As Robinson points out, although the eventual fate of Cantor's theory was a success story, it was not entirely so for its author. Despite the clear utility of Cantor's ideas, which arose in connection with his work on trigonometric series (later applied with great success by Lebesgue and others at the turn of the century), it was highly criticized by a spectrum of mathematicians, including, among the most prominent, Kronecker, Frege, and Poincaré. In addition to the traditional objection that the infinite should not be allowed in rigorous mathematics, Cantor's work was also questioned because of its abstract character. Ultimately, however, Cantor's ideas prevailed, despite criticism, and today set theory is a cornerstone, if not the major foundation, upon which much of modern mathematics rests.[52]

There was an important lesson to be learned, Robinson believed, in the eventual acceptance of new ideas of number, despite their novelty or the controversies they might provoke. Ultimately, utilitarian realities could not be overlooked or ignored forever. With an eye on the future of nonstandard analysis, Robinson was impressed by the fate of another theory devised late in the nineteenth century that also attempted, like those of Hamilton, Cantor, and Robinson, to develop and expand the frontiers of number.

In the 1890s, Kurt Hensel introduced a whole series of new number systems, his now familiar p-adic numbers. Hensel realized that he could use his p-adic numbers to investigate properties of the integers and other numbers. He also realized, as did others, that the same results could be obtained in other ways. Consequently, many mathematicians came to regard Hensel's work as a pleasant game; but, as Robinson himself observed, "Many of Hensel's contemporaries were reluctant to acquire the techniques involved in handling the new numbers and thought they constituted an unnecessary burden."[53]

The same might be said of nonstandard analysis, particularly in light of the transfer principle that demonstrates that theorems true in *$\mathbf{R}$ can also be proven for $\mathbf{R}$ by standard methods. Moreover, many mathematicians are clearly reluctant to master the logical machinery of model theory with which Robinson developed his original version of nonstandard analysis. This problem has been resolved by Keisler and Luxemburg, among others, who have presented nonstandard analysis in ways accessible to mathematicians without their having to take up the difficulties of mathematical logic as a prerequisite.[54] But for those who see nonstandard analysis as a fad that may be a currently pleasant game, like Hensel's p-adic numbers, the later history of Hensel's ideas should give skeptics an example to ponder. For today, p-adic numbers are regarded as coequal with the reals, and they have proven a fertile area of mathematical research.

The same has been demonstrated by nonstandard analysis. Its applications in areas of analysis, the theory of complex variables, mathematical physics, economics, and a host of other fields have shown the utility of Robinson's own extension of the number concept. Like Hensel's p-adic numbers, nonstandard analysis can be avoided, although to do so may complicate proofs and render the basic features of an argument less intuitive.

What pleased Robinson as much about nonstandard analysis as the interest it engendered from the beginning among mathematicians was the way it demonstrated the indispensability, as well as the power, of technical logic:

> It is interesting that a method which had been given up as untenable has at last turned out to be workable and that this development in a concrete branch of mathematics was brought about by the refined tools made available by modern mathematical logic.[55]

Robinson had begun his career as a mathematician by studying set theory and axiomatics with Abraham Fraenkel in Jerusalem, which eventually led to his Ph.D. from the University of London in 1949.[56] His early interest in logic was later amply repaid in his applications of logic to the development of nonstandard analysis. As Simon Kochen once put it in assessing the significance of Robinson's contributions to mathematical logic and model theory:

> Robinson, via model theory, wedded logic to the mainstreams of mathematics.... At present, principally because of the work of

Abraham Robinson, model theory is just that: a fully-fledged theory with manifold interrelations with the rest of mathematics.[57]

Kurt Gödel valued Robinson's achievement for similar reasons: it succeeded in uniting mathematics and logic in an essential, fundamental way. That union has proved to be not only one of considerable mathematical importance, but of substantial philosophical and historical content as well.

Notes

1. There is a considerable literature on the subject of the supposed crisis in mathematics associated with the Pythagoreans. See, for example, (Hasse and Scholz 1928). For a recent survey of this debate, see (Berggren 1984; Dauben 1984; Knorr 1975).
2. (Bishop 1975, 507).
3. (Bishop 1975, 513-14).
4. Robinson first published the idea of nonstandard analysis in a paper submitted to the Dutch Academy of Sciences (Robinson 1961).
5. (Cleave 1971; Edwards 1979; Laugwitz 1975, 1985; Luxemburg 1975).
6. (Lakatos 1978).
7. (Lakatos 1978, 43).
8. (Lakatos 1978, 44).
9. (Lakatos 1978, 49).
10. (Lakatos 1978, 50). Emphasis in original.
11. Cauchy offers his definitions of infinitely large and small numbers in several works, first in the *Cours d'analyse*, subsequently in later versions without substantive changes. See (Cauchy 1821, 19; 1823, 16; 1829, 265), as well as (Fisher 1978).
12. (Cauchy 1868).
13. (Lakatos 1978, 54).
14. For details of the successful development of the early calculus, see (Boyer 1939; Grattan-Guinness 1970, 1980; Grabiner 1981; Youshkevitch 1959).
15. (Newton 1727, 39), where he discusses the contrary nature of indivisibles as demonstrated by Euclid in Book X of the *Elements*. For additional analysis of Newton's views on infinitesimals, see (Grabiner 1981, 32).
16. See (Leibniz 1684). For details and a critical analysis of what is involved in Leibniz's presentation and applications of infinitesimals, see (Bos 1974-75; Engelsman 1984).
17. See (Robinson 1967, 35 [in Robinson 1979, 544]).
18. In (Lagrange 1784, 12-13; Dugac 1980, 12). For details of the Berlin Academy's competition, see (Grabiner 1981, 40-43; Youshkevitch 1971, 149-68).
19. (Lakatos 1978, 59). Emphasis added.
20. See (Grattan-Guinness 1970, 55-56), where he discusses "limit-avoidance" and its role in making the calculus rigorous.
21. (Robinson 1965b).
22. (Robinson 1965b, 184); also in (Robinson 1979, vol. 2, 87).
23. I am grateful to Stephan Körner and am happy to acknowledge his help in ongoing discussions we have had of Robinson and his work.
24. For a recent survey of the controversies surrounding the early development of the calculus, see (Hall 1980).
25. Borel in a letter to Hadamard, in (Borel 1928, 158).
26. (Körner 1979, xlii). Körner notes, however, that an exception to this generalization is to be found in Hans Vaihinger's general theory of fictions. Vaihinger tried to justify infinitesimals by "a method of opposite mistakes," a solution that was too imprecise, Körner suggests, to have impressed mathematicians. See (Vaihinger 1913, 511ff).

27. (Robinson 1965a, 230; Robinson 1979, 507). Nearly ten years later, Robinson recalled the major points of "Formalism 64" as follows: "(i) that mathematical theories which, allegedly, deal with infinite totalities do not have any detailed meaning, i.e. reference, and (ii) that this has no bearing on the question whether or not such theories should be developed and that, indeed, there are good reasons why we should continue to do mathematics in the classical fashion nevertheless." Robinson added that nothing since 1964 had prompted him to change these views and that, in fact, "well-known recent developments in set theory represent evidence favoring these views." See (Robinson 1975, 557).

28. (Robinson 1970, 45-49).

29. (Robinson 1970, 45-49).

30. (Birkhoff 1975, 504).

31. (Birkhoff 1975, 504).

32. (Bishop 1975, 507).

33. (Bishop 1975, 508).

34. For Cantor's views, see his letter to the Italian mathematician Vivanti in (Meschkowski 1965, 505). A general analysis of Cantor's interpretation of infinitesimals may be found in (Dauben 1979, 128-32, 233-38). On the question of rigor, see (Grabiner 1974).

35. (Bishop 1975, 514).

36. It should also be noted, if only in passing, that Bishop has not bothered himself, apparently, with a careful study of nonstandard analysis or its implications, for he offhandedly admits that he only "*gathers* that it has met with some degree of success" (Bishop 1975, 514; emphasis added).

37. (Bishop 1977, 208).

38. (Keisler 1976, 298), emphasis added; quoted in (Bishop 1977, 207).

39. (Bishop 1976, 207).

40. (Sullivan 1976, 370). Note that Sullivan's study used the experimental version of Keisler's book, issued in 1971. Bishop reviewed the first edition published five years later by Prindle, Weber and Schmidt. See (Keisler 1971, 1976).

41. (Sullivan 1976, 371).

42. (Sullivan 1976, 373).

43. (Sullivan 1976, 383-84).

44. (Bishop 1977, 208).

45. (Sullivan 1976, 373).

46. (Sullivan 1976, 375).

47. (Robinson 1966, 5).

48. See especially (Robinson 1972a, 1972b, 1974, 1975), as well as (Dresden 1976) and (Voros 1973).

49. (Robinson 1973, 14).

50. (Robinson 1973, 14).

51. For details, see (Dauben 1979).

52. See (Dauben 1979).

53. (Robinson 1973, 16).

54. (Luxemburg 1962, 1976; Keisler 1971).

55. (Robinson 1973, 16).

56. Robinson completed his dissertation, *The Metamathematics of Algebraic Systems*, at Birkbeck College, University of London, in 1949. It was published two years later; see (Robinson 1951).

57. (Kochen 1976, 313).

References

Berggren, J. L. 1984. History of Greek Mathematics: A Survey of Recent Research. *Historia Mathematica* 11: 394-410.

Birkhoff, Garrett. 1975. Introduction to "Foundations of Mathematics." *Proceedings of* the American Academy Workshop in the Evolution of Modern Mathematics. In *Historia Mathematica* 2: 503-5.

Bishop, Errett. 1975. The Crisis in Contemporary Mathematics. *Proceedings of the American Academy Workshop in the Evolution of Modern Mathematics.* In *Historia Mathematica* 2: 505-17.

———. 1977. Review of H. Jerome Keisler, "Elementary Calculus." *Bulletin of the American Mathematical Society* 83: 205-8.

Borel, E. 1928. *Leçons sur la théorie des ensembles.* Paris: Gauthier-Villars.

Bos, Hendrik J. M. 1974-75. Differentials, Higher-Order Differentials and the Derivative in the Leibnizian Calculus. *Archive for History of Exact Sciences* 14: 1-90.

Boyer, C. 1939. *The Concepts of the Calculus, a Critical and Historical Discussion of the Derivative and the Integral.* New York: Columbia University Press, 1939. Reprinted New York: Hafner, 1949. Reissued as *The History of the Calculus and Its Conceptual Development.* New York: Dover, 1959.

Cauchy, A.-L. 1821. *Cours d'analyse de l'école polytechnique.* Paris: Debure. In *Oeuvres complètes d'Augustin Cauchy,* vol. 3 (2d ser.). Paris: Gauthier-Villars, 1897, pp. 1-471.

———. 1823. *Résumé des leçons donées à l'Ecole Royale Polytechnique, sur le calcul infinitésimal.* Paris: Debure. In *Oeuvres complètes d'Augustin Cauchy,* vol. 4 (2d ser.). Paris: Gauthier-Villars, 1899, pp. 5-261.

———. 1829. *Leçons sur le calcul différentiel.* Paris: Debure. In *Oeuvres complètes d'Augustin Cauchy,* vol. 4 (2d ser.). Paris: Gauthier-Villars, 1899, pp. 265-609.

———. 1853. Note sur les séries convergentes dont les divers termes sont des fonctions continues d'une variable réelle ou imaginaire, entre des limites données. *Comptes Rendus* 36: 454-59. In *Oeuvres complètes d'Augustin Cauchy,* vol. 12 (1st ser.). Paris: Gauthier-Villars, 1900, pp. 30-36.

———. 1868. *Sept leçons de physique générale.* Paris: Gauthier-Villars.

Cleave, John P. 1971. Cauchy, Convergence and Continuity. *British Journal of the Philosophy of Science* 22: 27-37.

Dauben, J. 1979. *Georg Cantor: His Mathematics and Philosophy of the Infinite.* Cambridge, Mass.: Harvard University Press.

———. 1984. Conceptual Revolutions and the History of Mathematics: Two Studies in the Growth of Knowledge. In *Transformation and Tradition in the Sciences,* ed. E. Mendelsohn. Cambridge: Cambridge University Press, pp. 81-103.

Dresden, Max (with A. Ostebee and P. Gambardella). 1976. A "Nonstandard Approach" to the Thermodynamic Limit. *Physical Review A* 13: 878-81.

Dugac, Pierre. 1980. *Limite, point d'accumulation, compact.* Paris: Université Pierre et Marie Curie.

Edwards, C. H. 1979. *The Historical Development of the Calculus.* New York: Springer Verlag.

Engelsman, S. B. 1984. *Families of Curves and the Origins of Partial Differentiation.* Mathematical Studies 93. Amsterdam: North Holland.

Fisher, G. M. 1978. Cauchy and the Infinitely Small. *Historia Mathematica* 5: 313-31.

Grabiner, Judith V. 1974. Is Mathematical Truth Time-Dependent? *American Mathematical Monthly* 81: 354-65.

———. 1981. *The Origins of Cauchy's Rigorous Calculus.* Cambridge, Mass.: MIT Press.

Grattan-Guinness, Ivor. 1970. *The Development of the Foundations of Mathematical Analysis from Euler to Riemann.* Cambridge, Mass.: MIT Press.

Grattan-Guinness, Ivor, ed. 1980. *From the Calculus to Set Theory, 1630-1910. An Introductory History.* London: Duckworth.

Hall, A. R. 1980. *Philosophers at War: The Quarrel between Newton and Leibniz*. Cambridge: Cambridge University Press.

Hasse, H., and Scholz, H. 1928. Die Grundlagenkrisis der griechischen Mathematik. *Kant-Studien* 33: 4-34.

Keisler, H. J. 1971. *Elementary Calculus: An Approach Using Infinitesimals (Experimental Version)*. Tarrytown-on-Hudson, N.Y.: Bodgen and Quigley; photo-reproduced from typescript.

———. 1976. *Elementary Calculus: An Approach Using Infinitesimals*. Boston: Prindle, Weber and Schmidt.

Knorr, Wilbur R. 1975. *The Evolution of the Euclidean Elements*. Dordrecht: Reidel.

Kochen, S. 1976. Abraham Robinson: The Pure Mathematician. On Abraham Robinson's Work in Mathematical Logic. *Bulletin of the London Mathematical Society* 8: 312-15.

Körner, S. 1979. Introduction to Papers on Philosophy. In *Selected Papers of Abraham Robinson*, ed. H. J. Keisler et al. Vol. 2. New Haven: Yale University Press, pp. xli-xlv.

Lagrange, J. L. 1784. Prix proposés par l'Académie Royale des Sciences et Belles-Lettres pour l'année 1786. *Nouveaux mémoires de l'Académie Royale des Sciences et Belles-Lettres de Berlin*. Vol. 15. Berlin: G. J. Decker, pp. 12-14.

Lakatos, Imre. 1978. Cauchy and the Continuum: The Significance of Non-standard Analysis for the History and Philosophy of Mathematics. In *Mathematics, Science and Epistemology: Philosophical Papers*, ed. J. Worrall and G. Currie. Vol. 2. Cambridge: Cambridge University Press, pp. 43-60. Reprinted in *The Mathematics Intelligencer* 1: 151-61, with a note, Introducing Imre Lakatos, pp. 148-51.

Laugwitz, Detlef. 1975. Zur Entwicklung der Mathematik des Infinitesimalen und Infiniten. *Jahrbuch Überblicke Mathematik*. Mannheim: Bibliographisches Institut, pp. 45-50.

———. 1985. Cauchy and Infinitesimals. *Preprint 911*. Darmstadt: Technische Hochschule Darmstadt, Fachbereich Mathematik.

Leibniz, G. W. 1684. Nova methodus pro maxima et minima . . . . *Acta Eruditorum* 3: 467-73. Reprinted in (Leibniz 1848-63, vol. 3).

———. 1848-63. *Mathematische Schriften*, ed. G. I. Gerhardt. Berlin: Ascher (vol. 1-2); Halle: Schmidt (vol. 3-7).

Luxemburg, W. A. J. 1962. *Lectures on A. Robinson's Theory of Infinitesimals and Infinitely Large Numbers*. Pasadena: California Institute of Technology. Rev. ed. 1964.

———. 1975. Nichtstandard Zahlsysteme und die Begründung des Leibnizschen Infinitesimalkalküls. *Jahrbuch Überblicke Mathematik*. Mannheim: Bibliographisches Institut, pp. 31-44.

Luxemburg, W. A. J. and Stroyan, K. D. 1976. *Introduction to the Theory of Infinitesimals*. New York: Academic Press.

Meschkowski, Herbert. 1965. Aus den Briefbüchern Georg Cantors. *Archive for History of Exact Sciences* 2: 503-19.

Newton, Isaac. 1727. *Mathematical Principles of Natural Philosophy*. Trans. A. Motte, rev. ed. F. Cajori. Berkeley: University of California Press, 1934.

Robinson, Abraham. 1951. *On the Metamathematics of Algebra*. Amsterdam: North Holland Publishing Company.

———. 1961. Non-Standard Analysis. *Proceedings of the Koninklijke Nederlandse Akademie van Wetenschappen*, ser. A, 64: 432-40.

———. 1965a. Formalism 64. *Proceedings of the International Congress for Logic, Methodology and Philosophy of Science. Jerusalem, 1964*. Amsterdam: North Holland Publishing Company, pp. 228-46. Reprinted in (Robinson 1979, vol. 2, 505-23).

———. 1965b. On the Theory of Normal Families. *Acta Philosophica Fennica* 18: 159-84.

——. 1966. *Non-Standard Analysis*. Amsterdam: North-Holland Publishing Co.; 2d ed., 1974.

——. 1967. The Metaphysics of the Calculus. *Problems in the Philosophy of Mathematics*. Amsterdam: North-Holland Publishing Company, pp. 28-46.

——. 1970. From a Formalist's Point of View. *Dialectica* 23: 45-49.

——. (with P. Kelemen). 1972a. The Nonstandard $\lambda : \phi\,{}^4_2 : (x)$: Model: The Technique of Nonstandard Analysis in Theoretical Physics. *Journal of Mathematical Physics* 13: 1870-74. II. The Standard Model from a Nonstandard Point of View. Pp. 1875-78.

——. (with D. J. Brown). 1972b. A Limit Theorem on the Cores of Large Standard Exchange Economies. *Proceedings of the National Academy of Sciences* 69: 1258-60.

——. 1973. Numbers—What Are They and What Are They Good For? *Yale Scientific Magazine* 47: 14-16.

——. (with D. J. Brown). 1974. The Cores of Large Standard Exchange Economies. *Journal of Economic Theory* 9: 245-54.

——. (with D. J. Brown). 1975. Nonstandard Exchange Economies. *Econometrica* 43: 41-55.

——. 1975. Concerning Progress in the Philosophy of Mathematics. *Proceedings of the Logic Colloquium at Bristol, 1973* Amsterdam: North-Holland Publishing Company, pp. 41-52.

——. 1979. *Selected Papers of Abraham Robinson*. Ed. H. J. Keisler, S. Körner, W. A. J. Luxemburg, and A. D. Young. New Haven: Yale University Press. Vol. 1: *Model Theory and Algebra*; vol. 2: *Nonstandard Analysis and Philosophy*; vol. 3: *Aeronautics*.

Sullivan, Kathleen. 1976. The Teaching of Elementary Calculus Using the Nonstandard Analysis Approach. *American Mathematical Monthly* 83: 370-75.

Vaihinger, H. 1913. *Die Philosophie des Als Ob*. Berlin: Reuther and Reichard.

Voros, A. 1973. Introduction to Nonstandard Analysis. *Journal of Mathematical Physics* 14: 292-96.

Youshkevitch, Adolf P. 1959. Euler und Lagrange über die Grundlagen der Analysis. In *Sammelband der zu Ehren des 250. Geburtstages Leonhard Eulers der Deutschen Akademie der Wissenschaften zu Berlin vorgelegten Abhandlungen*, ed. K. Schröder. Berlin: Akademie Verlag, pp. 224-44.

——. 1971. Lazare Carnot and the Competition of the Berlin Academy in 1786 on the Mathematical Theory of the Infinite. In *Lazare Carnot Savant*, ed. C. C. Gillispie. Princeton: Princeton University Press, pp. 149-68.

How Can Mathematicians
and Mathematical Historians
Help Each Other?

1. Introduction

This paper could have a slightly different title with the word *How* dropped, but I could argue both sides of that question. The present title presumes the optimistic answer, and while we all hope that this is the correct answer, the present time may not be the right one for this answer.

The history of mathematics is not an easy field, and it takes a rare person to be good at it. Keynes supposedly said it took a rare person to be a great economist: one must be a second-rate historian, mathematician, and philosopher. For him second-rate was very good, but not great. The same is probably true about the history of mathematics, except one may not have to be a philosopher. Some great mathematicians have made important contributions to the history of mathematics, but very few have spent enough time and thought on the history of mathematics to be able to write a first-rate historical account of part of mathematics. One recent exception is A. Weil, who has added an excellent historical account of number theory before Gauss (Weil 1984) to the historical notes he wrote for the Bourbaki volumes (Bourbaki 1974). His paper from the Helsinki Congress (Weil 1980) should also be read by anyone interested in the history of mathematics.

Since my training is in mathematics, and my reading of history has been almost random, I have found it useful to think about what history is and how the history of mathematics differs from cultural or political history. Collingwood starts his book *The Idea of History* (1956) with the following four questions: What is history, what is it about, how does it proceed, and what is it for? Of the many answers that could be given to these questions, he gave general ones that others could probably agree with, although most would think the answers were incomplete. For the first question, he wrote that "history is a kind of research or inquiry" (1956, 9).

This is so general that it means little without a more specific statement about what *type* of research. He gave this, and I have nothing to add. Here there is no real difference between the history of mathematics and other types of history.

Collingwood rephrased the second question to ask: What kind of things does history find out? His answer: "actions of human beings that have been done in the past" (1956, 9). Here the answer for the history of mathematics would be slightly different. Beethoven and Shakespeare did unique work, and we would not have *Fidelio* or *King Lear* if they have not lived. However, symmetric functions would have been discovered by others if Newton had not lived. Most mathematicians are Platonists who believe that we discover mathematics rather than invent or create it. The American Revolution and the government that came after would have been significantly different if another group of people had been involved, but most of us believe that calculus would be essentially the same no matter when or where it was developed. I am not trying to claim that our way of thinking of the real numbers would always be the same, for that is clearly not true, but that the essence of calculus would be the same. The essence of calculus has not really changed since Euler's time, and many of us would be happy to teach from his texts if our students were significantly better. As Rota wrote about Mark Kac (Kac 1985, xi): "He warned them [students] that axioms will change with the whims of time, but an application is forever." The applications of calculus, and the facts in Euler, are the real essence of calculus.

Skipping the third question, where I have little to add, we are left with last: What is history for? Collingwood answered that it is "'for' human knowledge" (1956, 10). For the history of mathematics this could be changed to "for mathematical knowledge," except that does not distinguish the history of mathematics from doing mathematics. As a substitute, consider the related but different question raised by Weil in (1980, 227); as he remarked, it is a question that has been discussed by others in the past. The question is: Who is history for? Weil made a number of useful comments on this question for the history of mathematics. My view, which is fairly narrow, is summed up nicely by one of the great historians of early mathematics, O. Neugebauer (1956). More will be quoted than is necessary to try to entice the reader to look at his complete note: "I always felt that its total lack of mathematical competence as well as its moralizing and anecdotal attitude seriously discredited the history of mathematics in the eyes of mathematicians, for whom, after all, the history of mathematics has to be written.''

If the history of mathematics is to be written for mathematicians (exclusively, or even in part), then historians of mathematics need to know a lot of mathematics. One cannot form an adequate picture of what is really important on the basis of the current undergraduate curriculum and first-year graduate courses. In particular, I think there is far too much emphasis on the emergence of rigor and on the foundations of mathematics in much of what is published on the history of mathematics. In an interesting article, Jeremy Gray (1985, 18) wrote: "The foundations of analysis does not emerge as the central topic in mathematics that one might think it was from the historians of mathematics." He was writing about the period of one hundred years ago, and his conclusion was that the central topics then were the theory of differential equations and a variety of topics in geometry, including the theory of algebraic curves.

2. How Can Mathematicians Help?

Most mathematicians have little or no training in the ways of thought that historians have developed, so it is unrealistic to expect many of them to write papers or books that will satisfy mathematical historians. However, some mathematicians are tempted to write a paper on the history of a topic they have studied for years. I was tempted and did this over ten years ago. I had found a few series identities in papers that had been forgotten, and in one case an important result usually attributed to Saalschütz (1890) had been found by Pfaff (1797b) almost one hundred years earlier. Actually, I did not find this paper but read about it in (Jacobi 1848). The identity of Pfaff and Saalschütz is

$$(2.1) \quad \sum_{k=0}^{n} \frac{(-n)_k(a)_k(b)_k}{(c)_k(a+b+1-n-c)_k k!} = \frac{(c-a)_n(c-b)_n}{(c)_n(c-a-b)_n},$$

where the shifted factorial $(a)n$ is defined by

$$(2.2) \quad (a)_n = a(a+1) \ldots (a+n-1), \qquad n = 1,2, \ldots ,$$
$$= 1 \qquad , \qquad n = 0.$$

There were two reasons I wanted to call attention to Pfaff's paper. One is historical, and should have been of interest to historians. When $n \to \infty$

in (2.1), the result is

$$(2.3) \quad \sum_{k=0}^{\infty} \frac{(a)_k(b)_k}{(c)_k k!} = \frac{\Gamma(c)\Gamma(c-a-b)}{\Gamma(c-a)\Gamma(c-b)}$$

if Re$(c$-a-$b) > 0$ and $c \neq 0$, -1,.... This result was proved by Gauss in his published paper on hypergeometric series (1813). Notice the publication date: Gauss lived in Pfaff's home for a few months in 1797, and this was the year in which Pfaff published (1797b) and also published a book (1797a), the middle third of which contains the most comprehensive treatment of hypergeometric functions that appeared before Gauss's work. In addition to the published paper mentioned above, Gauss wrote a sequel that was only published posthumously (1866). Felix Klein was aware of Pfaff's book, and he raised the question of connection between Pfaff's work and Gauss's later work (Klein 1933). The sum (2.1) is not in Pfaff's book, so one can also ask if Gauss had seen this paper by Pfaff before he did not work in (1813). If he had, it is clear he was not then aware of the importance of hypergeometric functions, for he would have easily seen that (2.1) implies (2.3), and so not have thought of (2.3) as a new result. If he saw (2.1) before he appreciated the importance of hypergeometric functions, it is very likely he would have forgotten it. We will probably never know whether Gauss saw this work by Pfaff, but it is worth pointing out the possible influence, for a mathematical historian may find an annotated book or offprint and not appreciate the importance unless told why this influence is interesting.

The second reason interests me but might not interest others. There is a second sum that seems similar to (2.1) and that is attributed to Dixon (1903);

$$(2.4) \quad \sum_{k=0}^{\infty} \frac{(a)_k(b)_k(c)_k}{k!(a+1-b)_k(a+1-c)_k}$$

$$= \frac{\Gamma(1+\frac{a}{2})\Gamma(1+a-b)\Gamma(1+a-c)\Gamma(1+\frac{a}{2}-b-c)}{\Gamma(1+a)\Gamma(1+\frac{a}{2}-b)\Gamma(1+\frac{a}{2}-c)\Gamma(1+a-b-c)}.$$

A much more general result was given by Rogers (1895, sect. 8), and a few special cases of (2.4) were found earlier, but the earliest special case I know appeared in (Dixon 1891). This is almost one hundred years after Pfaff proved (2.1), but only one year after Saalschütz rediscovered it (1890). These two results are often given at the same time. For example,

Knuth (1968, 70, #31; 73, #62) has both of them as problems. He assigned 20 points to (2.1) and 38 points to (2.4), on a scale of 0 to 50, so it is clear he knew that (2.4) is deeper than (2.1). This depth is also illustrated by the approximately one hundred year difference in time when they were discovered. If the reader thought the time difference was one year (for the special case) or thirteen years (for the general case), these results would seem to be of the same depth. It took me a number of years before I appreciated the difference in depth.

I wanted to make one other point in this paper. Many people who write about (2.1) and (2.4) do not really understand what these identities say. For example, Knuth did not write either of these identities in the above form. He used binomial coefficients rather than shifted factorials. To explain the reason behind this difference, I will have to get technical.

In elementary calculus, the favorite test for convergence of an infinite series is the ratio test. It is easy to use and works on most power series that are given to students at this level. For these power series, the ratio between successive terms is a rational function of the index of the term. For example, for a power series about $x = 0$, if

$$f(x) = \sum_{n=0}^{\infty} c_n$$

then for $f(x) = \exp x$ the ratio is

$$\frac{c_{n+1}}{c_n} = \frac{x}{n+1},$$

for $f(x) = (1-x)^{-a}$ the ratio is

$$\frac{c_{n+1}}{c_n} = \frac{(n+a)}{(n+1)} x,$$

and for $\log(1+x)$ the ratio is

$$\frac{c_{n+1}}{c_n} = \frac{-n}{(n+1)} x, \qquad n = 1,2, \ldots.$$

A generalized hypergeometric series is a series

$$\sum c_n$$

with c_{n+1}/c_n a rational function of n. This rational function is usually factored as

$$(2.5) \quad \frac{c_{n+1}}{c_n} = \frac{(n+a_1)\dots(n+a_p)x}{(n+b_1)\dots(n+b_q)(n+1)}$$

and the series is written as

$$(2.6) \quad {}_pF_q\left[\begin{array}{c} a_1,\dots,a_p \\ b_1,\dots,b_q \end{array}; x\right] = \sum_{n=0}^{\infty} \frac{(a_1)_n \dots (a_p)_n}{(b_1)_n \dots (b_q)_n} \frac{x^n}{n!}.$$

The special case $p = 2$, $q = 1$ is often called the hypergeometric series, and the analytic continuation of this function is called the hypergeometric function. It is single valued on the plane cut on $[1,\infty)$ and on an appropriate Riemann surface.

Euler, Gauss, Kummer, Riemann, and many other mathematicians studied the general ${}_2F_1$, and facts about this function are important enough to be collected in many handbooks. Limiting or special cases such as Bessel functions, which are essentially ${}_0F_1$'s; confluent hypergeometric functions, which are ${}_1F_1$'s or linear combinations of two ${}_1F_1$'s; and Legendre functions, which are ${}_2F_1$'s with one or two parameters specialized in appropriate ways, have also been studied extensively. Facts about them are given in very large books, such as Watson's book on Bessel functions (1944) Robin's three volumes on Legendre functions (1957-59), and Hobson's book on spherical harmonics (1931), as well as the standard handbooks (Abramowitz and Stegun 1965; Erdélyi 1953-55). Thus one cannot claim this material is not well known, at least in some circles. However, there are many other circles where this work is almost completely unknown. The best example is combinatorics. One of the most important sets of mathematics books written in the last twenty years in Knuth's *The Art of Computer Programming* (1968, 1969, 1973). In this set of books, and in many others of a combinatorial nature, binomial coefficients occur regularly. The binomial coefficient $\left[\begin{array}{c} n \\ k \end{array}\right]$ is given by

$$\left[\begin{array}{c} n \\ k \end{array}\right] = \frac{n!}{k!(n-k)!}$$

and it counts the number of ways k identical objects can be put in n spots. Knuth wrote:

> There are literally thousands of identities involving binomial coefficients, and for centuries many people have been pleased to discover them. However, there are so many relations present that when someone finds a new identity, there aren't many people who get excited about

it any more, except the discoverer! In order to manipulate the formulas which arise in the analysis of algorithms, a facility for handling binomial coefficients is a must . . . . (1968, sect. 1.2.6, 52-53)

When a mathematician who is as good as Knuth write nonsense like the above (except for the last sentence, where he is probably right), then one must look seriously at what he wrote and try to understand why he missed the essence of what is really true. There are actually very few identities of the sort Knuth gave in this section—there just seem to be many because he does not know how to write them. For example, in (1968, sect. 1.2.6I) he gave six sums (21)-(26) and then wrote that (21) is by far the most important. What he did not point out is that five of these identities, (21)-(25), are all just disguised versions of

$$_2F_1\left[\begin{array}{c} -n,a \\ c \end{array} ; 1\right] = \frac{(c-a)_n}{(c)_n} .$$

In other words, they are all the same identity. Binomial coefficients are important, since they count things; but when one has a series of products of binomial coefficients, the right thing to do is to translate the sum to the hypergeometric series for (2.6). Translation is almost always easy (there can be some problems that require limits when division by zero arises), and it has been known for a long time that this is the right way to handle sums of products of binomial coefficients. Andrews spelled this out in detail in (1974, sect. 5), but the realization that hypergeometric series are just series with term ratio a rational function of n is very old. Horn (1889) used this as the definition of a hypergeometric series in two variables. R. Narasimhan told me that he found a definition of "comfortable" series in one of the late volumes of Euler's collected works. For Euler, a comfortable series is a power series whose term ratio is a rational function of n. When I asked Narasimhan to give me a specific reference, he was unable to find it again. I will be very pleased to pay $50 U.S. for this reference, for it would be worth that to know that Euler's insight was also good here. An even earlier place one might look for this insight would be in Newton's work. In any case, by the time of Kummer's early work (1836), some mathematicians started to look at higher hypergeometric series and write them as

$$1 + \frac{\alpha\beta\lambda}{1\cdot\gamma\nu} x + \frac{\alpha(\alpha+1)\beta(\beta+1)\lambda(\lambda+1)}{1\cdot2\cdot\gamma(\gamma+1)\nu(\nu+1)} x^2 + \ldots$$

which is (2.6) when $p = 3$, $q = 2$. Clausen (1828) used the same notation even earlier.

I wrote a short note (1975) mentioning many of the above facts and sent it to *Historia Mathematica*. It was sent to two referees, who disliked it because I had not written a history paper. The editor then sent it to two other referees, who also did not like it. One thought there were the makings of a reasonable paper if I only did some serious historical research, whereas the other thought the paper was silly. I had written that this material was treated badly in books of mathematical history and was not part of the standard curriculum. His view was essentially the following: if this work is not part of the curriculum, then it probably is not very important.

I would like to quote one paragraph from this technical report, changing it slightly to make it readable without including the earlier text, and then give the surprising sequel:

> The real reason for the obscurity of this material seems to be that it plays a minor role in most problems. Often this role is essential, but there is usually some other idea involved in the solution of a problem which seems to be more central (it usually is) and the explicit sum which is necessary remains a lemma. These sums are easy enough to derive so that a mathematician who has been able to come up with other ideas on how to solve a problem can also rediscover the required sum. But this has not always been true. For example, Good obtained the sum
>
> $$d_{0,2s} = \sum_{v=0}^{s} (-1)^v \binom{\beta}{v} \binom{\beta+s-v}{\beta} \frac{\alpha}{\alpha+s-v} .$$
>
> Then he says "(The sum of the series on the right must be non-negative if $\alpha > \beta$, an inequality that is not obvious directly.) If $\beta = 0$,
>
> $$d_{0,2s} = \frac{\alpha}{\alpha+s} . "$$
>
> See Good (1958). What he did not notice is that this series can be translated into hypergeometric form and summed by a special case of (2.1). The result is
>
> $$d_{0,2s} = \frac{(\alpha-\beta)_s}{(\alpha+1)_s} .$$
>
> Now the positivity for $\alpha > \beta$ is obvious. Recently some very complicated formulas for hypergeometric series have been used by Gasper (1975) to obtain some inequalities for integrals which have not been obtained by any other method. Askey and Gasper (1977) used some other deep facts about hypergeometric series to extend a result of Szegö (1933). If more complicated problems of this sort are to be solved in other areas then mathematicians are going to have to realize that even the subject

of explicit evaluation of sums will have to be looked at in a more systematic way. It will be interesting to see if the knowledge of these important results spreads more widely in the next few decades. If one judges by past history the prospects are poor.

This is probably the most prophetic paragraph I have ever written. In the winter of 1984, L. de Branges reduced the Bieberbach conjecture to showing that a certain integral of a $_2F_1$ is positive. He had a new idea about how to use the Loewner machine, and he reduced a generalization of the Bieberbach conjecture—the Milin conjecture—to the following inequality:

$$(2.7) \qquad \int_0^1 t^{n-k-1/2} {}_2F_1\left[\begin{matrix} -k, 2n-k+2 \\ 2n-2k+1 \end{matrix} ; tx\right] dt \geq 0,$$

$$0 < x < 1, \; k = 0, 1, \ldots n-1.$$

For the sake of those who do not know these conjectures, the Bieberbach conjecture is the following. A function $f(z)$ is univalent if it is one to one on its domain. Let $f(z)$ by analytic for $|z| < 1$ and normalized by

$$f(z) = z + \sum_{n=2}^{\infty} a_n z^n, \qquad |z| < 1.$$

If f is univalent, Bieberbach (1916) showed that $|a_2| \leq 2$ and conjectured that $|a_n| \leq n$. There is equality when $f(z) = z(1-z)^{-2}$. This had been proved for $n = 3, 4, 5, 6$ and for all n for some subclasses of univalent functions, such as those with real coefficients a_n, but the general case was open and thought to be very hard. The Milin conjecture is too technical to state here, but it was known to imply the Bieberbach conjecture, and many experts on univalent functions thought it was false.

The final step in the proof of these conjectures has been described by the participants (Askey 1986; de Branges 1986; Gautschi 1986). Briefly, it is as follows. De Branges asked Walter Gautschi for aid in seeing when (2.7) holds; Gautschi is a colleague of de Branges's and probably the best person in the world to approach when asking for numerical results on the integral in (2.7). De Branges had heard that many experts on univalent functions thought the Bieberbach conjecture was false for odd values of n starting at 17 or 19, and Gautschi was skeptical that this approach would work; so the two of them were very excited when the numbers Gautschi found seemed to say that (2.7) held for n up to 30. This strongly suggested that the Milin conjecture and the Bieberbach conjecture were true for these *n* and that it should be relatively easy to obtain error estimates to show that (2.7) held for those *n*, and probably a good deal higher. In fact, J. Hummel (personal communication) had independently done calculation on the integrated form of (2.7):

$$(2.8) \quad {}_3F_2\left[\begin{matrix}-k, 2n-k+2, n-k+\frac{1}{2} \\ n-k+\frac{3}{2}, 2n-2k+1\end{matrix}; x\right] > 0, \quad 0 < x < 1,$$

$$k = 0, 1, \ldots n-1,$$

and had shown that this is true for *n* up to about 20, and thus that the Milin and Bieberbach conjectures were true for these values of *n*.

The Bieberbach conjecture was a big problem, there was a nice new idea, and the final step dealt with hypergeometric functions, as in my outline quoted above. However, this time the hypergeometric function work was harder than just one identity. There is a relatively simple proof of (2.8), but it requires three identities, only one of which was well known. A second one is contained in the best handbooks. The third one is over a hundred years old and is contained in a few books, but not in the standard handbooks. Gautschi eventually called me and asked if I knew how to prove (2.7). I looked at it that evening, changed it to (2.8), and found this in the first place I looked (Askey and Gasper 1976). George Gasper and I had needed this inequality to prove a conjecture I had made, and Gasper had proved it.

De Branges's paper has now appeared (1985). After a version of de Branges's proof was available in preprint form, many mathematicians went through the details of his argument, gave talks on it, and some wrote their own accounts (Aharonov 1984; Anonymous [Maynooth] 1985; Fitz Gerald and Pommerenke 1985; Korevaar 1985; Milin 1984). However, only two of these accounts gave a complete proof of (2.7), and the general consensus was that the proof of (2.7) was magic and that it would be nice to have a more conceptual, less computational proof. One always wants simple proofs, but if one is willing to admit that Euler, Gauss, Kummer, Riemann et al. knew what they were doing when they studied hypergeometric functions, then this proof seems very natural. To prove that something is nonnegative, one tries to write it as a square or the sum of squares with nonnegative coefficients. The proof Gasper found is just that. What I am afraid of is that the last part of the quotation above will also be prophetic and that, even with this striking use of hypergeometric functions, knowledge of them will not spread to the mathematical community at large as it should.

My note was not the only paper written by a mathematician about some early mathematics of interest today that was turned down by the editors of *Historia Mathematica*. Another was written by George Andrews, and he published it elsewhere (1982). It is also not really a history paper by the standards of *Historia Mathematica*, but it deals with historical documents—in his case, two unpublished letters of L. J. Rogers; it is also of more interest to quite a few mathematicians than many of the papers in *Historia Mathematica*. I suggest it would be useful to publish an occasional article like this. If there were more articles of interest to mathematicians, then more mathematicians would read this journal. A few more papers by mathematicians would also tell historians what mathematicians consider important and would suggest topics that should be looked at in detail by historians. After my experience, the next time I had a topic of historical interest I wrote my comments for publication elsewhere. This topic was the orthogonal polynomials that generalize the classical polynomials of Hermite, Laguerre, and Jacobi. Most of these are older than is generally known. For example, a set of polynomials that is orthogonal on $x = 0, 1, \ldots, N$ with respect to the function $\left[\begin{smallmatrix} x+a \\ x \end{smallmatrix}\right]\left[\begin{smallmatrix} N-x+\beta \\ N-x \end{smallmatrix}\right]$ and that are known as Hahn polynomials were really discovered by Tchebychef (1875); see my comments to (Szegö 1968) in the reprinted version. There is a need for a historical treatment of orthogonal polynomials. Szegö (1968) wrote an outline, I added further comments, and a historical resource without equal exists in (Shohat et al. 1940). This bibliography is not complete, but when it was written it was probably the best bibliography of a part of mathematics, and I do not know of another that equals it in the coverage of the eighteenth and nineteenth centuries. My comments appeared in a place where some mathematicians will see them, but mathematical historians are unlikely to hear about them.

There should be some place where mathematicians can record historical observations that will be read by mathematical historians. For example, consider general history books, which mostly contain material copied by the author from other books or papers. Errors tend to be propagated from one book to another, some of the errors being historical and others mathematical. More frequently, they are errors of ignorance, where the real point of the work is not understood. There needs to be a place where these errors can be corrected so that they will not appear in future books. I will illustrate these errors by mentioning some in M. Kline's book (1972). It is used as an illustration because I agree with the following remark of Rota (1974): "It is easy to find something to criticize in a treatise 1,200 pages long and packed with information. But whatever we say for or against it, we had better treasure this book on our shelf, for as far as mathematical history goes, it is the best we have.'' If the best can have errors like the following, then it is clear that mathematical historians need all the help they can get.

First, a mathematical error. Kline attributed the following formula to Euler (Kline 1972, 489):

$$(2.9) \quad {}_2F_1\left[\begin{matrix} -n,b \\ c \end{matrix}; z\right] = \frac{n!}{c(c+1)\dots(c+n-1)} \cdot$$

$$\int_0^1 t^{-n-1}(1-t)^{c+n-1}(1-tz)^{-b}dt.$$

He does not say what n is, but by implication he has n a nonnegative integer, for he does not define what $c(c+1)\cdots(c+n\text{-}1)$ means when n is anything else. However, when $n = 0, 1, \cdots$, the integral diverges. This integral was copied from (Slater 1966, 3). Euler had a formula like (2.9) that is correct. It is

$${}_2F_1\left[\begin{matrix} a,b \\ c \end{matrix}; z\right] = \frac{\Gamma(c)}{\Gamma(a)\Gamma(c-a)} \int_0^1 t^{a-1}(1-t)^{c-a-1}(1-tz)^{-b}dt$$

when Re c > Re a > 0.

An example of a historical fact that is wrong is the following:

> The problem of solving ordinary differential equations over infinite interval or semi-infinite intervals and of obtaining expansions of arbitrary functions over such intervals was also tackled by many men during the second half of the century and such special functions as Hermite functions first introduced by Hermite in 1864 and Nikolai J. Sonine in 1880 serve to solve this problem. (Kline 1972, 714)

These functions were known more than a half century before Hermite. Laplace studied them in his work in probability theory (Laplace 1812) and gave as many facts about them as Hermite was to rediscover. The date of discovery is important, but there is a more important point that Kline could have illustrated here. The message he put across is that these functions were introduced to solve differential equations. That is not true. They arose for other reasons, such as their connection with Fourier transforms, and the differential equation was a minor fact to Laplace. Part of the power of mathematics is the surprising way in which mathematical ideas or results that arose in one area turn out to be useful in other areas. That is just as true of special functions as it is of other parts of mathematics.

Finally, Kline does not seem to be aware that the $_2F_1$ hypergeometric function is really the most important of the special functions that arise as solutions of differential equations that come from separation of variables. He even said that the Bessel equation is the most important ordinary differential equation that results from separation of variables (1972, 710). Historically it probably arose most frequently, but the hypergeometric equation is much more important, and it contains the Bessel equation and many others as limiting or special cases. Riemann was the first to give a structural reason why the hypergeometric equation is so important. He showed that it is the only linear, homogeneous, second-order differential equation with regular singularities at $z = 0$, 1, and ∞, with every other point an ordinary point. The Bessel equation has a regular singular point at $z = 0$ and an irregular singular point at $z = \infty$. Riemann's result is well over one hundred year old and should now be in a comprehensive history book such as Kline's. Kline's treatment of Riemann's problem and related work is considerably better than any other treatment in a general book on mathematical history—he just did not draw the right conclusions about the $_2F_1$ and its limits.

Kline mentions Gauss's published paper on the hypergeometric function twice but does not mention what I consider to be the major insight in this paper. In his first comment (p. 712), he mentions Gauss's sum (2.3) and the fact that Gauss recognized that special choices of the parameters will give many known elementary and higher transcendental functions. In the second comment (p. 962), he wrote about Gauss's work on convergence. Then he wrote the following, which I do not believe (p. 962): "The unusual rigor discouraged interest in the paper by mathematicians of the time."

The really new point of view that Gauss introduced in this paper was to treat

$$_2F_1\left[\begin{matrix} \alpha,\beta \\ \gamma \end{matrix}\ ; x\right]$$

as a function of four variables, α,β and γ as well as x. This is very important and eventually led to some orthogonal polynomials that have been used in the quantum theory of angular momentum, in the analysis of pea growth, in the first proof of the irrationality of $\zeta(3)$, and in discrete analogues of spherical harmonics that are used in coding theory. Vorsselman de Heer drew upon Gauss's paper and some neglected early work of Euler for his thesis in (1833), and Kummer read Gauss's paper very carefully before he wrote (1836). Twenty years is not an exceedingly long time for a piece of work to be appreciated and extended. The work that Gasper and I did on positive sums of Jacobi polynomials was not used for ten years, and no one has extended or used Gasper's deeper work (1977), or even considered the dual problems. Gauss pushed the field ahead quite a few ways, and it took time to understand his work well enough so that the next steps could be made. The most significant extension of (Gauss 1813) was done by Heine (1845) but that is another story that leads to the Rogers-Ramanujan identities in partition theory and eventually to some beautiful and deep work of Baxter in statistical mechanics (1982, chap. 14).

3. How Can Mathematical Historians Help Mathematicians?

The answer to this question is easy: they can write papers on the history of mathematics that mathematicians care about. The American Mathematical Society has periodic special sessions on the history of mathematics, and these are very well attended. Mathematicians care about mathematics outside their own field, but they find a lack of good papers that contain the essence of the subject—namely, the problems that led to the field and the ideas that were used to attack these problems. This is really mathematical history. Please help us to learn more about mathematics by writing such papers. Eventually it may be possible to write a book like Kline's (1972) that will allow us to get a good overview of mathematical development. Before it is possible to have a good global view, however, we must have much better local views than we now have. As an illustration of a good local treatment of some parts of mathematics, look at some of the papers in (Dieudonné 1978) and the excellent treatment of the arithmetic-geometric mean in (Cox 1984). We need more historical work like these.

I mentioned work on orthogonal polynomials by R.. A. Fisher and a co-worker of his, F. Allan, in my comments to (Szegö 1968), but I did not say how I found this. There was a cryptic comment in Box's biography of Fisher (Box 1978) that mentioned the existence of some discrete orthogonal polynomials. She did not appreciate the importance of the representation found by Allan, and there is no reason she should have, but this representation was very important. If I had not rediscovered it a few years earlier, it would have been a good illustration of how a work on mathematical history could have aided a mathematician. I never would have come across Fisher's paper in the *Journal of Agricultural Science*, and was unlikely to have seen Allan's paper in the *Proceedings of the Royal Society of Edinburgh* (in 1930). However, a comment in a biography alerted me to the existence of these very interesting papers. Mathematicians read mathematical history outside the areas where they work, so this can be a good way of helping to break down the barriers that we have created around our work. Historians can help in ways they do not know, but only if they treat significant mathematics.

4. Conclusion

One of the referees of my note objected that it only contained facts and that history is much more than just facts. That is clearly true, but history built on incorrect facts is justly suspect. There needs to be a place where facts can be recorded, or corrected, and this needs to be a place that will be read by mathematical historians.

Weil (1980) set very high standards for what one would like to have in a mathematical historian, and then he demonstrated that he was not writing about an empty set in this century by living up to his standards. To be realistic, it is unlikely that there will be many others who match his description of a great mathematical historian, so we are going to have to help each other. Together we may able to do some useful history. Separately, some will be done, but not as much as is needed, and the quality will often be lower than it should be.

References

Abramowitz, M., and Stegun, I. 1965. *Handbook of Mathematical Functions*. New York: Dover.

Aharonov, D. 1984. *The de Branges Theorem on Univalent Functions*. Technion-Israel Institute of Technology, Haifa.

Andrews, G. 1974. Applications of Basic Hypergeometric Series. *SIAM Review* 16: 441-84.

———. 1982. L. J. Rogers and the Rogers-Ramanujan Identities. *Mathematical Chronicle* 11(2): 1-15.

Anonymous. 1985. *The Proof of the Bieberbach Conjecture*. Maynooth Mathematics Seminar, Maynooth University.

Askey, R. 1975. *A Note on the History of Series*. Mathematical Research Center Technical Report 1532. University of Wisconsin, Madison.

———. 1986. My Reaction to de Branges's Proof of the Bieberbach Conjecture. In *The Bieberbach Conjecture: Proceedings of the Symposium on the Occasion of the Proof*, ed. A. Baernstein II, D. Drasin, P. Duren, and A. Marden. Providence, R. I.: American Mathematical Society, pp. 213-15.

Askey, R., and Gasper, G. 1976. Positive Jacobi Polynomial Sums. II. *American Journal of Mathematics* 98: 709-37.

——. 1977. Convolution Structures for Laguerre Polynomials. *Journal d'Analyse Mathématique* 31: 48-68.

Baxter, R. J. 1982. *Exactly Solved Models in Statistical Mechanics*. London: Academic Press.

Bieberbach, L. 1916. Über die Koeffizienten derjenigen Potenzreihen, welche eine schlichte Abbildung des Einheitskreises vermitteln. *Sitzungsberichte der Preussischen Akademie der Wissenschaften, physikalish-mathematische Klasse* 940-55.

Bourbaki, N. 1974. *Eléments d'histoire des mathématiques*. Paris: Hermann.

Box, J. F. 1978. *R. A. Fisher, the Life of a Scientist*. New York: Wiley.

Clausen, T. 1828. Ueber die Fälle, wenn die Reihe von der Form $y = 1 + (\alpha \cdot \beta)/(1 \cdot \gamma)x + (\alpha \cdot \alpha + 1)/(1 \cdot 2) \cdot (\beta \cdot \beta + 1)/(\gamma \cdot \gamma + 1)x^2 +$ etc. ein Quadrat von der Form $z = 1 + (\alpha' \beta' \gamma')/(1 \delta' \varepsilon')x +$ etc. hat. *Journal für die reine und angewandte Mathematik* 3: 89-91.

Collingwood, R. G. 1956. *The Idea of History*. New York: Oxford University Press.

Cox, D. A. 1984. The Arithmetic-Geometric Mean of Gauss. *L'Enseignement Mathématique* 30: 275-330.

de Branges, L. 1985. A Proof of the Bieberbach Conjecture. *Acta Mathematica* 154: 137-52.

——. 1986. The Story of the Verification of the Bieberbach Conjecture. In *The Bieberbach Conjecture: Proceedings of the Symposium on the Occasion of the Proof*, ed. A. Baernstein II, D. Drasin, P. Duren, and A. Marden. Providence, R.I.: American Mathematical Society, pp. 199-203.

Dieudonné, J. 1978. *Abrégé d'histoire des mathématiques, 1700-1900*. 2 vols. Paris: Hermann.

Dixon, A. C. 1891. On the Sum of the Cubes of the Coefficients in a Certain Expansion by the Binomial Theorem. *Messenger of Mathematics* 20: 79-80.

——. 1903. Summation of a Certain Series. *Proceedings of the London Mathematical Society* 35: 284-89.

Erdélyi, A. 1953-55. *Higher Transcendental Functions*. 3 vols. New York: McGraw-Hill.

FitzGerald, C., and Pommerenke, C. 1985. The de Branges Theorem on Univalent Functions. *Transactions of the American Mathematical Society* 290: 683-90.

Gasper, G. 1975. Positive Integrals of Bessel Functions. *SIAM Journal on Mathematical Analysis* 6: 868-81.

——. 1977. Positive Sums of the Classical Orthogonal Polynomials. *SIAM Journal on Mathematical Analysis* 8: 423-47.

Gauss, C. F. 1813. Disquisitiones generales circa seriem infinitam $1 + (\alpha\beta)/(1 \cdot \gamma)x + [\alpha(\alpha+1)\beta(B+1)]/[1 \cdot 2 \cdot \gamma(\gamma+1)]xx + [\alpha(\alpha+1)(\alpha+2)\beta(\beta+1)(\beta+2)]/[1 \cdot 2 \cdot 3 \cdot \gamma(\gamma+1)(\gamma+2)]x^3 +$ etc. Pars prior. *Commentationes societatis regiae scientiarum Gottingensis recentiores, B.* Vol. 2. Reprinted in *Carl Friedrich Gauss Werke*. Vol. 3: *Analysis*. Göttigen, 1866, pp. 123-62.

——. 1866. Determinatio seriei nostrae per aequationem differentialem secundi ordinis. In *Carl Friedrich Gauss Werke*. Vol. 3: *Analysis*. Göttingen, 1866, pp. 207-29.

Gautschi, W. 1986. Reminiscences of My Involvement in de Branges' Proof of the Bieberbach Conjecture. In *The Bieberbach Conjecture: Proceedings of the Symposium on the Occasion of the Proof*, ed. A. Baernstein II, D. Drasin, P. Duren, and A. Marden. Providence, R.I.: American Mathematical Society, pp. 205-11.

Good, I. J. 1958. Random Motion and Analytic Continued Fractions. *Proceedings of the Cambridge Philosophical Society* 54: 43-47.

Gray, J. 1985. Who Would Have Won the Fields Medals a Hundred Years Ago? *Mathematical Intelligencer* 7 (3): 10-19.

Heine, E. 1845. Untersuchungen über die Reihe $1 + [(1-q^\alpha)(1-q^\beta)]/[(1-q)(1-q^\gamma)]x$ *Journal für die reine und angewandte Mathematik* 34: 285-328.

Hobson, E. W. 1931. *The Theory of Spherical and Ellipsoidal Harmonics*. Cambridge: Cambridge University Press. Reprinted New York: Chelsea, 1955.

Horn, J. 1889. Ueber die Convergenz der hypergeometrischen Reihen zweier und dreier Veränderlichen. *Mathematische Annalen* 34: 544-600.

Jacobi, C. G. J. 1848. De seriebus ac differentiis observatiunculae. *Journal für die reine und angewandte Mathematik* 36: 135-42. Reprinted in *Gesammelte Werke*, vol. 6, New York: Chelsea, 1969, pp. 174-82.

Kac, M. 1985. *Enigmas of Chance*. New York: Harper and Row.

Klein, F. 1933. *Vorlesungen über die hypergeometrische Funktion*. Berlin: Springer.

Kline, M. 1972. *Mathematical Thought from Ancient to Modern Times*. New York: Oxford University Press.

Knuth, D. 1968. *The Art of Computer Programming*. Vol. 1: *Fundamental Algorithms*. Reading, Mass.: Addison Wesley.

——. 1969. *The Art of Computer Programming*. Vol. 2: *Seminumerical Algorithms*. Reading, Mass.: Addison Wesley. 2d ed., 1981.

——. 1973. *The Art of Computer Programming*. Vol. 3: *Sorting and Searching*. Reading, Mass.: Addison Wesley.

Korevaar, J. 1985. *Ludwig Bieberbach's Conjecture and Its Proof by Louis de Branges*. Report 85-08, Mathematics Department, University of Amsterdam.

Kummer, E. E. 1836. Über die hypergeometrische Reine $1 + (\alpha \cdot \beta)/(1 \cdot \gamma)x + [\alpha(\alpha+1) \beta(\beta+1)]/[1 \cdot 2 \cdot \gamma(\gamma+1)]x^2 + \ldots$. *Journal für die reine und angewandte Mathematik* 15: 39-83, 127-72.

Laplace, P. S. 1812. *Théorie analytique des probabilitiés*. 2d ed. Paris.

Milin, I. M. 1984. *L. de Branges' Proof of the Bieberbach Hypothesis* (in Russian). Leningrad seminar notes.

Neugebauer, O. 1956. A Notice of Ingratitude. *Isis* 47: 58.

Pfaff, J. F. 1797a. *Disquisitiones Analyticae*. Helmstadt.

——. 1797b. Observationes analyticae ad L. Euler Institutiones Calculi Integralis, vol. 4, suppl. 2 et 4. *Histoire de 1793. Nova acta academiae scientiarum Petropolitanae* 11: 38-57. (Note that the history section is paged separately from the scientific section of this journal.)

Robin, L. 1957-59. *Fonctions sphériques de Legendre et fonctions sphéroidales*. 3 vols. Paris: Gauthier-Villars.

Rogers, L. J. 1895. Third Memoir on the Expansion of Certain Infinite Products. *Proceedings of the London Mathematical Society* 26: 15-32.

Rota, G.-C. 1974. Review of "Mathematical Thought from Ancient to Modern Times" by M. Kline. *Bulletin of the American Mathematical Society* 80: 805-7.

Saalschütz, L. 1890. Eine Summationsformel. *Zeitschrift für Mathematik und Physik* 35: 186-88.

Shohat, J., Hille, E., and Walsh, J. 1940. *A Bibliography on Orthogonal Polynomials*. National Research Council, Bull. 103. National Academy of Sciences, Washington, D.C.

Slater, L. J. 1966. *Generalized Hypergeometric Functions*. Cambridge: Cambridge University Press.

Szegö, G. 1933. Über gewisse Potenzreihen mit lauter positiven Koeffizienten. *Mathematische Zeitschrift* 37: 674-88. Reprinted in *Collected Papers*, vol. 2. Boston: Birkhäuser Boston, 1982, pp. 480-94.

——. 1968. An Outline on the History of Orthogonal Polynomials. In *Orthogonal Expansions and Their Continuous Analogues*, ed. D. Haimo. Carbondale: Southern Illinois University Press, pp. 3-11. Reprinted in *Collected Papers*, vol. 3. Boston: Birkhäuser Boston, 1982, pp. 857-65. Comment on this paper is on pp. 866-69.

Tchebychef, P. L. 1875. Sur l'interpolation des valeurs équidistants. *Zapiski Imperatorskoi Akademii Nauk* (Russia), vol. 25, suppl. 5 (in Russian). French translation appears in *Oeuvres*, vol. 2. New York: Chelsea, 1961, pp. 219-42.

Vorsselman de Heer, P. O. C. 1833. *Specimen in augurale de fractionibus continuis*. Thesis, Utrecht.

Watson, G. N. 1944. *Theory of Bessel Functions*. 2d ed. Cambridge: Cambridge University Press.

Weil, A. 1980. History of Mathematics: Why and How. *Proceedings of the International Congress of Mathematicians, Helsinki, 1978*. Vol. 1, Helsinki, pp. 227-36.

——. 1984. *Number Theory*. Boston: Birkhäuser Boston.

III. CASE STUDIES IN
THE HISTORY AND PHILOSOPHY OF MATHEMATICS

——————— Lorraine J. Daston ———————

Fitting Numbers to the World: The Case of Probability Theory

By almost all current philosophical accounts, the success of applied mathematics is a perpetual miracle. Neither formalist nor logicist nor Platonist (at least of the latter-day dilute variety) can provide a plausible explanation of why numbers should fit the world. (Here I use *numbers* as a convenient shorthand for all elements of mathematics, since most— but not all—examples of applied mathematics involve numerical measurements.) Why should the relationships among undefined elements or logic or purely mental objects describe a multitude of phenomena with tolerable accuracy? Short of invoking preestablished harmony or a coincidence of mind-boggling improbability, we are philosophically at a loss to account for our great—and ever greater—good fortune in applied mathematics. We believe that pure mathematics is conceptually and for the most part historically prior to and independent of applied mathematics. Indeed, the very term *applied mathematics* tells all: in order to be applied, the mathematics must already exist in its own right, just as theory is "applied" to practice.

The situation was quite otherwise in the eighteenth century, which practiced "mixed" rather than "applied" mathematics. Enlightenment philosophers of mathematics certainly had their share of problems, but they were not our problems. They pondered, for example, why mathematics should enjoy greater certainty than various parts of physics and astronomy, but the triumphs of mixed mathematics, many of them barely fifty years old then, did not perplex them. An odd hybrid of Aristotelian, Cartesian, and Lockean views about the nature of mathematics had prepared them for such happy outcomes in astronomy, mechanics, optics, and acoustics, and they anticipated further mathematical breakthroughs in areas like pneumatics and what was then known as the art of conjecture, or the calculus of probabilities.

In this paper, I shall try to do three things: first, to briefly sketch the eighteenth-century distinction between "abstract" and "mixed" mathematics, and to contrast it with our "pure" versus "applied" distinction; second, to illustrate the significance of this philosophical distinction for mathematical practice, using examples drawn from the history of probability theory during the eighteenth and early nineteenth centuries; and third, to conclude with some reflections suggested by these examples about the preconditions for applying mathematics. Throughout, I shall be primarily concerned with the problem of how a mathematical theory gains—and sometimes loses—a domain of applications.

1. Mixed Mathematics

In the *Metaphysics*, Aristotle argues that mathematical entities cannot be truly separated from sensible things, but only abstracted from them (M.2.1077a 9-20; b 12-30). With some Lockean and Cartesian elaborations, this view continued to dominate philosophical accounts of mathematics throughout the eighteenth century. The most influential of these accounts was d'Alembert's *Preliminary Discourse* (1751) to the great *Encyclopédie*. Following Locke, d'Alembert asserts that all human knowledge derives from experience; borrowing from Descartes, he assumes analysis to be the fundamental intellectual operation turned upon the raw materials of sensation. Property by property, the mind strips away the tangle of particular features that compose any sensation until it arrives at the barest skeleton or "phantom" of the object, shaped extension. It is at this rarefied level that mathematics studies the objects of experience. Although these objects have been systematically denuded of all those traits that normally accompany them in perception, they are not, d'Alembert insists, thereby denatured. Mathematics may be the "farthest outpost to which the contemplation of properties can lead us," but it is nonetheless still anchored in the material universe of experience.

Once the limits of analysis have been reached at magnitude and extension, the mind reverses its path and begins to reconstitute perception, property by property, by the reciprocal operation of synthesis, until it ultimately arrives at its departure point, the concrete experience itself. The successive stages along this route demarcate the subject matter of the various sciences. For example, add impenetrability and motion to the magnitude and extension of mathematics, and the science of mechanics is created. All sciences study the same objects but embrace their perceptual complexity to a greater or lesser extent.[1]

That mathematics could describe the external world would thus have hardly puzzled d'Alembert: mathematics came from the world, and to the world it must return, for d'Alembert believed that mathematical abstractions, as he says, "are useful only insofar as we do not limit ourselves to them." Even the "abstract" mathematics of pure number and extension (i.e., of arithmetic and geometry) were simply the endpoints of a continuum along which mathematics was "mixed" with sensible properties in varying proportions. Of course, this schema brought its own quandaries: if mathematics was joined to the natural sciences by a continuous intellectual process, why did only mathematical results enjoy certainty, and exactly where did the boundary between mathematics and these sciences lie? D'Alembert was obliged to resort to Cartesian intuitions of clarity and simplicity to answer the first question, and he tendentiously drew the boundary to include rational mechanics within mathematics—even his devoted admirer Montucla could not follow him in that.[2] But d'Alembert's account of "abstract" and "mixed" mathematics *did* explain how, for example, geometric optics or celestial mechanics was possible, and it was echoed by many other in the next fifty years.

So far, so Aristotelian. We recognize the kinship of Aristotle's bronze isosceles triangle and d'Alembert's impenetrable moving body: both examples "mix" physical with abstract properties. But when we turn to d'Alembert's table depicting the actual contents of mixed mathematics, we see how much enlarged beyond the Aristotelian canon it had become by 1750 in the mind of one of Europe's leading mathematicians.[3] "Mixed" or "Physico-Mathematics" dwarfs "Pure Mathematics," even though the latter category has been swelled by the integral and differential calculus, and each of the headings of the Aristotelian "mixed" canon—mechanics, astronomy, optics, harmonics—has greatly expanded its range of subheadings. (Harmonics, for example, now comprises only one part of "Acoustics," and optics only one part of a broader division including dioptics, perspective, and catoptrics.) Moreover, the division of space in d'Alembert's table corresponds roughly to the division of labor among eighteenth-century mathematicians: Bossut's 1810 survey of mathematics from the origins of the calculus to the year 1800 devotes 177 pages to work in abstract mathematics and 319 pages to that in mixed mathematics.[4]

Not only did mixed mathematics preponderate; in the opinion of eighteenth-century mathematicians, it bid fair to expand still further. Montucla, in the 1758 edition of his celebrated *Histoire des mathématiques*, reflected on the open-ended nature of mixed mathematics, which grew apace as the natural sciences acquired greater stores of observations that could serve as "principles" of new branches such as pneumatics (the mathematical study of gases) and the art of conjecture (probability theory). Indeed, there seemed no more auspicious augury for the future of mixed mathematics than a calculus of chance, "this kind of Proteus, so difficult to get firm hold of, [and yet] the mathematician is on the verge of chaining it and of submitting it to his calculations."[5] If chance itself could be tamed, what phenomenon, however irregular, could withstand the forward march of mixed mathematics?

This, then, was the framework within which eighteenth- and early nineteenth-century probabilists worked. Mathematics about the world was "mixed" with its subject matter rather than "applied" to it, and pure or abstract mathematics was thought as likely to be enriched by physical insights as the other way around—the intimate, two-way links between the development of the calculus and the study of motion were an object lesson to eighteenth-century mathematicians. Mixed mathematics was more than just a *façon de parler*: all attempts to apply mathematics to the phenomena of experience presume some degree of analogy between the subject matter and the mathematics employed, but for eighteenth-century practitioners of mixed mathematics, the requisite degree of analogy bordered on congruence. Mathematical models were conceived not merely as analogues that shared certain key features with the phenomena they described, but rather as mathematical "portraits"—highly schematic ones, to be sure—of the phenomena and/or the underlying mechanisms that produced them. Moreover, mixed mathematics absorbed the greater part of mathematicians' energies in the eighteenth century. However, they revered pure mathematics as the only inexorably progressive part of the discipline. As Montucla noted, the very hybrid nature of mixed mathematics, partaking both of the "clarity and evidence" of abstract mathematics and the "uncertainty and shadows" of physics, put it at risk of error and (Montucla's term) "retrogression."[6] I shall now turn to the calculus of probabilities, which was at once emblematic for the distinctive character of mixed mathematics, of its vast horizon, and, ultimately, of the retrogression Montucla feared.

2. The Art of Conjecture

Probability theory has been described as simply the sum total of its applications. Until the late eighteenth century introduction of generating functions, it had no techniques of its own; until Kolmogorov's axiomatization of 1933, it had no independent standing as a mathematical theory. Early twentieth century mathematicians like Hilbert and Borel still narrowly identified probability theory with specific applications, and this was a fortiori true of the eighteenth century art of conjecture, which existed solely as mixed mathematics, like geometric optics and celestial mechanics. As such, it stood or fell with its treatment of the problems deemed specific to it—problems concerning rational belief or action under conditions of uncertainty—hence the "art of conjecture," after Jakob Bernoulli's 1713 treatise of that title. I shall outline three examples from the eighteenth-century corpus of probability theory: the St. Petersburg gambling problem; the probability of judgments; and insurance. I have deliberately chosen examples that are both familiar (insurance) and, to twentieth-century eyes, outlandish (the probability of judgments); examples that are still central to the probabilistic corpus (gambling and expectation) and others that have migrated to other disciplines (the St. Petersburg paradox and its solutions now belong properly to economics). Taken together, the examples illustrate the special constraints placed upon mixed mathematics to somehow "match" the phenomena, and also how the empire of numbers could contract as well as expand.

The St. Petersburg Problem

Although most modern expositions derive the concept of expectation from that of probability, the first published versions of the theory of the late seventeenth and early eighteenth centuries made expectation rather than probability fundamental.[7] In modern terms, expectation is defined as the product of the probability of an event m and its outcome value:

$$E = P(m) \cdot V(m), \quad \text{Where } P(m) = \text{probability of outcome } m$$
$$\text{and } V(m) = \text{value of outcome } m.$$

Early probabilists like Huygens and Jakob Bernoulli thought in terms of expectations rather than probabilities because the problems they posed were taken from contract law, and were concerned more with equity than with probabilities per se. Aleatory contracts were a long-recognized subdivision of Roman and canon law, and included all agreements involving an element of chance: gambling, annuities, maritime insurance, the prior purchase of the "cast of a net" from a fisherman—in short, any trade of here and present goods for future uncertain ones. Lawyers studied conditions of equal expectations rather than equal probabilities of gain or loss, and the first two generations of mathematical probabilists followed suit when they attempted to quantify these legal concerns.

The St. Petersburg paradox, posed by Nicholas Bernoulli in 1713,[8] sparked a long debate over the proper definition of the central concept of probabilistic expectation—a debate that threatened to undermine the foundations of eighteenth-century probability theory. Two players, A and B, play a coin-toss game. If the coin turns up heads on the first toss, A gives B \$1; if heads does not turn up until the second toss, B wins \$2; if not until the third toss; \$4; and so on. By conventional methods, B's expectation (and therefore the fair price to play the game) would be:

$$E = 1/2\ (\$1) + 1/4\ (\$2) + 1/8\ (\$4) + \ldots + (1/2^n)\ (2^{n-1}) + \ldots .$$

Therefore, B must pay A an infinite amount to play the game. But, as Nicholas Bernoulli and subsequent commentators were quick to point out, no reasonable person would pay even a small amount, much less a very large or infinite sum, for the privilege of playing such a game.

Why was this a paradox? Unlike most mathematical paradoxes, the contradiction lay not between discrepant mathematical results deduced from equally valid premises: there is nothing mathematically wrong with this answer. In order to understand the furor triggered by the St. Petersburg problem, we must return to eighteenth-century views on mixed mathematics and the peculiar mission of the calculus of probabilities. In essence, the branches of eighteenth-century mixed mathematics corresponded to what would now be termed mathematical models. If the mathematical description diverged significantly from the phenomena, it was incumbent upon the mixed mathematician to revise this theory. In other words, mathematical probability was as "corrigible" as the mathematical theory of lunar motion. In modern parlance, the mathematical theory had no existence independent of its applications. Thus the designated field of applications (the term is anachronistic here) played a critical role in the career of a branch of mixed mathematics, and such indeed was the case with mathematical probability. The distinctive field of applications to which classical probabilists attached their theory was the determination of standards for rational thought and conduct in civil society, a field pioneered by Jakob Bernoulli in the unfinished fourth book of the *Ars conjectandi* and taken up by his nephews Nicholas and Daniel Bernoulli, Condorcet, Lambert, Laplace, and others during the eighteenth century. By isolating and mathematizing the principles that underpinned the beliefs and actions of an elite of reasonable men, the probabilists hoped to make social rationality accessible to all. Since probability theory was meant to be a mathematical model of reasonableness, when its results clashed with the judgments of *hommes éclairés*, as in the case of the St. Petersburg problem, probabilists anxiously reexamined their premises and demonstrations for inconsistencies. Hence the paradox: the contradiction lay not between incompatible mathematical tenets, but rather between the unambiguous mathematical solution and good sense. This is why the St. Petersburg paradox, trivial in itself, exercised every first-rate probabilist from Daniel Bernoulli through Poisson.

I have discussed the numerous eighteenth-century solutions to this so-called paradox at length elsewhere, and I shall not review them here.[9] Here, I shall only point out that although mathematicians reached no consensus about the correct solution to the problem, they all agreed that the paradox was both real and dangerous to probability theory as a whole, that the definition of expectation was the nub of the problem, and that a satisfactory solution must realign the mathematical theory with the opinions of reasonable men. That is, their primary loyalty lay to the field of phenomena that the calculus of probabilities purported to describe, and they were willing to sacrifice the most fundamental definition of that calculus in order to bring about a better match between mathematical results and phenomena.

The Probability of Judgments

The preceding example of the St. Petersburg paradox and the animated debate it provoked was meant to show that (1) the eighteenth-century calculus of probabilities was, as a branch of mixed mathematics, inseparable from its designated field of applications; and (2) that field of applications was broadly conceived as rational decision-making under uncertainty—be it to buy a lottery ticket, sell an annuity, believe a witness, or accept a scientific hypothesis. For eighteenth- and early nineteenth-century practitioners, probability theory was, in Laplace's famous phrase, "nothing more than good sense reduced to a calculus."[10] We have seen how the failure of mathematical results to tally with good sense prompted a searching reevaluation of even the most fundamental mathematical premises—squarely in the descriptive tradition of mixed mathematics. My second example, the probability of judgments, illustrates just how the calculus of probabilities was made to seem "congruent" to good sense, and how rapidly shifting conceptions of that "good sense" eventually dissolved the bond between the calculus of probabilities and the phenomena it was originally intended to describe. It gained a measure of independence thereby, as an autonomous mathematical theory, but it also lost an entire field of applications.

The probability of judgments emerged in the late eighteenth and early nineteenth centuries largely through the work of Condorcet, Laplace, and Poisson.[11] The central problem was the optimal design of a jury or tribunal of judges so that, by adjusting the number of judges and the plurality required for conviction, one could minimize the probability of an erroneous decision. The mathematicians adopted both the mathematical format of the probability of causes based on Bernoulli's and Bayes's theorems, and the concomitant assumptions concerning the uniformity and independence of trials in their treatments of the probability of judgments. (Recall that Bernoulli's and Bayes's theorems are the inverse of one another: roughly speaking, Bernoulli's theorem tells you how to find the probability that *n* drawings from an urn with replacement will closely approximate the known ratio of black-to-white balls contained in the urn; Bayes's theorem tells you the probability that the unknown ratio of black-to-white balls will closely approximate the known results of *n* drawings with replacement.) In the probability of judgments, each judge or juror was likened to an urn containing so many balls marked "true," corresponding to a correct decision, and the rest marked "false," to denote an incorrect decision. The decisions of the judges were further assumed to be independent of one another, just as drawings from separate urns would be. Since the individual "truth" probabilities of the judges could not be ascertained a priori, the probabilists resorted to inverse probabilities, and, after the publication of French judicial statistics starting in 1825, to assumptions that the guilt or innocence of the defendants operated as an unknown cause of the observed acquittal and conviction rates, and that a certain prior probability of guilt obtained.

Why did late eighteenth- and early nineteenth-century probabilists accept the verisimilitude of this probabilistic model for judgment? Later critics such as the mathematician Louis Poinsot and the philosopher John Stuart Mill found it all but incomprehensible that thinkers of Laplace's stature could have compared judges to, in Poinsot's words, "so many dice, each of which has several sides, some for error, others for truth."[12] The nineteenth-century probabilist Joseph Bertrand objected that decision-making was intrinsically particular, governed by determinate but fluctuating factors: if a judge erred, it was for a specific reason, not because he had "put his hand in an urn" and made an unlucky draw.[13] Yet for the classical probabilists, these assumptions did not seem so outrageous. Condorcet and Laplace admitted that the conditions of independence, equality, and constancy for individual probabilities were simplifications, but they argued that all mixed mathematics involved idealizations (witness Newton's laws of dynamics), and further maintained that such approximations "founded on the data, indicated by good sense" were preferable to nonmathematical specious reasoning.

The eighteenth- and early nineteenth-century probabilists could be confident in assumptions that their successors found absurd because they subscribed to psychological theories that described mental operations in terms congenial to their mathematics. According to Locke, Hartley, Hume, and others, the sound mind reasoned by the implicit computation and comparison of probabilities.[14] Or as Montucla put it apropos of crack gamblers:

> The gambler's mind ["l'esprit du jeu"], that mind that seems to capture fortune, is nothing more than an innate or acquired talent for seeing at a glance all the chance combinations that could lead to gain or loss; human prudence is ultimately nothing other than the art of appreciating the probabilities of events, in order to act accordingly.[15]

The association of ideas in principle mirrored the regularity and frequency of events culled from experience: in an unbiased mind, associations of ideas corresponded to real connections between the events and objects represented by the ideas. The very workings of the human understanding, when undistorted by strong emotion or uncritical custom, imitated Bernoulli's theorem, which Hartley had claimed was "evident to attentive Persons, in a gross general way, from the Common Methods of Reasoning."[16] Associationist psychology also emphasized the combinatorial operations of the mind; indeed, all intellectual novelty owed to the mental combination and recombination of simple ideas by, as Condillac wrote, "a kind of calculus." Condorcet affirmed Condillac's claim that the best intellects were those that excelled in "uniting more ideas in memory and in multiplying these combinations."[17] If good minds worked by "a kind of calculus," then the combinatorial calculus of probabilities could be viewed as the mathematical expression and extension of the psychological processes that constituted right reasoning—in particular, the right reasoning of judges.

Eighteenth-century jurisprudence also supplied the probabilists with suggestive analogies to their mathematical techniques and concepts, thereby inviting new applications in that field. Continental jurists had developed an elaborate hierarchy of so-called legal proofs that assigned the evidence procured from both witnesses and things a fixed fractional value. These fractional "probabilities" (a usage that antedates the mathematical term) corresponded to degrees of assent in the mind of the judge and were summed to obtain the complete or "full" proof required for conviction.[18] Classical probabilists like Jakob Bernoulli, Condorcet, and Laplace adopted the legal interpretation of probability as a "degree of certainty" apportioned to the probative force of various types of evidence, and they also attempted to convert the legal "probabilities" into a mathematical probability of testimony and conjecture.

Hence there would have been no a priori objections to the probability of judgments, which computed the probability that a tribunal composed of a given number of members, each with a postulated probability of judging aright, would arrive at a correct decision by a certain majority, as an inappropriate application of mathematical probability. Despite the later criticisms of the probability of judgments as the "scandal of mathematics," classical probabilists viewed the theory as reasonable and indeed quite useful, given the urgent interest in judicial reform at the time. Their optimism stemmed in large part from the "congruence" between the phenomena of right reasoning, particularly the weighing of legal evidence, and the mathematics itself, a congruence prepared by contemporary psychology and jurisprudence. Geometric optics "fit" the phenomena of reflection and refraction because light rays were indeed like geometric lines; the art of conjecture "fit" the phenomena of rational decision-making because enlightened minds intuitively calculated probabilities. The probabilists' job was to make the tacit principles underlying good sense into an explicit calculus accessible to all.

Alas for classical probability theory, neither the phenomena of good sense nor the psychological and legal theories that linked them to mathematical probability survived the French Revolution unaltered. Both psychology and jurisprudence had changed considerably between the publication of Condorcet's and Poisson's treatises on the probability of judgments in 1785 and 1837, respectively. Although associationist theories were still influential, they emphasized the pathologies of reason created by habit, prejudice, self-interest, and ignorance rather than the smoothly functioning mental calculus of eighteenth-century psychologists. Good sense was no longer so clearly identified with computation and comparison of probabilities. In jurisprudence, the deliberately anti-formal, anti-analytic system of "free" proofs had replaced that of legal proofs, substituting an intuitive appeal to the "intimate conviction" of juror or judge for any formal reckoning.[19] Poinsot's objections to Poisson's work on the probability of judgments reveal how far the notion of good sense had diverged from the mental calculus of the associationists, and how completely the new free system of proofs had severed the older connection between legal and mathematical probabilities:

> It is the application of this calculus [of probabilities] to things of the moral order which offends the intellect . . . at the end of such calculations in which the numbers derive only from such hypotheses, to draw conclusions, which purport to guide a sensible man in his judgment of a criminal case . . . this is what seems to me a sort of aberration of the intellect, a false application of science, which it is only proper to discredit.[20]

Thus did probability theory lose, at least temporarily, one of its key domains of applications. Poinsot's verdict (so to speak) on the probability of judgments as an "aberration of the intellect" stuck. However, in detaching the probability of judgments from the corpus of respectable applications of mathematical probability, Poinsot was at pains to make clear that his attack on the probability of judgments in no way touched the certainty of the *theory* of mathematical probability, which he deemed to be as irrefragable as arithmetic. This careful distinction between theory and applications would have been foreign to mixed mathematics, and it shows that the opposition of "abstract" to "mixed" mathematics had been superseded by that of "pure" to "applied" mathematics by the time Poinsot took the floor of the Paris Académie des Sciences in 1835 to attack Poisson's work. The story of the ill-fated probability of judgments might serve as an object lesson in the need to exercise caution in the choice of a suitable set of phenomena to mathematize. The "good sense" of reasonable men turned out to be notoriously unstable, as the probabilists bent on mathematically describing it discovered to their chagrin. Faced with such a "retrogression," in Montucla's sense, mathematicians and philosophers protected the reputation of mathematics for certainty by sharply distinguishing untarnished pure mathematics from its sometimes dubious applications.

Insurance

My final example, actuarial mathematics, is by way of contrast an undisputed success story about applying probability theory to the world—or at least it became one after almost a century's worth of neglected mathematical efforts to make it so. The history of how mathematical probability eventually came to be applied to the insurance trade is an intricate one that I can barely sketch here.[21] My main point in doing so will be to make the converse point to my claim concerning mixed mathematics and the probability of judgments. If the probability of judgments enjoyed a bright if brief career, it was because peculiar circumstances conjoined first to suggest a striking analogy between legal judgments and mathematical probability, and later to destroy this analogy. Conversely, if eighteenth-century insurers were slow to make use of a mathematical technology that was in many ways tailor-made for them, it was because they perceived the conditions of their trade as downright disanalogous to those imposed by the mathematicians.

Why did the practitioners of risk—particularly insurers and sellers of annuities—fail to take advantage of mathematical techniques and collections of statistics that were devised for their purposes? There seems to have been, as the mystery writers put it, both opportunity and motive. Opportunity, for there existed a well-worked-out theory for pricing annuities and (potentially) life insurance using the new mathematics of probability and the new data of mortality statistics from circa 1700 on. Moreover, the mathematicians (De Moivre, Simpson, and others) had expressly translated this literature into elementary handbooks aimed at the innumerate clerk, with all algebra converted to verbal form and most technical material relegated to appendices, plus copious tables to obviate the need for onerous calculation. Motive, for both the Dutch and especially the English annuity and insurance markets were large and bustling enough by the turn of the eighteenth century to offer the mathematically based company a competitive edge. (This is not to say that such enterprises could not be profitable without mathematics—indeed, they were *too* profitable—but as the experience of the first mathematically based company was to show, even the loosest connection between calculation and premium could cut prices while still preserving a lavish margin of profit.)

Why, then, did the practice of risk taking lag so far behind the theory of risk taking? I shall argue, in summary form, that mathematically based insurance could not be sold until insurers and their customers came to believe that phenomena like human mortality, shipwrecks, fires, and other catastrophes were regular enough to make statistical and probabilistic approaches plausible. To make this point clear, I must briefly describe the practice of insurance *without* actuarial mathematics. Certain forms of risk taking—insurance (chiefly maritime), annuities, and gambling—were widely and successfully practiced in Europe long before the formulation of mathematical probability. Although experience no doubt honed the ability of the underwriter, dealer in annuities, or gambler to estimate odds, their approach to risk could hardly be described as statistical or probabilistic, even at an intuitive level. A sixteenth-century insurer might have found such a statistical approach impractical, for it assumes conditions that are stable over a long period as well as the homogeneity of categories. Insurance manuals and legal treatises of the period emphasized that the premium in any given case depends on a judicious weighting of the particular circumstances: the cargo, the season of the year, the route taken, the condition of the ship, the skill of the captain, the latest ''good or bad news'' concerning storms, warships, and privateers.[22] Moreover, in commercial centers populous enough to support whole markets of insurers, premium prices also reacted to levels of supply and demand as well as to the latest news about the Barbary pirates.

Thus, annuity rates and insurance premiums certainly reflected past experience, but it was a far more nuanced experience than a simple toting up of mortality and shipwreck statistics. It was an experience sensitive to myriad individual circumstances and their weighted interrelationships, not to mention market pressures: it was not simply astatistical, it was anti-statistical. Given the highly volatile conditions of both sea traffic and health in centuries notorious for warfare, plagues, and other unpredictable misfortunes, I am not persuaded that this was an unreasonable approach. In any case, it was the prevailing one—and it evidently turned a profit.

Although mathematicians and statisticians addressed insurance and annuities problems almost from the inception of mathematical probability in the late seventeenth century—Huygens, DeWitt, Leibniz, the Bernoullis, Halley, and De Moivre were all interested—and although by 1750 there existed an extensive literature in Dutch, English, Latin, and French on the subject, the impact of mathematical probability on practice was effectively nil prior to the establishment of the Equitable Society for the Assurance of Lives in 1762 (at the instance of a mathematician, James Dodson).[23] And even then, the dictates of mathematical theory were greatly tempered by other considerations. The Royal Exchange and London Assurance offices (established in 1720), both of which insured lives, charged a flat rate of 5 percent for every £100 insured, regardless of age.[24] The vast bulk of insurance trade remained maritime, and although premiums responded to decreases in risk (the disappearance of marauding Turks made insuring voyages to the Levant, Spain, and Portugal considerably cheaper), statistics played no role in pricing. Fire insurance was too new to be burdened with the weight of tradition, and clients were offered graduated premiums depending on the kind of building (brick versus wood) and trade housed therein (sugar bakers, for example, paid especially stiff rates). Yet fire offices apparently never collected statistics on the subject.[25]

Why was the practice of eighteenth-century insurance and annuities so resistant to the influence of mathematical theory? For maritime insurance and annuities, it might be argued that the inertia of an entrenched and successful practice based on nonmathematical estimates worked against the application of the new mathematical statistical methods. This is no doubt part of the answer, but it cannot explain why the new forms of insurance against fire and death, both inventions of the late seventeenth and early eighteenth centuries, did not take on a more statistical cast. The case of life insurance and annuities is particularly baffling because of the availability of mortality statistics drawn from several locales. The first mathematically based life insurance company, the Equitable, was founded by a mathematician, not an insurer, and was twice denied a Royal charter because, in the words of the Royal Council (1761), the company's proposed mathematical basis,

> Whereby the chance of mortality is attempted to be reduced to a certain standard . . . is a mere speculation, never yet tried in practice, and consequently subject, like all other experiments, to various chances in the execution. . . .[26]

And even the Equitable does not seem to have been entirely persuaded of the reliability of the new mathematical methods, for it was cautious in every respect: it calculated interest on investments at the lowest rates (3 percent); used the mortality table that gave the shortest lifespans; took the further precaution of insuring only healthy lives; added a flat 6 percent to all premiums, and surcharges of up to 22 percent in individual cases deemed especially risky. Nor could the company be persuaded to distribute dividends among its members, despite its almost embarrassing prosperity and the ever more regular contours of the Equitable's own mortality figures as membership increased. Year after year, William Morgan, the first mathematically trained actuary, held importunate stockholders at bay, warning that "extraordinary events or a season of uncommon mortality" might catch the Equitable unawares,[27] law of large numbers or no. Small wonder that the nineteenth-century mathematician Augustus De Morgan quipped: "We should write upon the door of every mutual office but one be *wary*; but upon that one should be written *be not too wary* and over it the *Equitable Society*."[28]

Thus, the confidence that the phenomena of human mortality revealed, in the words of German theologian and demographer Johann Süssmilch, "a constant, general, great, complete, and beautiful order"[29] stable enough to rest a trade upon was slow in coming. For some phenomena, that faith came later than for others, and the disparities are hard to explain. As early as 1662 some writers were confident enough of the regularities to collect data on mortality, but it is difficult to understand why mortality should have been assumed to be regular and other phenomena of equal practical interest like the incidence of fires not, in an age where both were subject to wild fluctuations: witness the plague and Great Fire of London in 1665-66. Whatever the reasons for such confidence that whole new realms of phenomena were indeed regular, it was a prerequisite for the transformation of insurance from an enterprise based on judgments of individual cases to one based on probability and statistics. In contrast to the case of the probability of judgments, here a new analogy was forged rather than an old one sundered.

3. Conclusion

These three examples drawn from the history of the classical theory of probability illustrate three cardinal points about applying mathematics to phenomena: (1) whether applications succeed or fail depends on the standards set, and in this respect eighteenth-century mixed mathematics differed significantly from nineteenth-century applied mathematics; (2) mathematical theories can lose as well as gain applications; and (3) the conditions that make a set of phenomena ripe for quantification—or the reverse—depend crucially on their *perceived* stability and degree of analogy with the mathematical techniques at hand. The mixed mathematics of the eighteenth century demanded a far closer fit between mathematics and the phenomena than did the applied mathematics of the nineteenth century—for example, the psychological and legal theories underlying the probability of judgments. As with applied mathematics, the proof of the pudding was in the eating, i.e. whether or not the mathematical results tallied with experience, but the consequences of a mismatch were far more dire for a branch of mixed mathematics: the St. Petersburg paradox was grave enough evidence against the calculus of probabilities for its practitioners to revise its most fundamental definitions or even to suggest abandoning it entirely. These bonds between subject matter and mathematics loosened by the end of the eighteenth century in large part because many mathematical techniques originally conceived in a rather restricted applied context (e.g., partial derivatives and hydrodynamics) had since acquired so large and varied a repertoire of applications that close analogy no longer seemed of the essence. This is certainly what occurred in probability theory: consider the diaspora of the normal distribution from astronomy to sociology to physics to psychology in the course of the nineteenth century.

As mixed mathematics became applied mathematics, probability theory shifted its characteristic domain of applications. The St. Petersburg paradox, once at the very center of probabilistic research, became a problem in economic theory that happened to employ probabilistic techniques. The probability of judgments disappeared altogether, and actuarial mathematics expanded steadily. Although mathematicians and philosophers continued to wrangle over the proper interpretation of probability, the theory achieved a measure of independence from both applications and, ultimately, interpretations. And finally, the configuration of stable phenomena was transformed: in the eighteenth century, good sense seemed uniform enough to be quantified, but not the frequency of hailstorms; in the nineteenth century, just the reverse. In a curious way, the expanding and contracting domain of applications belonging to probability theory—indeed, to mathematics in general—charts for us the changing landscape of what we believe to be the regular, the predictable, and the stable, for only at these points do we believe that numbers can fit the world.

Notes

1. Jean d'Alembert, "Discours préliminaire," *Encyclopédie, ou Dictionnaire raisonné des sciences, des arts et métiers* (Paris, 1751-80), vol. 1, pp. v ff.

2. J. F. Montucla, *Histoire des mathématiques* (Paris, 1758), vol. 1, p. 5.

3. D'Alembert, "Discours," Système figuré des connaissances humaines.

4. Charles Bossut, *Histoire générale des mathématiques* (Paris, 1802-10), vol. 2.

5. Montucla, *Histoire*, vol. 1, p. 6; 2d ed. (Paris, An VII-X [1798-1802]), vol. 3, p. 381.

6. Montucla, *Histoire*, vol. 1, p. xxv.

7. See Christiaan Huygens, *De ratiociniis in ludo aleae* (1657), in *Oeuvres complètes de Christian Huygens*, Société Hollandaise des Sciences (The Hague: Martinus Nijhoff, 1920), vol. 14, p. 60.

8. For a complete history of the paradox, see Gérard Jorland, "The St. Petersburg Paradox (1713-1937)," in *The Probabilistic Revolution. Vol. I: Ideas in History*, ed. Lorenz Krüger, Lorraine J. Daston, and Michael Heidelberger (forthcoming).

9. See Lorraine J. Daston, "Probabilistic Expectation and Rationality in Classical Probability Theory," *Historia Mathematica* 7 (1980): 234-60.

10. Pierre Simon de Laplace, *Essai philosophique sur les probabilités* (1814), in *Oeuvres completès*, Academie des Sciences (Paris, 1878-1912), vol. 7, p. cliii.

11. M.-J.-A.-N. Condorcet, *Essai sur l'applcation de l'analyse à la probabilité des décisions rendues à la pluralité des voix* (Paris, 1785); Laplace, *Théorie analytique des probabilités* (1812), in *Oeuvres*, vol. 7; Siméon Denis Poisson, *Recherches sur la probabilité des jugements en matière criminelle et en matière civile* (Paris, 1837). See also Lorraine J. Daston, "Mathematics and the Moral Sciences: The Rise and Fall of the Probability of Judgments," in *Epistemological and Social Problems of the Sciences in the Early Nineteenth Century*, ed. H. N. Jahnke and M. Otte (Dordrecht and Boston: Reidel, 1981), pp. 287-309.

12. John Stuart Mill, *A System of Logic* (1843), 8th ed. (New York, 1881), p. 382.

13. Joseph Bertrand, *Calcul des probabilités (Paris, 1889), p. 326.*

14. See John Locke, An Essay Concerning Human Understanding (1689; all editions dated 1690), ed. Alexander C. Fraser (New York: Dover, 1959), vol. 1, p. 529; David Hartley, *Observations on Man, His Frame, His Duty, and His Expectations* (London, 1749), vol. 1, pp. 335ff.; David Hume, *A Treatise of Human Nature* (1739), ed. L. A. Selby-Bigge (London: Oxford University Press, 1968), vol. 1, pp. 125ff.

15. Montucla, *Histoire*, 2d ed., vol. 3, p. 381.

16. Hartley, *Observations*, vol. 1, p. 331.

17. Étienne Condillac, *Essai sur l'origine des connaissances* (1746), in *Oeuvres de Condillac* (Paris, An VI [1798]), vol. 1, pp. 100 and 109; Condorcet, *Vie de Turgot* (1786), in *Oeuvres de Condorcet*, ed. F. Arago and A. Condorcet-O'Connor (Paris, 1847-49), vol. 1, p. 222.

18. See J. Gilissen, "La preuve en Europe du XVIe au début du XIXe siècle," *Recueils de la Société Jean Bodin pour l'Histoire Comparative des Institutions* 17(1965): 755-833.

19. From the Instruction of 21 October 1791, quoted in Gilissen, "Preuve," pp. 831-32. These instructions were reprinted in the Code of 1808.

20. Louis Poinsot, in *Comptes rendus de l'Académie des Sciences* 2 (1836): 399.

21. See Lorraine J. Daston, "The Domestication of Risk: Mathematical Probability and Insurance, 1650-1830," in Krüger et al., eds. *The Probabilistic Revolution*.

22. See, for example, Estienne Cleirac, *Les us et coutumes de la mer* (Rouen, 1671), pp. 271-72.

23. See Maurice Edward Ogborn, *Equitable Assurances* (London: George Allen and Unwin, 1962), for a detailed history of this company.

24. Barry Supple, *The Royal Exchange Assurance* (Cambridge: Cambridge University Press, 1970), p. 56; Royal Assurance Company, *Assurance Book on Lives*, vol. 1 (1733-57), MS. 8740 of the Guildhall Library, London.

25. H. A. L. Cockerell and Edwin Green, *The British Insurance Business 1547-1970* (London: Heinemann, 1976), p. 27.

26. Quoted in Ogborn, *Equitable*, p. 35.

27. Quoted in Ogborn, *Equitable*, p. 105.

28. Quoted in Ogborn, *Equitable*, p. 206. See also Augustus De Morgan, "Review of Théorie Analytique des Probabilités. Par M. le Marquis de Laplace," *Dublin Review* 2 (1837): 338-54, esp. pp. 341ff.

29. Johann Süssmilch, *Die göttliche Ordnung in den Veränderungen des menschlichen Geschlechts*, 3d. ed. (Berlin, 1775), p. 49.

Logos, *Logic, and* Logistiké:
Some Philosophical Remarks on
Nineteenth-Century Transformation
of Mathematics

I

Mathematics underwent, in the nineteenth century, a transformation so profound that it is not too much to call it a second birth of the subject—its first birth having occurred among the ancient Greeks, say from the sixth through the fourth century B.C. In speaking so of the first birth, I am taking the word *mathematics* to refer, not merely to a body of knowledge, or lore, such as existed for example among the Babylonians many centuries earlier than the time I have mentioned, but rather to a systematic discipline with clearly defined concepts and with theorems rigorously demonstrated. It follows that the birth of mathematics can also be regarded as the discovery of a capacity of the human mind, or of human thought—hence its tremendous importance for philosophy: it is surely significant that, in the semilegendary intellectual tradition of the Greeks, Thales is named both as the earliest of the philosophers and the first prover of geometric theorems.

As to the "second birth," I have to emphasize that it is *of the very same subject*. One might maintain with some plausibility that in the time of Aristotle there was no such science as that we call *physics*; that Plato and Aristotle were acquainted with mathematics in our own sense of the term is beyond serious controversy: a mathematician today, reading the works of Archimedes, or Eudoxos's theory of ratios in Book V of Euclid, will feel that he is reading a contemporary. Then in what consists the "second birth"? There was, of course, an enormous *expansion* of the subject, and that is relevant; but that is not quite it: The expansion was itself effected by the very same capacity of thought that the Greeks discovered; but in the process, something new was learned about the nature of that capacity—what it is, and what it is not. I believe that what has been learned, when properly understood, constitutes one of the greatest advances of philosophy—although it, too, like the advance in mathematics itself, has a close relation to ancient ideas (I intend by the word *logos* in the title an allusion to the philosophy of Plato). I also believe that, when properly understood, this philosophical advance should conduce to a certain modesty: one of the things we should have learned in the course of it is how much we do *not* yet understand about the nature of mathematics.

II

A few decades ago, the reigning cliché in the philosophy of mathematics was "the three schools": logicism (Frege-Russell-Carnap), formalism (Hilbert), intuitionism (Brouwer). Many people would now hold that all three schools have failed: Hilbert's program has not been able to overcome the obstacle discovered by Gödel; logicism was lamed by the paradoxes of set theory, impaired more critically by Gödel's demonstration of the dilemma: either paradoxes—i.e., inconsistency—or incompleteness, and (some would add) destroyed by Quine's criticisms of Carnap; intuitionism, finally, has limped along, and simply failed to deliver the lucid and purified new mathematics it had promised. In these negative judgments, there is a large measure of truth (which bears upon the lesson of modesty I have spoken of); although I shall argue later that none of the three defeats has been total, or necessarily final. But my chief interest here in the "three schools" will be to relate their positions to the actual developments in mathematics in the preceding century. I hope to show that in some degree the usual view of them has suffered from an excessive preoccupation with quasi-technical "philosophical"—or, perhaps better, ideological—issues and oppositions, in which perspective was lost of the mathematical interests these arose from. To this end, I shall be more concerned with the progenitors or foreshadowers of these schools than with their later typical exponents; more particularly, with Kronecker rather than Brouwer, and with Dedekind more than Frege. Hilbert is another matter: his program was very much his own creation. Yet in a sense, Hilbert himself, in his mathematical work before the turn of the century, stands as the precursor of his own later foundational program; and in the same sense, as we shall see, Dedekind is a very important precursor of Hilbert as well as of logicism. Since Kronecker and Frege, too, both contributed essential ingredients to Hilbert's program—Kronecker on the philosophical, Frege on the technical side—the web is quite intricate.

III

It is far beyond both the scope of this paper and the competence of its author to do justice, even in outline, to the complex of interrelated investigations and mathematical discoveries (or "inventions") by which mathematics itself was deeply transformed in the nineteenth century; I am going to consider only a few strands in the vast fabric. A good part of the interest will center in the theory of numbers—and in more or less related matters of algebra and analysis—but with a little attention to geometry too.

Let us begin with a very brief glance at the situation early in the century, when the transformation had just commenced. A first faint prefiguration of it can be seen in work of Lagrange in the 1770s on the problems of the solution of algebraic equations by radicals and of the arithmetical theory of binary quadratic forms. On both of these problems, Lagrange brought to bear new methods, involving attention to transformations or mappings and their invariants, and to classifications induced by equivalence-relations. (That Lagrange was aware of the deep importance of his methods is apparent from his remark that the behavior of functions of the roots of an equation under permutations of those roots constitutes "the true principles, and, so to speak, the *metaphysics* of the resolution of equations of the third and fourth degree." Bourbaki[1] suggests that one can see here a first vague intuition of the modern concept of *structure*.) This work of Lagrange was continued, on the number-theoretic side, by Gauss; on the algebraic side, by Gauss, Abel, and Galois; and, despite the failure of Galois's investigations to attract attention and gain recognition until nearly fifteen years after his death, the results attained by the early 1830s were enough to ensure that—borrowing Lagrange's word—the new "metaphysics" would go on to play a dominant role in algebra and number-theory. (I should not omit to remark here that a procedure closely related to that of classification was the introduction of new "objects" for consideration in general, and for calculation in particular. In the *Disquisitiones Arithmeticae* of Gauss, explicit introduction of new ideal "objects" is avoided in favor of the device of introducing new "quasi-equalities" [congruences]; Galois, on the other hand, went so far as to introduce "ideal roots" of polynomial congruences *modulo* a prime number, thus initiating the algebraic theory of finite fields.[2])

By the same period, the early 1830s, decisive developments had also taken place in analysis and in geometry: in the former, the creation—

privately by Gauss, publicly by Cauchy—of the theory of complex-analytic functions; in the latter, besides the emergence of projective geometry, above all the discovery of the existence of at least one geometry alternative to that of Euclid. For although Bolyai-Lobachevskyan geometry made no great immediate impression on the community of mathematicians, Gauss—who, again privately, had discovered it for himself—well understood its importance; and Gauss's influence upon several of the main agents in this story was profound and direct.

IV

As pivotal figures for the history now to be discussed, I would name Dirichlet, Riemann, and Dedekind—all closely linked personally to one another, and to Gauss. Among these, Dirichlet is perhaps the poet's poet—better appreciated by mathematicians, more especially number-theorists, with a taste for original sources, than by any wide public. It is not too much to characterize Dirichlet's influence, not only upon those who had direct contact with him—among those in our story: Riemann and Dedekind, Kummer and Kronecker—but upon a later generation of mathematicians, as a spiritual one (the German *geistig* would do better). Let me cite, in this connection, Hilbert and Minkowski. In his Göttingen address of 1905, on the occasion of Dirichlet's centenary, Minkowski, naming a list of mathematicians who had received from Dirichlet "the strongest impulse of their scientific aspiration," refers to Riemann: "What mathematician could fail to understand that the luminous path of Riemann, this gigantic meteor in the mathematical heaven, had its starting-point in the constellation of Dirichlet"; and he then remarks that although the assumption to which Riemann gave the name "Dirichlet's Principles"— we may recall that this had just a few years previously been put on a sound basis by Hilbert—was in fact introduced not by Dirichlet but by the young William Thomson, still "the modern period in the history of mathematics" dates from what he calls "the other Dirichlet Principle: to conquer the problems with a minimum of blind calculation, a maximum of clear-seeing thoughts."[3] Just four years later, in his deeply moving eulogy of Minkowski, Hilbert says of his friend: "He strove first of all for simplicity and clarity of thought—in this Dirichlet and Hermite were his models."[4] And for perhaps the strongest formulation of Minkowski's "other Dirichlet Principle," consider this passage quoted by Otto Blumenthal, in his biographical sketch of Hilbert, from a letter of Hilbert to Minkowski:

"In our science it is always and only the reflecting mind *der überlegende Geist*], not the applied force of the formula, that is the condition of a successful result."[5]

V

I am going to make a sudden jump here: Why did Dedekind write his little monograph on continuity and irrational numbers? To be sure, that work is in no need of an excuse; but I have long been struck by these circumstances: (1) Dedekind himself tells us in his foreword that when, some dozen years earlier, he first found himself obliged to teach the elements of the differential calculus, he "felt more keenly than ever before the lack of a really scientific foundation of arithmetic." He goes on to express his dissatisfaction with "recourse to the geometrically evident" for the principles of the theory of limits.[6] (Note that Dedekind speaks of a "foundation of *arithmetic*," rather than of analysis.) (2) In the foreword to the first edition of his monograph on the natural number, Dedekind invokes, as something "self-evident," the principle that every theorem of algebra and of the higher analysis can be expressed as a theorem about the natural numbers: "an assertion," he says, "that I have also heard repeatedly from the mouth of Dirichlet."[7] (3) From antiquity (e.g., Aristotle, Euclid) through the late eighteenth and early nineteenth centuries (e.g., Kant, Gauss), a prevalent view was that there are two distinct sorts of "quantity": the discrete and the continuous, represented mathematically by the theories of *number* and of *continuous magnitude*. Gauss, for instance, excludes from consideration in the *Disquisitiones Arithmeticae* "fractions for the most part, surds always."[8] But Dirichlet, in 1837, succeeded in proving that there are infinitely many prime numbers in any arithmetic progression containing two relatively prime terms, by an argument that makes essential use of continuous variables and the theory of limits. This famous investigation was the beginning of analytic number theory; and Dirichlet himself signalizes the importance of the new methods he has introduced into arithmetic: "The method I employ seems to me above all to merit attention by the connection it establishes between the infinitesimal Analysis and the higher Arithmetic [*l'Arithmétique transcendante*];[9]

> I have been led to investigate a large number of questions concerning numbers [among them, those related to the number of classes of binary quadratic forms and to the distribution of primes] from an entirely new point of view, which attaches itself to the principles of infinitesimal

analysis and to the remarkable properties of a class of infinite series and infinite products.[10]

From all this it seems reasonable—I would even say, inevitable—to conclude that a part at least of Dedekind's motive was that of clearing "arithmetic" in the strict sense, i.e. the theory of the rational integers, from any taint of reliance upon principles drawn from a questionable source. In any case, this motive is explicitly stressed by Dedekind's great rival, Kronecker, in the first lecture of his course of lectures on number theory (published by Hensel in 1901). He there quotes from Gauss's preface to the *Disquisitiones Arithmeticae*: "The investigations contained in this work belong to that part of mathematics which is concerned with the whole numbers—fractions excluded for the most part, and surds always." But, says Kronecker, "the Gaussian dictum... is only then justified, if the quantities he wishes to exclude are borrowed from geometry or mechanics...." Then—after referring to Gauss's own treatment, in the *Disquisitiones*, of cyclometry (thus, irrational numbers) and of forms (thus, "algebra" or *Buchstabenrechnung*)—he cites elementary examples (Leibniz's series for $\pi/4$; partial fraction expansion of $z \tan(z)$) to support the claim that analysis has its roots in the theory of the whole numbers (with, he says, the sole exception of the concept of the limit—as to how this is to be dealt with, he unfortunately leaves us in the dark); and goes on to conclude:

> Thus arithmetic cannot be demarcated from that analysis which has freed itself from its original source of geometry, and has been developed independently on its own ground; all the less so, as Dirichlet has succeeded in attaining precisely the most beautiful and deep-lying arithmetical results through the combination of methods of both disciplines.[11]

Finally, the epistemological connection is made explicit by Kronecker in another place—his essay "Über den Zahlbegriff"—again with a reference to Gauss:

> The difference in principles between geometry and mechanics on the one hand and the remaining mathematical disciplines, here comprised under the designation "arithmetic," consists according to Gauss in this, that the object of the latter, Number, is *solely* the product of our mind, whereas Space as well as Time have also a *reality, outside* our mind, whose laws we are unable to prescribe completely a priori.[12]

VI

It would be tempting to spend much time on Dedekind's theory of continuity—on its relation to Eudoxos and to geometry, and on the difficulty his contemporaries had in understanding its point. Let me just refer to the extracts from Dedekind's correspondence with Rudolf Lipschitz, given by Emmy Noether in vol. III of Dedekind's *Gesammelte Mathematische Werke*, and particularly to the fact that Lipschitz objected to Dedekind that the property the latter calls "completeness" or "continuity"—in the terminology now standard, *connectedness*—is self-evident and doesn't need to be stated—that no man can conceive of a line without that property. Dedekind replies that this is incorrect, since *he himself* can conceive of all of space and each line in it as entirely discontinuous (adding that Herr Professor Cantor in Halle is evidently another man of the same sort). He refers to §3 of his monograph, where he had said, "If space has a real existence at all, it does *not* have necessarily to be continuous."[13] But it is in the foreword to the first edition of the monograph on the natural numbers that Dedekind returns to this point and indicates the particular grounds for his claim: namely, the existence (as we would say) of *models* of Euclid's geometry in which all ratios of lengths of straight segments are *algebraic numbers*. For the latter concept, he refers the reader to Dirichlet's *Vorlesungen über Zahlentheorie*, §159 of the second, §160 of the third edition. The second edition of Dirichlet's lectures—of which, of course, Dedekind was the editor—was published the year before *Stetigkeit und irrationale Zahlen*; and the section indicated[14] is part of the famous Supplement written by Dedekind in which the theory of algebraic number fields and algebraic integers was developed for the first time.

I have the impression that the central importance of that very great work of Dedekind for the *entire* subsequent development of mathematics has not been generally appreciated. It is certainly well known—to those who know such things—that this Supplement was of the first importance for algebraic number theory and for what is now called "commutative algebra." It is perhaps less well known that this is also the place in which Galois's theory was developed for the first time in its modern form—as a theory of field extensions and their automorphisms, rather than of substitutions in formulas and of functions invariant under substitutions. But that new perspective upon Galois's achievement is itself only one manifestation of a general principle that permeates the work—one that could be summed up in Minkowski's phrase expressing the "other Dirichlet

Principle": "a minimum of blind calculation, a maximum of clear-seeing thoughts." Here is how Dedekind puts it, in the version of his theory that he published in French in 1877:

> A theory based upon calculation would, as it seems to me, not offer the highest degree of perfection; it is preferable, as in the modern theory of functions, to seek to draw the demonstrations, no longer from calculations, but directly from the characteristic fundamental concepts, and to construct the theory in such a way that it will, on the contrary, be in a position to predict the results of the calculation (for example, the composition of decomposable forms of all degrees). Such is the aim that I shall pursue in the following Sections of this Memoir.[15]

The reference to "the modern theory of functions" is, unmistakably, to Riemann; but the methods by which Dedekind pursues his stated aim are distinctly his own. They may be summed up, in a word, as *structural*: Emmy Noether, in her notes to the *Mathematische Werke*, recognizes plainly and with evident enthusiasm the strong kinship of Dedekind's point of view with her own (and I hope it may be presumed that the influence of Noether's point of view upon the mathematics of our century is itself well known).

VII

Perhaps, however, a distinction should be made: One theme associated with the name of Emmy Noether is that of the "abstract axiomatic approach" to algebra; a second is that of attention to *entire algebraic structures and their mappings*, rather than to, say, just numerical attributes of those structures (as a notable example, in topology, attention to homology groups and their induced mappings, not just to Betti numbers and torsion coefficients). One might consider that, of these themes, the second links Noether to Dedekind, the first rather to Hilbert. I want to consider both themes; but I begin with the second, which certainly is present in full measure in Dedekind. And I shall here take up this theme in connection with the monograph *Was sind und was sollen die Zahlen*?

In the foreword to the first edition of that work, Dedekind speaks of "the simplest science, namely that part of logic which treats of numbers."[16] Arithmetic, then, is a "part of logic"; but what is logic?

The same question can be asked of other philosophers who claim that mathematics is, or belongs to, logic; the inquiry tends to be frustrating. For example, Frege says[17] that his claim that arithmetical theorems are analytic can be conclusively established "only by a gap-free chain of inferences, so that no step occurs that does not conform to one of a small number of modes of inference recognized as logical''; but he quite fails to tell us how such recognition occurs—or what its content is. The trouble lies in the reigning presumption that to recognize a proposition as a ''logical truth'' is to identify its *epistemological basis*; and this is a trouble because—if I may be forgiven for pontificating on the point—no cogent theory of the ''epistemological basis'' of *any* kind of knowledge has ever been formulated.

On the other hand, some things can be said about Dedekind's claim—and even, I think, some epistemological things. In a general way, we should remember, ''logic'' in the period in question was taken to be concerned with the ''laws of thought.'' I have said in my opening remarks that the discovery of mathematics was also the discovery of a capacity of the human mind. Dedekind tells us what, in his opinion, this capacity is. His earliest published statement dates from 1879 (by coincidence, the year of Frege's *Begriffsschrift*) and occurs in §161 of the Eleventh Supplement to Dirichlet's *Zahlentheorie*, third edition:[18] in the text he introduces the notion of a *mapping*; in a footnote he remarks, and repeats the statement in the foreword to the first edition of *Was sind und was sollen die Zahlen?*[19] that the whole science of numbers rests upon *this capacity of this mind*— the capacity to envisage mappings—*without which no thinking at all is possible*. So the claim that arithmetic belongs to logic is the claim that the principles of arithmetic are essentially involved in all thought—without anything said about an epistemological basis. Moreover, the principles involved are indeed those employed explicitly in Dedekind's algebraic-number-theoretic investigations: the formation of what he calls ''systems''— fields, rings (or ''orders''), modules, ideals—and mappings.

But there is another aspect of Dedekind's view that should not be overlooked. What is his answer to the question posed by the title of his monograph? But first, what is the question? The version given by the English translator of the work, *The Nature and Meaning of Numbers*, is (setting aside its abandonment of the interrogative form) quite misleading: the German idiom ''Was soll [etwas]?'' is much broader than the English ''What does [something] mean?'' What it connotes is always, in some sense, ''intention''; but with the full ambiguity of the latter term. We can, however, see the precise sense of Dedekind's question from the quite explicit answer he gives:

My general answer [or ''principal answer'': *Hauptantwort*] to the question posed in the title of this tract is: numbers are free creations of the human mind; they serve as a means for more easily and more sharply conceiving the diversity of things.[20]

The title, therefore, asks: What are numbers, and what are they for? (What is their use, their function?)

The answer is not very satisfying. I think Dedekind makes a mistake by assuming that numbers are "for" some one thing. But it is a venial mistake that lies really only on the surface of his formulation. It is, indeed, part of the great discovery of the nineteenth century that mathematical constructs may have manifold "uses," and uses that lie far from those envisaged at their "creation." On the other hand, this formulation of Dedekind's is so general and vague that perhaps it could be stretched to cover absolutely any application of the concept of number.

What I think is more interesting in all this is that it is not what numbers "are" *intrinsically* that concerns Dedekind. He is not concerned, like Frege, to identify numbers as particular "objects" or "entities"; he is quite free of the preoccupation with "ontology" that so dominated Frege, and has so fascinated later philosophers. Dedekind's general answer to his first question, "Numbers are free creations of the human mind," later takes the following specific form: He defines—or, equivalently, axiomatizes—the notion of a "simply infinite system"; and then says (in effect: this is my own free rendering) *it does not matter what numbers are; what matters is that they constitute a simply infinite system.* He adds—and here I translate literally—"In respect of this freeing of the elements from any further content (abstraction), one can justly call the numbers a free creation of the human mind."[21] Of course it follows from this characterization that numbers are "for" *any use to which a simply infinite system can be put*; it is because this answer does follow, and because it is the right one, that I described Dedekind's mistake about this as venial.

It should be noted that a very similar point applies to Dedekind's analysis of real numbers. Here again the contrast with Frege is instructive. In the second volume of his *Grundgesetze der Arithmetik*, Frege moves slowly toward a definition of the real numbers.[22] He does not quite reach it—that was reserved for the third volume, which never appeared; but what it would have been is pretty clear. Frege's idea was that real numbers are "for" representing ratios of measurable magnitudes (as, for him, whole numbers—*Anzahlen*—are "for" representing sizes of sets— in his terminology, sizes of "extensions of concepts"); and he wants the real numbers to be "objects" *specifically adapted to that function*. That is why he rejects Dedekind's, Cantor's, and Weierstrass's constructions. But Dedekind says, again, that *it does not matter what real numbers "are."* In particular, he does *not* define them as cuts in the rational line. He says, rather, "Whenever a cut (A_1, A_2) is present that is induced by no rational number, we *create* a new, an *irrational* number α, which we regard as completely defined by the cut (A_1, A_2)."[23] In a letter to his friend and collaborator Heinrich Weber,[24] he defends this point, together with the related one concerning the integers, in a rather remarkable passage. Weber has evidently suggested that the natural numbers be regarded primarily as cardinal rather than ordinal numbers, and that they be defined as Russell later did define them. Dedekind replies:

> If one wishes to pursue your way—and I would strongly recommend that this be carried out in detail—I should still advise that by number . . . there be understood not the *class* (the system of all mutually similar finite systems), but rather something *new* (corresponding to this class), which the mind *creates*. We are of divine species [*wir sind göttlichen Geschlechtes*] and without doubt possess creative power not merely in material things (railroads, telegraphs), but quite specially in intellectual things. This is the same question of which you speak at the end of your letter concerning my theory of irrationals, where you say that the irrational number is nothing else than the cut itself, whereas I prefer to create something *new* (different from the cut), which corresponds to the cut . . . . We have the right to claim such a creative power, and besides it is much more suitable, for the sake of the homogeneity of all numbers, to proceed in this manner.

Dedekind continues with essentially the points made in a well-known paper of Paul Benacerraf: there are many attributes of cuts that would sound very odd if one applied them to the corresponding numbers; one will say many things about the class of similar systems that one would be most loath to hang—as a burden—upon the number itself. And he concludes with a reference to algebraic number theory: "On the same grounds I have always held Kummer's *creation* of the ideal numbers to be entirely justified, if only it is carried out rigorously"—a condition that Kummer, in Dedekind's opinion, had not fully satisfied.

In contrast with this last remark of Dedekind's, no less cultivated a mathematician than Felix Klein, as late as the 1910s (when his invaluable lectures on the development of mathematics in the nineteenth century were delivered), felt it necessary to demystify Kummer's "creation" by insisting on the fact that one can—although not in a uniquely distinguished (as one would now say, "canonical") fashion—identify Kummer's "ideal divisors" with actual complex algebraic numbers in the ordinary sense.[25] There can be no doubt in the mind of anyone acquainted with the later development of the subject that—at least in point of actual practice (but I would argue, also in principle)—it was Dedekind in 1888, not Klein thirty years later, who had this right.

VIII

Dedekind's term "*free* creation" also deserves some attention. (The theme has, again, some Dirichletian resonance, since Dirichlet in his work on trigonometric series played a significant role in legitimating the notion of an "absolutely arbitrary function," unrestricted by any necessary reference to a formula or "rule.") It is very characteristic of Dedekind to wish to open up the possibilities for developing concepts, and to wish also that alternative, and new, paths be explored. We have just seen him urging Weber to develop his own views on the natural numbers; he repeatedly urged Kronecker to make known his way of developing algebraic number theory; in the foreword to his fourth (and last) edition of Dirichlet's *Vorlesgunen über Zahlentheorie*, containing his final revision of the Eleventh Supplement, he expresses the hope that one of Kronecker's students may prepare a complete and systematic presentation of Kronecker's theory—and also recommends the attempt to simplify the foundations of his own theory to younger mathematicians, who enter the field without preconceived notions, and to whom therefore such simplification may be easier than to himself.[26] (This was in September 1893. Kronecker had recently died; Hilbert had just begun to work on algebraic number theory.)

If this has come to sound too much like a panegyric on Dedekind, I can only say that that is because he does seem to be a great and true prophet of the subject—a genuine philosopher, of and in mathematics.

IX

In his brief account of Dirichlet, Felix Klein mentions[27] as "a particular characteristic" of Dirichlet's number-theoretic investigations the type of proof, which he was the first to employ, that establishes the existence of something without furnishing any method for finding or constructing it. (One recalls that when, some forty-five years later, Hilbert published proofs of a similar type in the algebraic theory of invariants, they were regarded as unprecedented; and that Paul Gordan is said to have pronounced, "This is not mathematics, it is theology!"[28]) The nonconstructive character of Dirichlet's theorem on the arithmetic progression is noted by Kronecker in his lectures, where he is able to tell his students, with legitimate pride, that he had himself succeeded in the year 1885 in repairing this defect and that the more complete result is to be presented for the first time in the course of those lecture.[29]

What is perhaps most notable about this is the absence, in these published lectures of Kronecker's, of any polemical tone in his comments on constructive vs. nonconstructive methods, and—in the passage I quoted earlier—his positive emphasis upon the role of irrational numbers and limiting processes in number theory. It is of course possible that the restrained tone of these comments, which contrast so markedly with what is generally reported about Kronecker (and also with the more explosive reaction of Gordan to Hilbert), is partly conditioned by his reverence for Dirichlet (it is a rather pleasing fact that the two great number-theoretic rivals and philosophical opponents, Kronecker and Dedekind, each edited publications of work of Dirichlet), and partly by the exigencies of the subject itself (it is clear enough that Kronecker hoped to reduce everything to a constructive and finite basis, but clear also that he was far from having any definite idea of how to do this for the theory of limits).[30] It is possible also, since the work as published was assembled from a variety of manuscript sources,[31] that Hensel in editing this material exercised some moderating influence. But the foreword by Hensel does provide us with a more substantive clue to Kronecker's philosophical stand on the nonconstructive in mathematics:[32]

> He believed that one can and must in this domain formulate each definition in such a way that its applicability to a given quantity can be assessed by means of a finite number of tests. Likewise that an existence proof for a quantity is to be regarded as entirely rigorous only if it contains a method by which that quantity can really be found. Kronecker was far from the position of rejecting entirely a definition or proof that did not meet those highest demands, but he believed that in that case something remained lacking, and held a completion in this direction to be an important task, through which our knowledge would be advanced in an essential point.

No mathematician could quarrel with the statement that a constructive definition or proof adds something to our knowledge beyond what is contained in a nonconstructive one. However, Kronecker's public stand on the work of others was certainly a more repressive one than Hensel's characterization suggests. I think the issue concerns definitions rather more crucially than proofs; but let me say, borrowing a usage from Plato, that it concerns the mathematical *logos*, in the sense both of "discourse" generally, and of definition—i.e., the *formation of concepts*—in particular.

Kronecker's vehement polemic against the ideas and methods of Cantor is more or less notorious. So far as I am aware, that polemic is not represented in the published writings. The first place I know of in which Kronecker published strictures against nonconstructive definition is a paper of 1886, "Über einige Anwendungen der Modulsysteme auf elementare algebraische Fragen"; and here it is in the first instance the apparatus of Dedekind's algebraic number-theory—"jene Dedekind'sche Begriffsbildungen wie 'Modul', 'Ideal', u.s.w"—that comes under attack.[33] Unfortunately, this attack of Kronecker's contains *no* hint that nonconstructive definitions can be accepted at least provisionally (or as a Platonic "second best")—even where, as in the case of analysis, Kronecker has in fact nothing to offer that meets his "higher" demands. It is possible, therefore, that Hensel's reading is off on this point: it is possible, and seems to accord with the fact, that whereas Kronecker was willing to acknowledge the provisional value of a nonconstructive *argument* like Dirichlet's, he meant to exclude more rigidly any introduction of *concepts* by nonconstructive *logoi*. (It must be confessed, again, that this seems to leave analysis, for Kronecker, in limbo.)[34]

X

On the other hand, Kronecker appears uniformly to exempt "geometry and mechanics from his stringent requirements.[35] Are these still to be considered parts of mathematics? Kronecker does not seem to want to exclude them; but perhaps he regards them (e.g., on the basis of Gauss's remarks) as in some measure empirical sciences.

From this a Dedekindian question arises: "Was ist und was soll die Mathematik?" As in the case of number, I immediately repudiate the idea of seeking a definite formula to state "what mathematics is for"—it is "for" whatever it proves to be useful for. Still, what may that be?

Riemann's wonderful habilitation-lecture begins with a characterization of an "*n*-tuply extended magnitude" in terms that it would not be unreasonable to describe as belonging to "logic":[36] the points of such a manifold are "modes of determination [or "specification"] of a general concept"; examples of concepts whose modes of specification constitute such a manifold are found both in "ordinary life" places in sensible objects; colors) and within mathematics (e.g., in the theory of analytic functions). In the concluding section of that paper, Riemann considers the question of the bearings of his great generalization of geometry upon our understanding of ordinary physical space; *this*, he says, is an empirical question, and must remain open, subject to what developments may occur in physics itself: "Investigations which, like that conducted here, proceed from general concepts, can serve only to ensure that this work shall not be hindered by a narrowness of conceptions, and that progress in the knowledge of the connections of things shall not be hampered by traditional prejudices."[37]

If I may paraphrase: Geometry is not an empirical science, or a part of physics; it is a part of mathematics. *The role of a mathematical theory is to explore conceptual possibilities*—to open up the scientific *logos* in general, in the interest of science in general. One might say, in the language of C. S. Peirce, that mathematics is to serve, according to Riemann, among other interests (e.g., that of facilitating calculation), the interest of "abduction"—of providing the means of *formulating* hypotheses or theories for the empirical sciences.

The requirement of constructiveness is the requirement that all mathematical notions be effectively computable; that mathematics be fundamentally reduced to what the Greeks called *logistiké*: to processes of calculation. Kronecker's concession to geometry and mechanics of freedom from this requirement is tacit recognition that there is no reason to assume a priori that structural relationships in nature are necessarily all of an effectively computable kind. But then, when one sees, with Riemann, the usefulness of elaborating *in advance*, that is, independently of empirical evidence (and, in this sense, a priori), a theory of structures that need not but may prove to have empirical application, Kronecker's limitation to "geometry and mechanics" of the license he offers to those sciences loses much of its plausibility.

Another aspect of this point: Kant made a distinction between *logic*, which, concerned exclusively with *rules of discursive thought* in abstraction from all "content," is a "canon" but not an "organon"—not an instrument for gaining knowledge; and *mathematics*, which *is* an organon, because it is not purely discursive but has a definite content, given in the a priori intuition of space and time.[38] But the geometry of Riemann is quite freed from any such specific content: Kant's view that the creative or productive power of geometry rested upon its concrete "intuitive" spatial content was simply mistaken, and mathematics is seen to be an *abstract* and a priori "organon" of knowledge (whether one chooses to call it a "logic," or a "dialectic," or whatever). And again, in view of this its *general* capacity, there is no more reason to hamper it by restrictions to the effectively computable than to hamper it by restrictions to the "spatially intuitive."

But then—so *this* dialectic goes—why restrict the license at all (e.g., to mathematical disciplines that one thinks *might* serve the ends of some empirical science)? Why not complete freedom of conceptual elaboration in mathematics? Then, *within* any *logos* so freely developed, one can pay appropriate attention to the distinction between what is and what is not constructive.

XI

This point of view—which is really not very far from the moderate position attributed to Kronecker by Hensel—is the one that Hilbert so vigorously championed. Let me call attention, in briefest outline, to a few salient points.

Notice of Hilbert's interest in the foundations of geometry dates back to the earliest days of his work in algebraic number-theory, or even somewhat before: it was in 1891, according to Otto Blumenthal,[39] that Hilbert, in a mathematical discussion in a Berlin railway waiting room, made his famous statement that in a proper axiomatization of geometry "one must always be able to say, instead of 'points, straight lines, planes', 'tables, chairs, beer mugs'." This view—that the basic terms of an axiomatized system must be "meaningless"—is often misconstrued as "formalism." But the very same requirement was stated, fifteen years earlier, by the "logicist" Dedekind, in his letter to Lipschitz already cited: "All technical expressions [are to be] replaced by arbitrary newly invented (heretofore nonsensical) words; the edifice must, if it is rightly constructed, not collapse.'[40]

Hilbert uses as epigraph to his *Grundlagen der Geometrie* a well-known Kantian aphorism:[41] "Thus all human knowledge begins with intuitions, proceeds to concepts, and ends with ideas." For Hilbert, axiomatization, with all basic terms replaced by "meaningless" symbols, is *the elimination of the Kantian "intuitive"*—the "proceeding to concepts"; hence the axioms taken together constitute a *definition*. Of what? Of a *species of structure*. Now, by this, we need not mean "a set." It is a point of interest, and a very useful one, that the notion of set proves so serviceable both as a tool for concept-formulation *within* a mathematical discipline (ideals, etc.) and as affording a *general framework* for many or all mathematical disciplines; but set theory is a tool, not a foundation. For Hilbert, it is the axiomatized discourse itself that constitutes the mathematical *logos*; and the only restriction upon it that he recognizes is that of formal consistency.

As to "ideas": Kant associates ideas with *completeness* or *totality*. For Hilbert, I believe, this connoted the survey of the general characteristics of the structures of a species and of related species—more explicitly, in the geometrical case, the study of the problems of consistency and categoricity of the axioms, their independence, and the sorts of alternatives one gets by changing certain axioms; in short, something very much like model-theory. (Of course, when one now gets down to brass tacks, sets are pretty much indispensable; and because of the essential incompletability of set theory, the "absolute completeness or totality" fails: Kantian "dialectical illusion" has not after all been avoided.)

In the correspondence of Frege and Hilbert,[42] it is amusing or exasperating, depending upon one's mood, to see Frege, wondering what Hilbert can mean by calling his axioms "definitions," come in his ponderous but thorough way to the conclusion that *if* they are definitions, they must define what he calls a "concept of the second level"—and then more or less drop this notion as implausible or uninteresting to pursue; and to see Hilbert, not very interested in Frege's terminology or his point of view, fail to understand what Frege's undervalued insight really was: for a Fregean "second-level concept" simply *is* the concept of a species of structure. So: a tragically or comically missed chance for a meeting of minds.

But now, what of Hilbert's "program"? I think it is unfortunate that Hilbert, in his later foundational period, insisted on the formulation that ordinary mathematics is "meaningless" and that only finitary mathematics has "meaning." Hilbert certainly never abandoned the view that mathematics is an organon for the sciences: he states this view very strongly in the last paper reprinted in his *Gesammelte Abhandlungen*, called

"Naturerkennen und Logik" (1930);[43] and he surely did not think that physics is meaningless, or its discourse a play with "blind" symbols. His point is, I think, this rather: that the mathematical *logos* has no responsibility to any imposed *standard* of meaning: not to Kantian or Brouwerian "intuition," not to finite or effective decidability, not to anyone's metaphysical standards for "ontology"; its *sole* "formal" or "legal" responsibility is to be consistent (of course, it has also what one might call a "moral" or "aesthetic" responsibility: to be useful, or interesting, or beautiful; but to this it cannot be constrained—poetry is not produced through censorship).

In proceeding to his "program," however, Hilbert set as his goal the *mathematical investigation of the mathematical* logos *itself*, with the principal aim of establishing its consistency (or "their" consistency—for he did not envisage a single canonical axiom-system for all of mathematics). And here he made essential use both of Frege and of Kronecker. For Frege's extremely careful and minute regimentation of logical language or "concept-writing" did not, as Frege himself thought it would, serve as a guarantee of consistency; but the techniques he used for that regimentation did render the formal languages of mathematical theories, and their formal rules of derivation, subject to mathematical study in their own right, regarded as purely "blind" symbolic systems. Moreover, the regimentation was itself of such a kind that the play with symbols was a species of calculation—of *logistiké*. Without Frege, proof-theory in Hilbert's sense would have been impossible.

Of course, the final irony of this story, and the collapse of Hilbert's dream of establishing the consistency of the logic of the *logos* by means restricted to *logistiké*, lies in the discovery by Gödel, Post, Church, and Turing that there is a *general theory* of *logistiké*, and that this theory is nonconstructive; in particular, that neither the notion of consistency nor that of provability is (in general) effective; and further that all sufficiently rich consistent systems fall short of the Kantian "ideal"—are incomplete.

This leaves us with a mystery—a subject of "wonder," in which, according to Aristotle, philosophy begins. He says it *ends* in the contrary state; I am inclined to believe, as I think Aristotle's master Plato did, that philosophy does not "end," but that mysteries become better understood—and deeper. The mysteries we now have about mathematics are certainly better understood—and deeper—than those that confronted Kant or even Gauss. To attempt, however, to survey what they are, even in the sketchiest way, although enticing, is a task that would require another paper, and is certainly beyond the scope of this one.[44]

Notes

1. N. Bourbaki, *Éléments d'histoire des mathématiques*, 2d ed. (Paris: Hermann, 1974), p. 100.

2. Ibid., pp 108-9.

3. Hermann Minkowski, *Gesammelte Abhandlungen*, vol. 2, (Leipzig, 1911; reprinted New York: Chelsea 1967), pp. 460-61.

4. Ibid., vol. 1, p. xxix; David Hilbert, *Gesammelte Abhandlungen*, vol. 3, (Berlin: Springer, 1935), p. 362.

5. Ibid., p. 394.

6. Richard Dedekind, *Stetigkeit und irrationale Zahlen*, Vorwort; in his *Gesammelte Mathematische Werke*, vol. 3, ed. R. Fricke, E. Noether, and O. Ore (Braunschweig: F. Vieweg, 1932), pp. 315-16; translation in Dedekind, *Essays on the Theory of Numbers*, trans. W. W. Beman (Open Court, 1901; reprinted New York: Dover, 1963), p. 1.

7. Dedekind, *Was sind und was sollen die Zahlen?*, Vorwort zur ersten Auflage; *Werke*, vol. 3, p. 338; *Essays*, p. 35.

8. Carl Friedrich Gauss, *Disquisitiones arithmeticae*, preface; quoted by Leopold Kronecker, *Vorlesungen über Zahlentheorie*, vol. 1, ed. Kurt Hensel (Leipzig: Teubner, 1901), p. 2.

9. Peter Gustav Lejeune Dirichlet, "Sur l'usage des séries infinies dans la théorie des nombres," in *G. Lejeune Dirichlet's Werke*, vol. 1, ed. L. Kronecker (Berlin, 1889; reprinted New York: Chelsea, 1969), p. 360.

10. Dirichlet, "Recherches sur diverses applications de l'analyse infinitésimale à la théorie des nombres," ibid., p. 411.

11. Kronecker, *Vorlesungen über Zahlentheorie*, vol. 1, pp. 2-5 (italics added).

12. Kronecker, *Werke*, vol. 3, 1st half-volume, ed. K. Hensel (Leipzig: Teubner, 1899), p.253 (emphasis in original; Kronecker quotes, in a footnote, a letter from Gauss to Bessel, 9 April 1830).

13. Dedekind, *Werke*, vol. 3, p. 478.

14. See Dedekind, *Werke*, vol. 3, pp. 223ff. (§159, in Supplement X to the second edition of Dirichlet's lectures), and ibid., pp. 2ff. (§160, in Supplement XI to the third and fourth editions of that work).

15. Dedekind, "Sur la théorie des Nombres entiers algébrique," in his *Werke*, vol. 3, p. 296.

16. Dedekind, *Werke*, vol. 3, p. 335; *Essays*, p. 31.

17. Gottlob Frege, *Die Grundlagen der Arithmetik* (Breslau: Verlag von W. Koebner, 1884; edition with parallel German and English texts, trans. J. L. Austin, reprinted Evanston, Ill.: Northwestern University Press, 1968), p. 102; cf. also his *Grundgesestze der Arithmetik*, vol. 1 (Jena, 1893; reprinted (two volumes in one) Hildesheim: Georg Olms Verlagsbuchhandlung, 1962), p. vii.

18. Dedekind, *Werke*, vol. 3, p. 24.

19. Ibid., p. 336; *Essays*, p. 32.

20. Dedekind, *Werke*, vol. 3, p. 335; *Essays*, p. 31.

21. Dedekind, *Was sind und was sollen die Zahlen?*, §73; *Werke*, vol. 3, p. 360, *Essays*, p. 68.

22. Frege, *Grundgesetze der Arithmetik*, vol. 2 (Jena, 1903; reprinted, together with vol. 1, as cited in n. 17 above), pp. 155-243; for general discussion, see §§156-65, 173, 175, 197, and 245—pp. 154-63, 168-69, 170-72, 189-90, 243.

23. Dedekind, *Werke*, vol. 3, p. 325; *Essays*, p. 15.

24. Dedekind, *Werke*, vol. 3, pp. 489-90 (letter dated 24 January 1888).

25. Felix Klein, *Vorlesungen über die Entwicklung der Mathematik im 19ten Jahrhundert*, vol. 1, ed. R. Courant and O. Neugebauer (reprinted New York: Chelsea 1956), p. 322.

26. Dedekind, *Werke*, vol. 3, p. 427. Another passage is worth quoting as an illustration both of Dedekind's generous attitude toward alternative constructions and of the Dirichletian ideal that governs his own preferences; it comes, again, from the foreword to *Was sind und was sollen die Zahlen?*, but refers back to his procedure in constructing the real numbers. Dedekind recognizes that the theories of Weierstrass and of Cantor are both entirely adequate—possess complete rigor; but of his own theory, he says that "it seems to me somewhat simpler, I might say *quieter*," than the other two. (See *Werke*, vol. 3, p. 339. Italics added—Dedekind's word is *ruhiger*; the published English translation, "easier," loses the point of the expression.)

27. Klein, *Entwicklung der Mathematik*, vol. 1, p. 98.

28. See Hilbert, *Gesammelte Abhandlungen*, vol. 3, pp. 394-95.

29. Kronecker, *Vorlesungen über Zahlentheorie*, vol. 1, p. 11.

30. Kronecker had, of course, a generally workable technique for dealing with algebraic irrationals; but when he speaks in Lecture 1 of the definitions that occur in analysis, he remarks that "from the entire domain of this branch of mathematics, only the concept of limit or bound has thus far remained alien to number theory" (ibid., vol. 1, pp. 4-5).

31. Ibid., p. viii.

32. Ibid., p. vi.

33. Dedekind, *Werke*, vol. 3, p. 156 n. What Kronecker criticizes in the *second* instance here is "the various concept-formations with the help of which, in recent times, it has been attempted from several sides (first of all by Heine) to conceive and to ground the 'irrational' in general." That Kronecker mentions Heine—who in fact adopted, with explicit acknowledgment, the method introduced by Cantor—as "first" is striking; was his antipathy to Cantor so great that he refused to mention the latter's name at all—or, indeed, suppressed his recollection of it? The suspicion that some degree of pathological aversion is involved here is increased by an example cited by Fraenkel in his biographical sketch of Cantor appended to Zermelo's edition of Cantor's works (Georg Cantor, *Abhandlungen mathematischen und philosophischen Inhalts*, ed. Ernst Zermelo [Berlin, 1932; reprinted Hildesheim: Georg Olms Verlagsbuchhandlung, 1966], p. 455, n. 3). Cantor published in 1870 his famous theorem on the uniqueness of trigonometric series (see his *Abhandlungen*, pp. 80ff.). In the following year, he published the first stage in his extension of that result (allowing a finite number of possibly exceptional points in the interval of periodicity); this paper also contains a simplification of his earlier proof, which he acknowledges as due to "a gracious oral communication of Herr Professor Kronecker" (ibid., p. 84). What Fraenkel reports—with an exclamation point—is that in his posthumously published lectures on the theory of integrals, Kronecker cites the problem of the uniqueness of trigonometric expansions as still open! It is worth remembering that Cantor achieved his broadest generalization of his uniqueness theorem in 1872, in the same paper (ibid.) pp. 92ff.) in which he first presented his construction of the real numbers and initiated the study of the topology of point sets. As to Dedekind, he responds, in his own quiet style, to Kronecker's footnote attack in a footnote of his own to §1 of *Was sind und was sollen die Zahlen?* (see Dedekind, *Werke*, vol. 3, p. 345).

34. This point deserves a little further emphasis. Kronecker had a *program* for the elimination of the nonconstructive from mathematics. Such a program was of unquestionably great interest. It continues to be so, in the double sense that a radical elimination—showing how to render constructive a sufficient body of mathematics, and at the same time so greatly simplifying that body, as to make it plausible that this constructive part is really *all* that deserves to be studied and that the constructive point of view is the clearly most fruitful way of studying it—would be a stupendous achievement; whereas in the absence of such radical elimination, it remains always important to investigate how far constructive methods may be applied. But Kronecker's own published ideas for constructivization appear to have extended only to the *algebraic*, as indeed the remark cited in n. 28 above explicitly acknowledges (for when Kronecker says that "the concept of the limit . . . has thus far remained *alien to number theory*," that last phrase has to be taken to mean *irreducible to*

the finitary theory of the natural numbers—which reducibility is what Kronecker's constructive program aimed at). Kronecker, therefore, in his *documented* strong statements against "the various concept-formations with the help of which . . . it has been attempted . . . to conceive and ground the 'irrational' in general," is in the position of arguing that, in the absence of a constructive definition, nonconstructive ones are nevertheless to be avoided; that it is better to have *no definition at all*. This is the position that seems to me to follow from Kronecker's statements, although it conflicts with the report of Hensel.

35. See Kronecker, *Werke*, vol. 3, pp. 252-53; *Vorlesungen über Zahlentheorie*, vol. 1, pp. 3, 5.

36. *Bernhard Riemann's Gesammelte Mathematische Werke*, ed. Heinrich Weber with the assistance of Richard Dedekind, 2d ed. (1892; reprinted New York: Dover, 1953), pp. 273-74.

37. Ibid., p. 286.

38. See *Logik nach den Vorlesungen des Herrn Prof. Kant im Sommerhalbenjahre 1792*, in *Die philosophischen Hauptvorlesungen Immanuel Kants*, ed. Arnold Kowalewski (Munich, 1924; reprinted Hildesheim: Georg Olms Verlagsbuchhandlung, 1965), pp. 393-95.

39. Hilbert, *Gesammelte Abhandlungen*, vol. 3, pp. 402-3.

40. Dedekind, *Werke*, vol. 3, p. 479.

41. Kant, *Critique of Pure Reason*, A702/B730.

42. Gottlob Frege, *On the Foundations of Geometry and Formal Theories of Arithmetic*, trans. Eike-Henner W. Kluge (New Haven: Yale University Press, 1971), pp. 6-21. See esp. p. 19 (where Frege says quite distinctly, "It seems to me that you really intend to define second-level concepts"); but cf. also Frege's subsequent remark in the first of his two papers, "On the Foundations of Geometry" (ibid., p. 36): "If any concept is defined by means of [Hilbert's axioms], it can only be a second-level concept. It must of course be doubted whether any concept is defined at all, since not only the word 'point' but also the words 'straight line' and 'plane' occur." In the latter passage, Frege's notion seems to be that Hilbert must have set out to define the (second-level) concept of a *concept of a point*; and similarly for line, etc. The earlier passage, in his letter to Hilbert, was preceded by this: "The characteristics which you state in your axioms . . . do not provide an answer to the question, 'What property must an object have to be a point, a straight line, a plane, etc.?' Instead they contain, for example, second-level relations, such as that of the concept of 'point' to the concept 'straight line'." It is not far from this to the conclusion that what Hilbert's axioms *do* define is a *single* second-level relation, namely, that relation among a whole system of first-level concepts—corresponding to the undefined terms of the axiomatization, 'point', 'line', 'plane', 'between', and 'congruent'—which qualifies the whole as a Euclidean geometry. But however near that conclusion lies, Frege clearly failed to draw it.

43. Hilbert, *Gesammelte Abhandlungen*, vol. 3, pp. 378-87.

44. As I have suggested earlier in this paper, the role that may yet be played in further clarification of the nature of mathematics by the views of the "three schools" cannot be regarded as certain. My own opinion is that all three views (if they can really be called "three": within each "school," there have been quite significant differences) have made useful contributions to our understanding, but also that each will remain as merely partial and not fully satisfactory. Yet, as I have also remarked (above, n. 34), one cannot exclude the possibility that a genuine constructivization of mathematics will yet be achieved (although I consider the arguments of section X above as counting rather strongly against this). In the same way, it remains possible that a recognizably constructive proof of the consistency of analysis, or even of (some version of) set theory, will be found; and such an event would afford great impetus to the revival of (a modified form of) the Hilbert program. As to "logicism," as it seems to me the vaguest of the three doctrines, I find it hardest to envisage prospects for it, and am most strongly inclined to see its positive contribution as exhausted by the insight that mathematics is in some sense "about" conceptual possibilities or conceptual structure—an insight, as I have remarked, already clearly present in Riemann, but much more fully worked out after the stimulus provided by Dedekind, Frege, Russell, and

Whitehead. Nevertheless it is (barely, I think) conceivable that progress in understanding the structure of "knowledge" will succeed in isolating a special kind of knowledge that reasonably deserves to be considered "logical," thus conferring new interesting content upon the question whether mathematics does or does not belong to logic.

— Michael J. Crowe —

Ten Misconceptions about Mathematics and Its History

For over two decades, one of my major interests has been reading, teaching, and writing history of mathematics. During those decades, I have become convinced that ten claims I formerly accepted concerning mathematics and its development are both seriously wrong and a hindrance to the historical study of mathematics. In analyzing these claims, I shall attempt to establish their initial plausibility by showing that one or more eminent scholars have endorsed each of them; in fact, all seem to be held by many persons not fully informed about recent studies in history and philosophy of mathematics. This paper is in one sense a case study; it has, however, the peculiar feature that in it I serve both as dissector and frog. In candidly recounting my changes of view, I hope to help newcomers to history of mathematics to formulate a satisfactory historiography and to encourage other practitioners to present their own reflections. My attempt to counter these ten claims should be prefaced by two qualifications. First, in advocating their abandonment, I am not in most cases urging their inverses; to deny that all swans are white does not imply that one believes no swans are white. Second, I realize that the evidence I advance in opposition to these claims is scarcely adequate; my arguments are presented primarily to suggest approaches that could be taken in more fully formulated analyses.

1. The Methodology of Mathematics Is Deduction

In a widely republished 1945 essay, Carl G. Hempel stated that the method employed in mathematics "is the method of mathematical demonstration, which consists in the logical deduction of the proposition to be proved from other propositions, previously established." Hempel added the qualification that mathematical systems rest ultimately on axioms and postulates, which cannot themselves be secured by deduction.[1] Hempel's claim concerning the method of mathematics is widely shared;

I accepted it as a young historian of mathematics, but was uneasy with two aspects of it. First, it seemed to make mathematicians unnecessary by implying that a machine programmed with appropriate rules of inference and, say, Euclid's definitions, axioms, and postulates could deduce all 465 propositions presented in his *Elements*. Second, it reduced the role of historians of mathematics to reconstructing the deductive chains attained in the development of mathematics.

I came to realize that Hempel's claim could not be correct by reading a later publication, also by Hempel; in his *Philosophy of Natural Science* (1966), he presented an elementary proof that leads to the conclusion that deduction cannot be the sole method of mathematics. In particular, he demonstrated that from even a single true statement, an infinity of other true statements can be validly deduced. If we take "or" in the nonexclusive sense and are given a true proposition p, Hempel asserted that we can deduce an infinity of statements of the form "p or q" where q is any proposition whatsoever. Note that all these propositions are true because with the nonexclusive meaning of "or," all propositions of the form "p or q" are true if p is true. As Hempel stated, this example shows that the rules of logical inference provide only tests of the validity of arguments, not methods of discovery.[2] Nor, it is important to note, do they provide guidance as to whether the deduced propositions are in any way significant. Thus we see that an entity, be it man or machine, possessing the deductive rules of inference and a set of axioms from which to start, could generate an infinite number of true conclusions, none of which would be significant. We would not call such results mathematics. Consequently, mathematics as we know it cannot arise solely from deductive methods. A machine given Euclid's definitions, axioms, and postulates might deduce thousands of valid propositions without deriving any Euclidean theorems. Moreover, Hempel's analysis shows that even if definitions, axioms, and postulates could be produced deductively, still mathematics cannot rely solely on deduction. Furthermore, we see from this that historians of mathematics must not confine their efforts to reconstructing deductive chains from the past of mathematics. This is not to deny that deduction plays a major role in mathematical methodology; all I have attempted to show is that it cannot be the sole method of mathematics.

2. Mathematics Provides Certain Knowledge

In the same 1945 essay cited previously, Hempel stated: "The most distinctive characteristic which differentiates mathematics from the various branches of empirical science . . . is no doubt the peculiar certainty and necessity of its results." And, he added: "a mathematical theorem, once proved, is established once and for all. . . ."[3] In noting the certainty of mathematics, Hempel was merely reasserting a view proclaimed for centuries by dozens of authors who frequently cited Euclid's *Elements* as the prime exemplification of that certainty. Writing in 1843, Philip Kelland remarked: "It is certain that from its completeness, uniformity and faultlessness, . . . and from the universal adoption of the completest and best line of argument, Euclid's 'Elements' stand preeminently at the head of all human productions."[4] A careful reading of Hempel's essay reveals a striking feature; immediately after noting the certainty of mathematics, he devoted a section to "The Inadequacy of Euclid's Postulates." Here in Hilbertian fashion, Hempel showed that Euclid's geometry is marred by the fact that it does not contain a number of postulates necessary for proving many of its propositions. Hempel was of course correct; as early as 1892, C. S. Peirce had dramatically summarized a conclusion reached by most late-nineteenth-century mathematicians: "The truth is, that elementary [Euclidean] geometry, instead of being the perfection of human reasoning, is riddled with fallacies. . . ."[5]

What is striking in Hempel's essay is that he seems not to have realized the tension between his claim for the certainty of mathematics and his demonstration that perhaps the most famous exemplar of that certainty contains numerous faulty arguments. Hempel's claim may be construed as containing the implicit assertion that a mathematical system embodies certainty only after all defects have been removed from it. What is problematic is whether we can ever be certain that this has been done. Surely the fact that the inadequacy of some of Euclid's arguments escaped detection for over two millennia suggests that certainty is more elusive than usually assumed. Moreover, in opposition to the belief that certainty can be secured for formalized mathematical systems, Reuben Hersh has stated: "It is just not the case that a doubtful proof would become certain by being formalized. On the contrary, the doubtfulness of the proof would then be replaced by the doubtfulness of the coding and programming."[6] Morris Kline has recently presented a powerful demonstration that the certainty purportedly present throughout the development of mathematics is an illusion; I refer to his *Mathematics: The Loss of Certainty*, in which he states: "The hope of finding objective, infallible laws and standards has faded. The Age of Reason is gone."[7] Much in what follows sheds further light on the purported certainty of mathematics, but let us now proceed to two related claims.

3. Mathematics Is Cumulative

An elegant formulation of the claim for the cumulative character of mathematics is due to Hermann Hankel, who wrote: "In most sciences one generation tears down what another has built and what one has established another undoes. In Mathematics alone each generation builds a new story to the old structure."[8] Pierre Duhem made a similar claim: "Physics does not progress as does geometry, which adds new final and indisputable propositions to the final and indisputable propositions it already possessed...."[9] The most frequently cited illustration of the cumulative character of mathematics is non-Euclidean geometry. Consider William Kingdon Clifford's statement: "What Vesalius was to Galen, what Copernicus was to Ptolemy, that was Lobatchewsky to Euclid."[10] Clifford's claim cannot, however, be quite correct; whereas acceptance of Vesalius entailed rejection of Galen, whereas adoption of Copernicus led to abandonment of Ptolemy, Lobachevsky did not refute Euclid; rather he revealed that another geometry is possible. Although this instance illustrates the remarkable degree to which mathematics is cumulative, other cases exhibit opposing patterns of development. As I wrote my *History of Vector Analysis*,[11] I realized that I was also, in effect, writing *The Decline of the Quaternion System*. Massive areas of mathematics have, for all practical purposes, been abandoned. The nineteenth-century mathematicians who extended two millennia of research on conic section theory have now been forgotten; invariant theory, so popular in the nineteenth century, fell from favor.[12] Of the hundreds of proofs of the Pythagorean theorem, nearly all are now nothing more than curiosities.[13] In short, although many previous areas, proofs, and concepts in mathematics have persisted, others are now abandoned. Scattered over the landscape of the past of mathematics are numerous citadels, once proudly erected, but which, although never attacked, are now left unoccupied by active mathematicians.

4. Mathematical Statements Are Invariably Correct

The most challenging aspect of the question of the cumulative character of mathematics concerns whether mathematical assertions are ever refuted.

The previously cited quotations from Hankel and Duhem typify the widespread belief that Joseph Fourier expressed in 1822 by stating that mathematics "is formed slowly, but it preserves every principle it has once acquired...."[14] Although mathematicians may lose interest in a particular principle, proof, or problem solution, although more elegant ways of formulating them may be found, nonetheless they purportedly remain. Influenced by this belief, I stated in a 1975 paper that "Revolutions never occur in mathematics."[15] In making this claim, I added two important qualifications: the first of these was the "minimal stipulation that a necessary characteristic of a revolution is that some previously existing entity (be it king, constitution, or theory) must be overthrown and irrevocably discarded"; second, I stressed the significance of the phrase "in mathematics," urging that although "revolutions may occur in mathematical nomenclature, symbolism, metamathematics, [and] methodology...," they do not occur *within* mathematics itself.[16] In making that claim concerning revolutions, I was influenced by the widespread belief that mathematical statements and proofs have invariably been correct. I was first led to question this belief by reading Imre Lakatos's brilliant *Proofs and Refutations*, which contains a history of Euler's claim that for polyhedra $V - E + F = 2$, where V is the number of vertices, E the number of edges, and F the number of faces.[17] Lakatos showed not only that Euler's claim was repeatedly falsified, but also that published proofs for it were on many occasions found to be flawed. Lakatos's history also displayed the rich repertoire of techniques mathematicians possess for rescuing theorems from refutations.

Whereas Lakatos had focused on a single area, Philip J. Davis took a broader view when in 1972 he listed an array of errors in mathematics that he had encountered.[18] Philip Kitcher, in his recent *Nature of Mathematical Knowledge*, has also discussed this issue, noting numerous errors, especially from the history of analysis.[19] Morris Kline called attention to many faulty mathematical claims and proofs in his *Mathematics: The Loss of Certainty*. For example, he noted that Ampère in 1806 proved that every function is differentiable at every point where it is continuous, and that Lacroix, Bertrand, and others also provided proofs until Weierstrass dramatically demonstrated the existence of functions that are everywhere continuous but nowhere differentiable.[20] In studying the history of complex numbers, Ernest Nagel found that such mathematicians as Cardan, Simson, Playfair, and Frend denied their existence.[21] Moreover,

Maurice Lecat in a 1935 book listed nearly 500 errors published by over 300 mathematicians.[22] On the other hand, René Thom has asserted: "There is no case in the history of mathematics where the mistake of one man has thrown the entire field on the wrong track.... Never has a significant error slipped into a conclusion without almost immediately being discovered."[23] Even if Thom's claim is correct, the quotations from Duhem and Fourier seem difficult to reconcile with the information cited above concerning cases in which concepts and conjectures, principles and proofs *within* mathematics have been rejected.

5. The Structure of Mathematics Accurately Reflects Its History

In recent years, I have been teaching a course for humanities students that begins with a careful reading of Book I of Euclid's *Elements*. That experience has convinced me that the most crucial misconception that students have about mathematics is that its structure accurately reflects its history. Almost invariably, the students read this text in light of the assumption that the deductive progression from its opening definitions, postulates, and common notions through its forty-eight propositions accurately reflects the development of Euclid's thought. Their conviction in this regard is reinforced by the fact that most of them have earlier read Aristotle's *Posterior Analytics*, in which that great philosopher specified that for a valid demonstration "the premises ... must be ... better known than and prior to the conclusion...."[24] My own conception is that the development of Euclid's thought was drastically different. Isn't it plausible that in composing Book I of the *Elements*, Euclid began not with his definitions, postulates, and common notions but rather either with his extremely powerful 45th proposition, which shows how to reduce areas bounded by straight lines to a cluster of measurable triangular areas, or with his magnificent 47th proposition, the Pythagorean theorem, for which he forged a proof that has been admired for centuries. Were not these two propositions the ones he knew best and of which he was most deeply convinced? Isn't it reasonable to assume that it was only after Euclid had decided on these propositions as the culmination for his first book that he set out to construct the deductive chains that support them? Is it probable that Euclid began his efforts with his sometimes abstruse and arbitrary definitions— "a point is that which has no parts"—and somehow arrived forty-seven propositions later at a result known to the Babylonians fifteen centuries earlier? An examination of Euclid's 45th and 47th propositions shows that they depend upon the proposition that if two coplanar straight lines meet at a point and make an angle with each other equal to two right angles, then those lines are collinear. Should it be seen as a remarkable coincidence that thirty-one propositions earlier Euclid had proved precisely this result, but had not used it a single time in the intervening propositions? It seems to me that accepting the claim that the history and deductive structures of mathematical systems are identical is comparable to believing that Saccheri was surprised when after proving dozens of propositions, he finally concluded that he had established the parallel postulate.

Is not the axiomatization of a field frequently one of the last stages, rather than the first, in its development? Recall that it took Whitehead and Russell 362 pages of their *Principia Mathematica* to prove that $1 + 1 = 2$. Calculus texts open with a formulation of the limit concept, which took two centuries to develop. Geometry books begin with primary notions and definitions with which Hilbert climaxed two millennia of searching. Second-grade students encounter sets as well as the associative and commutative laws—all hard-won attainments of the nineteenth century. If these students are gifted and diligent, they may years later be able to comprehend some of the esoteric theorems advanced by Archimedes or Apollonius. When Cauchy established the fundamental theorem of the calculus, that subject was nearly two centuries old; when Gauss proved the fundamental theorem of algebra, he climaxed more than two millennia of advancement in that area.[25] In teaching complex numbers, we first justify them in terms of ordered couples of real numbers, a creation of the 1830s. After they have magically appeared from this process, we develop them to the point of attaining, say, Demoive's theorem, which came a century before the Hamilton-Bolyai ordered-couple justification of them. In presenting a theorem, first we name it and state it precisely so as to exclude the exceptions it has encountered in the years since its first formulation; then we prove it; and, finally, we employ it to prove results that were probably known long before its discovery. In short, we reverse history. Hamilton created quaternions in 1843 and simultaneously supplied a formal justification for them, this being the first case in which a number system was discovered and justified at the same time; half a century later Gibbs and Heaviside, viewing the quaternion method of space analysis as unsatisfactory, proposed a simpler system derived from quaternions by a process now largely forgotten.

Do not misunderstand: I am not claiming that the structure of mathematics, as a whole or in its parts, is in every case the opposite of its history. Rather I am suggesting that the view that students frequently have, implicitly or explicitly, that the structure in which they encounter areas of mathematics is an adequate approximation of its history, is seriously defective. Mathematics is often compared to a tree, ever attaining new heights. The latter feature is certainly present, but mathematics also grows in root and trunk; it develops as a whole. To take another metaphor, the mathematical research frontier is frequently found to lie not at some remote and unexplored region, but in the very midst of the mathematical domain. Mathematics is often compared to art; yet reflect for a moment. Homer's *Odyssey*, Da Vinci's *Mona Lisa*, and Beethoven's Fifth Symphony are completed works, which no later artist dare alter. Nonetheless, the latest expert on analysis works alongside Leibniz and Newton in ordering the area they created; a new Ph.D. in number theory joins Euclid, Fermat, and Gauss in perfecting knowledge of the primes. Kelvin called Fourier's *Théorie analytique de la chaleur* a "mathematical poem,"[26] but many authors shaped its verses. Why did some mathematicians oppose introduction of complex or transfinite numbers, charging that they conflicted with the foundations of mathematics? Part of the reason is that, lacking a historical sense, they failed to see that foundations are themselves open to alteration, that not only premises but results dictate what is desirable in mathematics.

6. Mathematical Proof in Unproblematic

Pierre Duhem in his *Aim and Structure of Physical Theory* reiterated the widely held view that there is nothing problematic in mathematical proof by stating that geometry "grows by the continual contribution of a new theorem demonstrated once and for all and added to theorems already demonstrated...."[27] In short, Duhem was claiming that once a proposition has been demonstrated, it remains true for all time. Various authors, both before and after Duhem, have taken a less absolutist view of the nature and conclusiveness of proof. In 1739, David Hume observed:

> There is no ... Mathematician so expert ... as to place entire confidence in any truth immediately upon his discovery of it, or regard it as any thing, but a mere probability. Every time he runs over his proofs, his confidence encreases; but still more by the approbation of his friends; and is rais'd to its utmost perfection by the universal assent and applauses of the learned world."[28]

G. H. Hardy concluded in a 1929 paper entitled "Mathematical Proof" that "If we were to push it to its extreme, we should be led to rather a paradoxical conclusion: that there is, strictly, no such thing as mathematical proof; that we can, in the last analysis, do nothing but *point*; that proofs are what Littlewood and I call *gas*, rhetorical flourishes designed to affect psychology...."[29] E. T. Bell in a number of his writings developed the point that standards of proof have changed dramatically throughout history. For example, in his *Development of Mathematics* (1940), he challenged the assertion of an unnamed "eminent scholar of Greek mathematics" that the Greeks, by their "'unerring logic,' had attained such perfect mathematical results that 'there has been no need to reconstruct, still less to reject as unsound, any essential part of their doctrine....'" Bell responded that among, for example, Euclid's proofs, "many have been demolished in detail, and it would be easy to destroy more were it worth the trouble."[30] Raymond Wilder, who also discussed the process of proof in various writings, asserted in 1944 that "we don't possess, and probably will never possess, any standard of proof that is independent of the time, the thing to be proved, or the person or school of thought using it." Over three decades later, he put this point most succinctly: "'proof' in mathematics is a culturally determined, relative matter."[31]

That research in history and philosophy of mathematics has contributed far more toward understanding the nature of proof than simply showing that standards of proof have repeatedly changed can be illustrated by briefly examining the relevant writings of Imre Lakatos. In his *Proofs and Refutations* (1963-64), Lakatos, proceeding from his conviction (derived from Karl Popper) that conjectures play a vital role in the development of mathematics and his hope (derived from George Pólya) that heuristic methods for mathematics can be formulated, reconstructed the history of Euler's conjecture concerning polyhedra so as to show that its history ill accords with the traditional accumulationist historiography of mathematics. Whereas some had seen its history as encompassing little more than Euler's formulation of the conjecture and Poincaré's later proof for it, Lakatos showed that numerous "proofs" had been advanced in the interim, each being falsified by counterexamples. Fundamental to this essay is Lakatos's definition of proof as "*a thought-experiment—or 'quasi-experiment'—which suggests a decomposition of the original conjecture into subconjectures or lemmas,* thus *embedding it* in a possibly quite distant body of knowledge."[32] On this basis, Lakatos, in opposition to the belief that the proof or refutation of a mathematical claim is final, argued forcefully that on the one hand mathematicians should seek counterexamples to proved theorems (pp. 50ff.) and on the other hand be cautious in abandoning refuted theorems (pp. 13ff.). Moreover, he warned of the dangers involved in recourse, if counterexamples are found, to the techniques he called "monster-barring," "monster-adjustment," and "exception-barring" (pp. 14-33). Lakatos also wrote other papers relevant to the nature of mathematical proof; for example, in his "Infinite Regress and the Foundations of Mathematics" (1962), he provided insightful critiques of the "Euclidean programme" as well as of the formalist conception of mathematical method. In his "A Renaissance of Empiricism in Recent Philosophy of Mathematics" (1967), he stressed the importance of empirical considerations in mathematical proof, while in his "Cauchy and the Continuum . . . ," he urged that Abraham Robinson's methods of nonstandard analysis could be used to provide a radically new interpretation of the role of infinitesimals in the creation of the calculus.[33] The sometimes enigmatic character of Lakatos's writings and the fact that his interests shifted in the late 1960s toward the history and philosophy of science—to which he contributed a "methodology of scientific research programmes"—left, after his death in 1974, many unanswered questions about his views on mathematics. Various authors have attempted to systematize his thought in this regard,[34] and Michael Hallett has advanced and historically illustrated the thesis that "mathematical theories *can* be appraised by criteria like those of [Lakatos's] methodology of scientific research programmes. . . ."[35]

7. Standards of Rigor Are Unchanging

Writing in 1873, the Oxford mathematician H. J. S. Smith repeated a conclusion often voiced in earlier centuries; Smith stated: "The methods of Euclid are, by almost universal consent, unexceptionable in point of rigour."[36] By the beginning of the present century, Smith's claim concerning Euclid's "perfect rigorousness" could no longer be sustained. In his *Value of Science* (1905), Henri Poincaré asked: "Have we finally attained absolute rigor? At each stage of the evolution our fathers . . . thought they had reached it. If they deceived themselves, do we not likewise cheat ourselves?" Surprisingly, Poincaré went on to assert that "in the analysis of today, when one cares to take the trouble to be rigorous, there can be nothing but syllogisms or appeals to this intuition of pure number, the only intuition which can not deceive us. It may be said that today absolute rigor is attained."[37] More recently, Morris Kline remarked: "No proof is final. New counterexamples undermine old proofs. The proofs are then revised and mistakenly considered proven for all time. But history tells us that this merely means that the time has not yet come for a critical examination of the proof."[38]

Not only do standards of rigor intensify, they also change in nature; whereas in 1700 geometry was viewed as providing the paradigm for such standards, by the late nineteenth century arithmetic-algebraic considerations had assumed primacy, with these eventually giving way to standards formulated in terms of set theory. Both these points, as well as a number of others relating to rigor, have been discussed with unusual sensitivity by Philip Kitcher. For example, in opposition to the traditional view that rigor should always be given primacy, Kitcher has suggested in his essay "Mathematical Rigor—Who Needs It?" the following answer: "Some mathematicians at some times, but by no means all mathematicians at all times."[39] What has struck me most forcefully about the position Kitcher developed concerning rigor in that paper and in his *Nature of Mathematical Knowledge* are its implications for the historiography of mathematics. I recall being puzzled some years ago while studying the history of complex numbers by the terms that practitioners of that most rational discipline, mathematics, used for these numbers. Whereas their inventor Cardan called them "sophistic," Napier, Girard, Descartes, Huygens, and Euler respectively branded them "nonsense," "inexplicable," "imaginary," "incomprehensible," and "impossible." Even more mysteriously, it seemed, most of these mathematicians, despite the invective implied in thus naming these numbers, did not hesitate to use them. As Ernest Nagel observed, "for a long time no one could defend the 'imaginary numbers' with any plausibility, except on the logically inadequate ground of their mathematical usefulness." He added: "Nonetheless, mathematicians who refused to banish them . . . were not fools . . . as subsequent events showed."[40]

What I understand Kitcher to be suggesting is that the apparent irrationality of the disregard for rigor found in the pre-1830 history of both complex numbers and the calculus is largely a product of unhistorical, present-centered conceptions of mathematics. In particular, if one recognizes that need for rigor is a relative value that may be and has at times been *rationally* set aside in favor of such other values as usefulness, then one will be less ready to describe various periods of mathematics as ages of unreason and more prone to undertake the properly historical task of understanding why mathematicians adopted such entities as "impossible numbers" or infinitesimals. As Kitcher suggests, it may be wise for historians of mathematics to follow the lead of historians of science who long ago became suspicious of philosophic and historiographic systems that entail the reconstruction of scientific controversies in terms of such categories as irrationality, illogicality, and stubbornness.[41]

8. The Methodology of Mathematics Is Radically Different from the Methodology of Science

The quotations previously cited from Duhem's *Aim and Structure of Physical Theory* illustrate a subtheme running through that book: that the methodology of mathematics differs greatly from that of physics. In other passages, Duhem lamented that physics had not achieved "a growth as calm and as regular as that of mathematics" (p. 10) and that physics, unlike mathematics, possesses few ideas that appear "clear, pure, and simple" (p. 266). Moreover, largely because Duhem believed that "these two methods reveal themselves to be profoundly different" (p. 265), he concluded that, whereas history of physics contributes importantly to understanding physics, "The history of mathematics is, [although] a legitimate object of curiosity, not essential to the understanding of mathematics" (p. 269). The position developed in this and the next section is that important parallels exist between the methods employed in mathematics and in physics.

The first author who explicitly described the method that, according to most contemporary philosophers of science, characterizes physics was Christiaan Huygens. He prefaced his *Treatise on Light* by stating that in presenting his theory of light he had relied upon "demonstrations of those kinds which do not produce as great a certitude as those of Geometry, and which even differ much therefrom, since whereas the Geometers prove their Propositions by fixed and incontestable Principles, here the Principles are verified by the conclusions to be drawn from them...."[42] What I wish to suggest is that, to a far greater extent than is commonly realized, mathematicians have employed precisely the same method—the so-called hypothetico-deductive method. Whereas the pretense is that mathematical axioms justify the conclusions drawn from them, the reality is that to a large extent mathematicians have accepted axiom systems on the basis of the ability of those axioms to bring order and intelligibility to a field and/or to generate interesting and fruitful conclusions. In an important sense, what legitimized the calculus in the eyes of its creators was that by means of its methods they attained conclusions that were recognized as correct and meaningful. Although Hamilton, Grassmann, and Cantor, to name but a few, presented the new systems for which they are now famous in the context of particular philosophies of mathematics (now largely discarded), what above all justified their new creations, both in their own eyes and among their contemporaries, were the conclusions drawn from them. This should not be misunderstood; I am not urging that only utilitarian criteria have determined the acceptability of mathmatical systems, although usefulness has undoubtedly been important. Rather I am claiming that characteristics of the results attained—for example, their intelligibility—have played a major role in determining the acceptability of the source from which the results were deduced. To put it differently, calculus, complex numbers, non-Euclidean geometries, etc., were in a sense hypotheses that mathematicians subjected to test in ways comparable in logical form to those used by physicists.

My claim that mathematicians have repeatedly employed the hypothetico-deductive method is not original; a number of recent authors have made essentially the same suggestion. Hilary Putnam began a 1975 paper by asking how we would react to finding that Martian mathematicians employ a methodology that, although using full-blown proofs when possible, also relies upon quasi-empirical tests; for example, his Martians accept the four-color conjecture because much empirical evidence supports and none contradicts it. Putnam proceeded to claim that we should not see this as resulting from some bizarre misunderstanding of the nature of mathematics; in fact, he asserted that "we have been using quasi-empirical and even empirical methods in mathematics all along. . . ."[43] The first example he used to illustrate this claim is Descartes's creation of analytical geometry, which depends upon the possibility of a one-to-one correspondence between the real numbers and the points on a line. The fact that no justification for this correspondence, let alone for the real numbers, was available in Descartes's day did not deter him or his contemporaries; they proceeded confidently ahead. As Putnam commented on his Descartes illustration: "This is as much an example of the use of hypothetico-deductive methods as anything in physics is" (p. 65). Philip Kitcher, who has stressed the parallels between the evolution of mathematics and of science, has advocated a similar view. In 1981, he stated:

> Although we can sometimes present parts of mathematics in axiomatic form, . . . the statements taken as axioms usually lack the epistemological features which [deductivists] attribute to first principles. Our knowledge of the axioms is frequently less certain than our knowledge of the statements we derive from them. . . . In fact, our knowledge of the axioms is sometimes obtained by nondeductive inference from knowledge of the theorems they are used to systematize.[44]

Finally, statements urging that mathematical systems are, like scientific systems, tested by their results occur in the writings of Haskell Curry, Willard Van Orman Quine, and Kurt Gödel.[45]

9. Mathematical Claims Admit of Decisive Falsification

In the most widely acclaimed section of his *Aim and Structure of Physical Theory*, Duhem attacked the view that crucial experiments are possible in physics. He stated: "Unlike the reduction to absurdity [method] employed by geometers, experimental contradiction does not have the power to transform a physical hypothesis into an indisputable truth" (p. 190). The chief reason he cited for this inability is that a supposed crucial experiment can at most decide "between two sets of theories each of which has to be taken as a whole, i.e., between two entire systems . . ." (p. 189). Because physical theories can be tested only in clusters, the physicist, when faced with a contradiction, can, according to Duhem, save a particular theory by modifying one or more elements in the cluster, leaving the particular theory of most concern (for example, the wave or particle theory) intact. In effect, Duhem was stating that individual physical theories can always be rescued from apparent refutations. Having criticized a number of Duhemian claims, I wish now to pay tribute to him by urging that a comparable analysis be applied to mathematics. In particular, I suggest that in history of mathematics one frequently encounters cases in which a mathematical claim, faced with an apparent logical falsification, has been rescued by modifying some other aspect of the system. In other words, mathematical assertions are usually not tested in isolation but in conjunction with other elements in the system.

Let us consider some examples. Euclid brought his *Elements* to a conclusion with his celebrated theorem that "no other figure, besides [the five regular solids], can be constructed which is contained by equilateral and equiangular figures equal to one another." How would Euclid respond if presented with a contradiction to this theorem—for example, with a hexahedron formed by placing two regular tetrahedra face to face? It seems indisputable that rather than rejecting his theorem, he would rescue it by revising his definition of regular solid so as to exclude polyhedra possessing noncongruent vertices. For centuries, complex numbers were beset with contradictions; some charged that they were contradicted by the rules that every number must be less than, greater than, or equal to zero and that the square of any number be positive. Moreover, others urged that no geometric interpretation of them is possible.[46] Complex numbers survived such attacks, whereas the cited rules and the traditional definition of number did not. Many additional cases can be found; in fact, Lakatos's *Proofs and Refutations* is rich in examples of refutations that were themselves rejected. Of course, mathematicians do at times choose to declare apparent logical contradictions to be actual refutations; nonetheless, an element of choice seems present in many such cases.

10. In Specifying the Methodology Used in Mathematics, the Choices Are Empiricism, Formalism, Intuitionism, and Platonism

For decades, mathematicians, philosophers, and historians have described the alternative positions concerning the methodology of mathematics as empiricism, formalism, intuitionism, and Platonism. This delineation of the options seems ill-conceived in at least two ways. First, it tends to blur the distinction between the epistemology and methodology of mathematics. Although related in a number of complex ways, the two areas can and should be distinguished. Epistemology of mathematics deals with how mathematical knowledge is possible, whereas methodology of mathematics focuses on what methods are used in mathematics. The appropriateness of distinguishing between these two areas is supported by the fact that the history of the philosophy of mathematics reveals that different epistemological positions have frequently incorporated many of the same methodological claims. Second, this fourfold characterization tends to blur the distinction between normative and descriptive claims. To ask what methods mathematicians should use is certainly different from asking what methods they have in fact used. Failure to recognize this distinction leads not only to the so-called naturalistic fallacy—the practice of inferring from "is" to "ought"—but also to the unnamed opposite fallacy of inferring from "ought" to "is" (or to "was").

It has been my experience that both mathematicians and historians of mathematics are primarily interested in issues of the methodology rather than of the epistemology of mathematics. When they turn to writings in the philosophy of mathematics, they usually find these composed largely in terms of one or more of the four primarily epistemological positions mentioned previously. Unfortunately, these categories seem relatively unilluminating in exploring methodological issues. Reuben Hersh very effectively discussed this problem in a 1979 paper in which he asked: "Do we really have to choose between a formalism that is falsified by our everyday experience, and a Platonism that postulates a mythical fairyland where the uncountable and the inaccessible lie waiting to be observed...?" Hersh proposed a different and more modest program for those of us interested in investigating the nature of mathematics; he suggested that we attempt "to give an account of mathematical knowledge as it really is— fallible, corrigible, tentative and evolving.... That is, reflect honestly on what we do when we use, teach, invent or discover mathematics—by studying history, by introspection, and by observing ourselves and each other...."[47] If Hersh's proposal is taken seriously, new categories will probably emerge in the philosophy and historiography of mathematics, and these categories should prove more interesting and illuminating than the traditional ones. Moreover, increased study of the descriptive methodology of mathematics should itself shed light on epistemological issues. This paper not only concludes with an endorsement of Hersh's proposed program of research, but has been designed to serve as an exemplification of it.

Notes

1. Carl G. Hempel, "Geometry and Empirical Science," in *Readings in the Philosophy of Science*, ed. Philip P. Wiener (New York: Appleton-Century-Crofts, 1953), p. 41; reprinted from *American Mathematical Monthly* 52 (1945): 7-17.

2. Carl G. Hempel, *Philosophy of Natural Science* (Englewood Cliffs, N.J.: Prentice-Hall, 1966) pp. 16-17.

3. Hempel, "Geometry," pp. 40-41.

4. As quoted in *On Mathematics and Mathematicians*, ed. Robert Edouard Moritz (New York: Dover, 1942), p. 295.

5. Charles S. Peirce, "The Non-Euclidean Geometry," in *Collected Papers of Charles Sanders Peirce*, vol. 8, ed. Arthur W. Burks (Cambridge, Mass.: Harvard University Press, 1966), p. 72.

6. Reuben Hersh, "Some Proposals for Reviving the Philosophy of Mathematics," *Advances in Mathematics* 31 (1979): 43.

7. Morris Kline, *Mathematics: The Loss of Certainty* (New York: Oxford University Press, 1980), p. 7.

8. As quoted in Moritz, *Mathematics*, p. 14.

9. Pierre Duhem, *The Aim and Structure of Physical Theory*, trans. Philip P. Wiener

(Princeton, N.J.: Princeton University Press, 1954), p.177.

10. As quoted in Moritz, *Mathematics*, p. 354.

11. Michael J. Crowe, *A History of Vector Analysis: The Evolution of the Idea of a Vectorial System* (Notre Dame, Ind.: University of Notre Dame Press, 1967); reprinted New York: Dover Publications, 1985.

12. Charles S. Fisher, "The Death of a Mathematical Theory: A Study in the Sociology of Knowledge," *Archive for the History of the Exact Sciences* 3 (1966-67): 137-59.

13. For a compilation of 362 proofs, see Elisha Scott Loomis, *The Pythagorean Proposition* (Ann Arbor, Mich.: National Council of Teachers of Mathematics, 1940).

14. Joseph Fourier, *Analytical Theory of Heat*, trans. Alexander Freeman (New York: Dover, 1955), p. 7.

15. Michael J. Crowe, "Ten 'Laws' Concerning Patterns of Change in the History of Mathematics," *Historia Mathematica* 2 (1975): 161-66. For discussions of my claim concerning revolutions, see I. Bernard Cohen, *Revolution in Science* (Cambridge, Mass.: Harvard University Press, 1985) pp. 489-91, 505-7; Joseph Dauben, "Conceptual Revolutions and the History of Mathematics," in *Transformation and Tradition in the Sciences*, ed. Everett Mendelsohn (Cambridge: Cambridge University Press, 1984), pp. 81-103; Herbert Mehrtens, "T. S. Kuhn's Theories and Mathematics: A Discussion Paper on the 'New Historiography' of Mathematics," *Historia Mathematica* 3 (1976): 297-320; and Raymond L. Wilder, *Mathematics as a Cultural System* (Oxford: Pergamon, 1981), pp. 142-44.

16. Crowe, "Patterns," pp. 165-66.

17. Imre Lakatos, *Proofs and Refutations: The Logic of Mathematical Discovery*, ed. John Worrall and Elie Zahar (Cambridge: Cambridge University Press, 1976). This study first appeared in 1963-64 in *British Journal for the Philosophy of Science*.

18. Philip J. Davis, "Fidelity in Mathematical Discourse: Is One and One Really Two?" in *American Mathematical Monthly* 79 (1972): 252-63; see esp. pp. 260-62.

19. Philip Kitcher, *The Nature of Mathematical Knowledge* (New York: Oxford University Press, 1983), pp.155-61, 178-85, 236-68

20. Kline, *Mathematics*, pp. 161, 177.

21. Ernest Nagel, "'Impossible Numbers': A Chapter in the History of Modern Logic," *Studies in the History of Ideas* 3 (1935): 427-74; reprinted in Nagel's *Teleology Revisited* (New York: Columbia University Press, 1979), pp.166-94.

22. Maurice Lecat, *Erreurs de mathématiciens des origines à nos jours* (Brussels: Castaigne, 1935), p. viii.

23. René Thom, "'Modern Mathematics': An Educational and Philosophic Error?" *American Scientist* 59 (1971): 695-99.

24. Aristotle, *Posterior Analytics*, in *The Basic Works of Aristotle*, ed. Richard McKeon (New York: Random House, 1941), book 1, chap. 2, lines 20-22. In the same chapter, Aristotle admitted that "prior" can be understood in two senses, but he apparently saw no significance in this fact.

25. Morris Kline, *Mathematical Thought from Ancient to Modern Times* (New York: Oxford University Press, 1972), pp. 958, 595.

26. As quoted in Silvanus P. Thompson, *The Life of William Thomson, Baron Kelvin of Largs*, vol. 2 (London: Macmillan, 1910), p. 1139.

27. Duhem, *Aim*, p. 204.

28. David Hume, *Treatise of Human Nature*, ed. Ernest C. Mossner (Baltimore: Penguin, 1969), p. 231.

29. G. H. Hardy, "Mathematical Proof," *Mind* 38 (1929): 18.

30. Eric Temple Bell, *The Development of Mathematics*, 2d ed. (New York: McGraw-Hill, 1945), p. 10. See also Bell's "The Place of Rigor in Mathematics," *American Mathematical Monthly* 41 (1934): 599-607.

31. Raymond L. Wilder, "The Nature of Mathematical Proof," *American Mathematical Monthly* 51 (1944): 319. See Also Wilder, *Mathematics*, e.g. pp. 39-41, and his "Relativity of Standards of Mathematical Rigor," *Dictionary of the History of Ideas*, ed. Philip P. Wiener, vol. 3 (New York: Charles Scribner's, 1973), pp. 170-77.

32. Lakatos, *Proofs*, p. 9.

33. Imre Lakatos, "Infinite Regress and the Foundations of Mathematics," "A Renaissance of Empiricism in Recent Philosophy of Mathematics," and "Cauchy and the Continuum: The Significance of the Non-Standard Analysis for the History and Philosophy of Mathematics," in Lakatos's *Mathematics, Science and Epistemology. Philosophical Papers*, vol. 2, ed. John Worrall and Gregory Currie (Cambridge: Cambridge University Press, 1978), pp. 3-23, 24-42, 43-60.

34. See, for example, Hugh Lehman, "An Examination of Imre Lakatos' Philosophy of Mathematics," *Philosophical Forum* 12 (1980): 33-48; Peggy Marchi, "Mathematics as a Critical Enterprise," in *Essays in Memory of Imre Lakatos*, ed. R. S. Cohen, P. K. Feyerabend, and M. W. Wartofsky (Dordrecht and Boston: Reidel, 1976), pp. 379-93.

35. Michael Hallett, "Towards a Theory of Mathematical Research Programmes," *British Journal for the Philosophy of Science* 30 (1979): 1-25, 135-59.

36. H. J. S. Smith, "Opening Address by the President . . . ," *Nature* 8 (25 Sept. 1873): 450. Smith in that year was president of Section A of the British Association for the Advancement of Science.

37. Henri Poincaré, *The Value of Science,* trans. George Bruce Halsted (New York: Dover, 1958), pp. 19-20.

38. Kline, *Mathematics*, p. 313; see also his *Why Johnny Can't Add* (New York: Vintage, 1974), p. 69.

39. Philip Kitcher, "Mathematical Rigor—Who Needs It?" *Nous* 15 (1980): 490.

40. Nagel, "Numbers," pp. 435, 437.

41. Kitcher, *Mathematical Knowledge*, pp. 155-58. For his "rational reconstruction" of the history of calculus, see chap. 10.

42. Christiaan Huygens, *Treatise on Light*, trans. Silvanus P. Thompson (New York: Dover, n.d.), p. vi.

43. Hilary Putnam, "What Is Mathematical Truth?" in Putnam, *Mathematics, Matter and Method. Philosophical Papers*, vol. 1, 2d ed. (Cambridge: Cambridge University Press, 1979), p. 64.

44. Kitcher, "Rigor," p. 471; see also Kitcher, *Mathematical Knowledge*, pp. 217-24, 271.

45. See quotations in Kline, *Mathematics*, pp. 330-31.

46. Nagel, "Numbers," pp. 434, 437, 441.

47. Hersh, "Proposals," p. 43. See also Philip J. Davis and Reuben Hersh, *The Mathematical Experience* (Boston: Birkhäuser, 1981), p. 406.

Mathematics and the Sciences

1. Introduction

The principal thrust of this essay is to describe the current state of interaction between mathematics and the sciences and to relate the trends to the historical development of mathematics as an intellectual discipline and of the sciences as they have developed since the seventeenth century. This story is interesting in the context of the history of present-day mathematics because it represents a shift in the preconceptions and stereotypes of both mathematicians and scientists since World War II.

The notion that significant mathematical and scientific advances are closely interwoven is not particularly new. The opposing notion (associated, with whatever degree of justice, with the name of Bourbaki) was never as fashionable, at least among working mathematicians, as in the two decades immediately after World War II. The situation has changed significantly during the past decade and had begun turning even earlier. It turned not only among mathematicians, but even more significantly in such sciences as physics. The frontier of mathematical advance was seen again to be in forceful interaction with the basic problems and needs of scientific advance.

This is an essay on significant trends in *mathematical practice*. The relations between the history and philosophy of mathematics as usually conceived and mathematical practice have often been very ambiguous. In part, this has resulted from the efforts of some historians and philosophers to impose a framework of preconceptions upon mathematical practice that had little to do with the latter. In part, however, it resulted from the diversity of mathematical practice, to lags in its perception, and to the complexity of viewpoints embedded in that practice.

Let me preface this account with two statements by great American mathematicians of an earlier period who put the case in a sharp form.

The first is John von Neumann in a 1945 essay titled "The Mathematician":[1]

> Most people, mathematicians and others, will agree that mathematics is not an empirical science, or at least that it is practiced in a manner which differs in several decisive respects from the techniques of the empirical sciences. And, yet, its development is very closely linked with the natural sciences. One of its main branches, geometry, actually started as a natural, empirical science. Some of the best inspirations of modern mathematics (I believe, the best ones) clearly originated in the natural sciences. The methods of mathematics pervade and dominate the "theoretical" divisions of the natural sciences. In modern empirical sciences it has become more and more a major criterion of success whether they have become accessible to the mathematical method or to the near-mathematical methods of physics. Indeed, throughout the natural sciences an unbroken chain of successive pseudomorphoses, all of them pressing toward mathematics, and almost identified with the idea of scientific progress, has become more and more evident. Biology becomes increasingly pervaded by chemistry and physics, chemistry by experimental and theoretical physics, and physics by very mathematical forms of theoretical physics.

The second is Norbert Wiener in a 1938 essay titled "The Historical Background of Harmonic Analysis:"[2]

> While the historical facts in any concrete situation rarely point a clear-cut moral, it is worth while noting that the recent fertility of harmonic analysis has followed a refertilization of the field with physical ideas. It is a falsification of the history of mathematics to represent pure mathematics as a self-contained science drawing inspiration from itself alone and morally taking in its own washing. Even the most abstract ideas of the present time have something of a physical history. It is quite a tenable point of view to urge this even in such fields as that of the calculus of assemblages, whose exponents, Cantor and Zermelo, have been deeply interested in problems of statistical mechanics. Not even the influence of this theory on the theory of integration, and indirectly on the theory of Fourier series, is entirely foreign to physics. The somewhat snobbish point of view of the purely abstract mathematician would draw but little support from mathematical history. On the other hand, whenever applied mathematics has been merely a technical employment of methods already traditional and jejune, it has been very poor applied mathematics. The desideratum in mathematical as well as physical work is an attitude which is not indifferent to the extremely

instructive nature of actual physical situations, yet which is not dominated by these to the dwarfing and paralyzing of its intellectual originality. Viewed as a whole, the theory of harmonic analysis has a very fine record of this sort. It is not a young theory, but neither is it yet in its dotage. There is much more to be learned and much more to be proved.

2. Mathematics and the Natural Sciences

One of the most striking features of recent developments in the physical sciences has been the convergence of focal theoretical problems with major themes in mathematical research. Mathematical concepts and tools that had arisen in an autonomous way in relatively recent research have turned out to be important components in the description of nature. At the same time, this use of novel mathematical tools in the sciences has reacted upon the development of mathematical areas having no direct connection with scientific applications to yield new and surprising mathematical consequences. We ask whether this kind of interaction will continue in a serious way in the foreseeable future, and, if so, what the consequences will be for the future development of mathematics and the sciences.

Let us begin our analysis by examining the different ways in which novel, relatively sophisticated mathematical tools have been applied in recent scientific developments. We may classify them into five relatively broad modes of attack.

(1) *The use of sophisticated mathematical concepts in the formulation of new, basic physical theories on the most fundamental level.* At the present moment, this takes the form of the the *superstring theory*, which has as its objective the total unification of all the basic physical forces and interactions—electromagnetic, weak, strong, and gravitational. This new phase of physical theory, which is the culmination of the earlier development of gauge field theories and of theories of supersymmetry, exhibits the use of a wide variety of relatively new mathematical tools developed in the past two decades; examples are Kac-Moody algebras and their representations, the existence of Einstein metrics on compact Kahlerian manifolds satisfying simple topological restrictions, and representations of exceptional Lie groups. The body of techniques and mathematical arguments embodied here includes the theory of Lie groups and algebras, their generalizations and representation theory; differential geometry in its modern global form in terms of vector bundles; the study of the ex-

istence of solutions of manifolds of highly nonlinear partial differential equations; differential and algebraic topology; and the whole melange of analysis, algebra, and geometry on manifolds that has been called *global analysis*. The implementation of this program may involve still other major directions of mathematical research—for example, noncommutative geometries based on rings of operators.

A similar pattern of use of sophisticated mathematical tools in the development of fundamental physical theories appeared earlier in the context of the study of *instantons* in gauge field theories and the study of singularities in the equation of general relativity in connection with *black holes*. What must be strongly emphasized is that the role assumed by sophisticated mathematics was not the result of a willful act by either physicists or mathematicians, but of the intrinsic necessities of the development of the physical theory. Physicists, no matter how sophisticated mathematically they may be, are not free ad libitum to choose the mathematical tools they wish to use. Certainly, the mathematicians have no power to prescribe such uses to the physicists. We are very far from the decades after World War II when it was a commonplace among physicists that all the mathematics they would ever need had been completely worked out (at least as far as the involvement of research mathematicians was concerned) before World War I. It is the radical transformation of fundamental physics in the past decades that has caused the disappearance of this commonplace, and not any basic transformation in the sociology of the relations between physicists and mathematicians.

(2) *A focal interest in the complex mathematical consequences of simple physical laws*. One sees major examples in the modeling of turbulence in terms of bifurcation, of the asymptotic properties of differential equations, and of iteration of simple nonlinear transformations (Hopf bifurcation, the Lorenz equation, strange attractors, and Feigenbaum cascades). Simple causal mechanics can be shown to lead to disorderly regimes (*chaos*), but in relatively simple and classifiable forms. A historically earlier example of an attack on turbulence in the 1930s to 1950s used models in terms of stochastic processes where disorder was directly injected into the premises of the theory. Another current example is the use of fractal models (self-similarity under changes of scale, fractional Hausdorff dimensions) to describe complex phenomena in the study of materials.

(3) *Mathematical models of pattern formation and symmetry breaking*

as paradigms for structured systems developing out of apparently unstructured regimes. We might think of this mode of attack (which goes back to Turing in 1952) as the converse of (2). Stable structures arise from mathematical models of differential equations or stochastic games of an apparently structureless nature in the presence of noise and possible disorder. The most striking paradigm is the oscillating chemical reaction of the Belousov-Zhabotinskii type. The objective here is eventually to model phenomena in such areas as developmental biology and brain function.

(4) *Soliton theories, involving the existence in nonlinear differential equations of stable structures (solitons) arising from complete integrability.* The now classical paradigm is the Korteweg-De Vries equation of shallow wave theory, rediscovered by Kruskal and his collaborators in the late 1950s after an earlier partial rediscovery in computer experiments by Fermi-Pasta-Ulam. New models of a similar kind have been found and extensively analyzed as a possible way of describing a broad range of physical and engineering phenomena.

(5) *The need to develop a usable and fruitful mathematical theory of complex systems whose elements might well be simple to an extreme but whose complexity arises from the interaction of these elements, whether linear or nonlinear, local or global.* It is abundantly clear that every mode of analysis in science or in practice will eventually get to the stage where this theme is dominant.

After this summary of major themes in present-day scientific investigations to which sophisticated mathematical tools are being applied, we may ask whether this is really a new situation. A careful answer to this question demands an analysis with a historical and philosophical focus, which I present in the final section of the paper.

3. Mathematics and the Computer

The observant reader is already aware that the description in section 2 of the mathematical component of important themes in contemporary scientific research omitted explicit mention of the high-speed digital computer, one of the most conspicuous objects of our age. In the context of the present kind of discussion, this might seem to many like a performance of Hamlet without the Noble Dane. Yet we must segregate the discussion of the computer and its interrelation with the development of contemporary mathematics, both because of its important and distinctive role and because of the prevalence and intensity of myths in this domain that prevent realistic assessment of the situation.

We are all very conscious of the role of the high-speed digital computer as one of the decisive facts of the present epoch and for the foreseeable future. We all know of the tremendous impact it has had on the structure of processes in industrial society that depend on calculation, communication, and control. In practice, this excludes very few domains of human existence in modern society, whether technological, economic, social, political, or military. The sciences and mathematics have not been immune from this impact. Indeed, the scope and nature of scientific and mathematical instrumentation and practice in our society have already been radically changed by the existence of high-speed digital computation and its continual decrease in cost during recent decades. I have deliberately used the unusual phrase *mathematical instrumentation* to point up the radically new fact that such a phenomenon now exists and is an important component of our situation.

At the same time, although we are all conscious of the importance of the digital computer (sometimes to the point of hysteria), and indeed are inundated with advertising hyperbole from the most diverse quarters about all the wonders that supercomputers will accomplish, many are much less conscious of what is ultimately an even more important fact: the computer is as much a *problem* as it is a *tool*. We must understand the nature and limitations of this most powerful of all human tools. It is important to know what *cannot be computed* and the dangers of what can be miscomputed.

These limitations can be seen most plainly in the context of mathematical and scientific practice. Perhaps the most significant use of the computer in this context is as an experimental tool, sometimes even displacing the laboratory experiment altogether. One translates a scientific or mathematical problem into a simpler mathematical model and then uses the computational power of the computer to study particular cases of the general model. This approach has turned out to be very useful, particularly when the conditions for experiment in the usual sense or of precise calculation become impossibly difficult. The mystique of such practices has grown to such an extent that some speak of replacing Nature, an analog computer, by a newer and better model of a digitalized nature.

The drawbacks and dangers of such practices without a background of thorough critical analysis are equally clear. We must ask about the adequacy of the model, about the accuracy (not to say the meaningfulness) of the computational process, and, last but not least, about the representative character of the particular cases that one computes. Without serious cross-checks on these factors, we are left with yet another case of the zeroth law of the computer: *garbage in, garbage out*, particularly with serious scientific and mathematical problems that cannot be solved by computation as they stand. One replaces them by manageable problems, and the validity of the replacement is precisely the crucial question. It is the importance of this question that has led to pointed comments about the adjective *scientific* in the currently fashionable emphasis on programs for scientific computation on supercomputers.

These critical questions do not mean that we should neglect the computer as a tool in science and mathematics. They do point up a sometimes neglected fact—namely, that the computer is a difficult tool whose use must be studied and refined. Computers are *brute force* instruments; their effective use depends vitally on human insight and ingenuity. I intend here to emphasize the importance of the intellectual arts and insights that are or can be connected with the digital computer and its uses. These intellectual arts have a vital relation to the mathematical enterprise. They constitute a specialized and different way of applying classical mathematical ideas and techniques with radically new purposes in mind. Their vitality, both intellectual and practical, depends in an essential way upon a continuing contract with the central body of mathematical activity.

There is an interesting and slightly ironic aspect to the relationship between computer science and the central body of mathematics. Since the mid-nineteenth century, mathematicians and physical scientists have tended to see a dichotomy between mathematics that is *applicable* to the uses of physical modeling and calculation and another kind that is *not applicable*. The rules for this division have changed in recent years, with an ever-increasing diversity of mathematical themes and theories falling into the first category. Even so, the stereotype tends to persist, and some areas of active mathematical research—like algebraic number theory or mathematical logic—tend to be relegated to the second category. Yet it is precisely these areas, grouped together with various forms of combinatorics under the general label of *discrete mathematics*, that have turned out to be most vital in major areas of advance in computer science. The basic theoretical framework of computer science and the development of complexity of computation rest upon the foundation of mathematical logic.

The development of algorithms depends essentially upon combinatorics, number theory, and, most recently, on probabilistic models of a combinatorial type. The practical area of coding and cryptology is vitally dependent upon sharp results in number theory and algebraic number theory.

We should reject efforts to oppose the natural sciences to the "artificial" sciences. Human art and artifice are part of all the sciences, as is the confrontation with the objective realities beyond human will and control that we personify under the figure of Nature. Indeed, computer science in its necessary advance, seen today under such perspectives as parallel processing, artificial intelligence, and expert systems, and the whole family of problems subsumed under the label of computer systems and structure, can be seen as part of the general perspective of *complexity of organization*.

4. The Core of Mathematics

There is a danger in any form of discussion of the role of mathematics that emphasizes (as I have done) the active participation of new mathematical concepts and tools in the development of other scientific disciplines. Despite the strong emphasis on the *new*, it is far too easy to use such an approach as a prescription for future mathematicians simply to facilitate the interactions that I have described. Such a prescription would be a recipe for a massive failure, not only for the development of mathematics itself but also for the sciences. Such prescriptions are based upon the unconscious principles that *creativity* and *newness* in conceptual advance are always a matter of the past. The *autonomy* of mathematical research, in the sense of its freedom from any strong dependence upon the current processes of research in other disciplines and upon their rhythm of activity, has been one of the principal components of its creativity throughout its lengthy history since the Greeks. There are commonsense reasons why this creativity is important for the sciences: for example, when the advance of scientific understanding needs mathematical concepts, theories, or methods of calculation and argument, it is often essential that they be already available in a reasonably usable form.

Once mathematical problems are solved in any context, the solutions can be digested and turned to new uses in other contexts. Yet new mathematics (concepts, solutions, theorems, algorithms, proofs, calculations), if it is genuinely new, must be created by someone, and whoever does the job is a *mathematician* by definition. The task of the practitioner of another scientific discipline with respect to mathematics is to use it to understand and analyze the subject matter of that discipline, to *see through* the mathematics to the structure of the subject matter. From the fact that mathematics from the latter point of view *ought to be transparent*, one cannot draw the false conclusion that mathematics does not exist and needs no process of development in its own right.

It may seem like a paradox to some that I should introduce this strong affirmation of the essential autonomy of mathematics into a paper devoted to the interaction of mathematics and the sciences. This paradox is superficial. Any affirmation of interaction is only significant if the two sides of the interaction have a full-fledged separate existence and meaningfulness. In particular, we must affirm a central autonomous core of meaning in the mathematical enterprise if our thesis of strong interaction is to have full significance.

What is this core of meaning? I shall give a number of related answers in the form of programmatic definitions of *mathematics*. Each of these definitions points to important characteristics of mathematical practice, and each program leads to a slightly different perspective on that practice. It would take me too far afield to describe the interrelation of these perspectives and the tension between them. Suffice it to say that I am among those who believe in an essential unity of mathematics, though rejecting some of the dogmatic and oversimplified programs for achieving that unity by putting mathematics in a Procrustean bed and cutting off some of its limbs.

(1) Mathematics is the science of significant forms of order and relation.
(2) Mathematics is the science of the structure of possible worlds.
(3) Mathematics is the science of infinity.
(4) Mathematics is the science of the structure of complex systems.
(5) Mathematics is the study of the modeling of reality in symbolic form.

Each of these definitions taken by itself is a deep truth in the sense of Niels Bohr: its negation is also a deep truth. Taken jointly, they give us a reasonable perspective on the broad range of mathematics since the Renaissance. (Definitions 1 and 2 are due to Descartes and Leibniz, combined under the term *mathesis*, whereas definition 3, which was originated by Leibniz, was revived in modern times by Poincaré and Weyl.)

5. Applicable Mathematics

Mathematical research in its various forms is an enterprise of great vitality in the present-day world (although it is invisible to some outsiders). Despite its fundamental autonomy, the enterprise of advanced mathematical research has interacted strongly in the last two decades with various advances in the sciences. For the purposes of the present discussion, I present two kinds of evidence.

The first consists of taking a conventional breakdown of the principal active branches of contemporary mathematical research and inquiring in general terms whether these branches have interactions of the type described with the sciences. In the table of organization for the International Congress of Mathematicians in Berkeley, California, in the summer of 1986, we have such a breakdown in the division of the Congress into nineteen sections. Of these nineteen sections, we may set aside two (history of mathematics, teaching of mathematics) and ask about the applicability of the seventeen mathematical areas in this classification. Five (probability and mathematical statistics, mathematical physics, numerical methods and computing, mathematical aspects of computer science, applications of mathematics to nonphysical sciences) relate directly to the sciences and technology. Eight have direct relation in contemporary practice to theory and practice in the natural sciences (geometry, topology, algebraic geometry, complex analysis, Lie groups and representations, real and functional analysis, partial differential equations, ordinary differential equations and dynamical systems). The remaining four (mathematical logic and foundations, algebra, number theory, discrete mathematics and combinatorics) have an equally vital relation to computer science. There is no residue of mathematics that is fundamentally not applicable on this list.

The second kind of evidence is illustrated by the study of the soliton theory of the Korteweg-De Vries equation in the periodic case. The applications of algebraic geometry and complex analysis to the study of the Korteweg-De Vries equation under periodic boundary conditions not only contributed to the understanding of the physical model involved but reacted upon the disciplines involved. New ideas and methods in both mathematical disciplines arose from this interaction, resulting in the solution of classical problems in algebraic geometry and function theory. In an even more striking case, the young Oxford mathematician Simon Donaldson observed that if one combined the mathematical techniques developed for the study of the mathematical theory of gauge fields by Karen Uhlenbeck with the penetrating geometric attack upon the structure of four dimensional manifolds of Michael Freedman one could obtain a new and totally surprising geometric result in four dimensions. The result in question asserts that, unlike Euclidean spaces in every other dimension, four-dimensional Euclidean space possesses two systems of coordinates that are fundamentally different from one another.

These two cases illustrate a possibility that has turned into a current reality—that the strong mathematical attack upon mathematical problems raised in the context of development of scientific research can provide the occasion and stimulus for major conceptual advances in mathematics itself.

6. Historical Perspectives

To close, let us turn to the question of the future relation of mathematics and the sciences. We may recall another well-known saying of Niels Bohr: Prediction is difficult, especially of the future. Attempts to predict the future are hypotheses about the past and present. Let us formulate a hypothesis that we can check for coherence and accuracy against the past and present and then try to gauge its consequences for the future.

Our scientific tradition is inherited from the civilization of the ancient Greeks. It was there that the concept of science as a self-conscious structuring of objective lawful knowledge of the world (or, more strictly, of the hidden processes of the world) first arose. Although the Greeks investigated the full range of human experience, their achievement in creating permanent scientific knowledge was primarily in the mathematical sciences, in mathematics itself and in such highly mathematical disciplines as mathematical planetary astronomy, musical theory, and the mathematical treatment of statics. The Greeks created a highly perfected form of sophisticated mathematical theory treating of whole number, geometry, ratio, and geometrical measure. In this theory, they also perfected a fully mature concept of mathematical argument, of logical deduction. On the basis of these achievements, Plato might argue in his celebrated dialogue *Timaeus* for a mathematical *myth* of the cosmos and its formation on the basis of geometric elements, and Aristotle could formulate the logical principles of deduction while rejecting the possibility of mathematical laws for the phenomena of terrestrial physics.

It is fashionable to talk of scientific revolutions. On the most fundamental level, there has been only one scientific revolution—that of the seventeenth century, in which modern science was formed. The concept of science that this century produced gave a description of the cosmos, the physical universe, in terms of the geometry of space and numerical relations, a description that applied to both the skies and the earth. It saw this cosmos as a realm of objective lawful relations, devoid of human agency or affect. Reality was separated after Descartes into two completely distinct parts, the physical universe and a separate world of human consciousness and spirit. In this framework, it made total sense for human consciousness to try to determine the secrets of natural processes not by passive observation but by transforming nature by experiment.

There was a mathematical counterpart of the new physical science, which served both as its precursor and principal tool. This was the mathematics of the new algebra and of the analytic movement of Vieta and Descartes, a mathematics that substituted calculation and manipulation of symbolic expressions for the deductive sophistication of the Greeks. It substituted the analysis of complex phenomena into simple elements for the Greek emphasis on deduction. In the seventeenth century, this new mathematics had two overwhelming triumphs. The first was the creation of an analytic geometry through which the geometric structure of space could be transformed by coordinatization into the subject matter of algebraic analysis. The second was the invention of the great analytic engine of the differential and integral calculus, by which the sophisticated and difficult arguments by exhaustion of Eudoxos and Archimedes for handling infinite processes were replaced by much simpler and more manageable algebraic formulae or *calculi*. This was the tool with which Newton built his great mathematical world-machine, the central paradigm for the scientific world pictures of all succeeding ages.

There are essentially two forms in which objective human knowledge can be formulated: in words and in mathematical forms. Aristotle opted for the first and created a systematic description of the world in which the subject-predicate form of the sentence was transformed into the pattern of the individual object or substance possessing a certain quality. From the seventeenth century on, modern science has rejected this form of description and replaced it by descriptions in various mathematical forms. These forms have altered as the stock of mathematical forms has increased and become richer and more sophisticated. The original forms were geometric, in the style of the Greeks. In the Renaissance, a new and more flexible concept of number, the 'real' number in the present-day sense, came into being as the common measure of lengths, areas, volumes, masses, and so forth, without the precise distinction between these measures in terms of geometric form to which the Greeks had held. In the ensuing development of algebra, new kinds of 'numbers' appeared as the solutions of algebraic equations. Since they were not numbers in the old sense, some were called 'imaginary', and mixtures of the two types were called 'complex'. It was not until the end of the eighteenth century that these 'complex' numbers were fully naturalized as members of the commonsense mathematical realm by being identified in a simple way with the points of a Euclidean plane, the complex plane.

Since the seventeenth century, the enterprise of the scientific description of nature has continued to develop within this mathematical medium, which was dimly foreshadowed by Plato's mathematical world myth. New scientific disciplines developed, and they entered the same framework of numerical relationship, geometric form in space, and formulation of basic principles in mathematically expressed laws. As Kant put it in a well-known aphorism from his *Metaphysical Foundations of Natural Sciences*: "In every special doctrine of nature, only so much science proper can be found as there is mathematics in it."

In the nearly four centuries that have elapsed since Galileo began the seventeenth century scientific revolution, the curious relationship of autonomy and mutual dependence between the natural sciences and mathematics has taken ever more complex and sophisticated forms. The mathematical medium in which the various sciences live has continued to develop and take on new shapes. In the early nineteenth century, the intuitive concept of symmetry that was applied to the study of the roots of algebraic equations gave rise to the concept of group. The group concept, passing through the medium of its application to geometry and differential equations, became in the twentieth century the most essential building block of the fundamental description of the physical universe. The concept of space, enriched by the insights of Gauss and Riemann, gave rise to the richer geometric concepts of Riemannian manifold and of curvature, through which the theory of general relativity of Einstein was to describe the cosmos. Through the analysis of integral equations and differential equations in the early twentieth century, the concept of an infinite-dimensional vector space was born, and the especially rich concept of a Hilbert space. Hermitian operators on a Hilbert space with their spectral theory served as the eventual underpinning of the the formal structure of quantum mechanics. These are three important examples of a very broad phenomenon.

New concepts and theories arise in mathematical research through the pressure of the need to solve problems and create intellectual tools through which already existing mathematical theories and structures can be extended and applied. Once the new concepts and theories are established, they themselves become the focus of intensive investigation. The new is achieved by mathematical imagination, applied through the medium of mathematical constructions through which the new concepts and structures are given definite form. Although the imaginative process is free in some ultimate sense, its result once produced becomes a new objective realm of relationship of a determinate character. Classical tools like deduction and calculation are used to establish its properties, leading to new technical problems that may eventually demand new concepts and constructions for their solution. The jump of insight and imagination that leads to new mathematical breakthroughs belies the stereotypes of mathematical activity as an automatic, machinelike process of mechanical application of formal rules.

Mathematical research as a whole balances the *radical* process of generation of new concepts and theories with the *conservative* tendency to maintain in existence all those domains, problems, and conceptual themes that once became established as foci of significant mathematical research. The balance between these opposing tendencies gives rise to the striking fact that, at the same moment, one can find active research programs of apparently equal vitality bearing on two themes, one of which is two thousand years old and the other perhaps only a decade old. Yet the two-thousand-year-old problem might well be solved with tools and concepts of relatively recent vintage.

The richer the repertoire of modern mathematical research, the broader the arsenal of concepts and tools available for the use of the mathematicized sciences. The difficulty lies in the problem of communication, of the scientific practitioners being able to penetrate through the difficulties of translation between the languages of different disciplines, of knowing what is relevant in the concepts and techniques that are available.

As the concerns and principal foci of scientific interest move into domains ever further from the classical ones of theory and experience, the role of mathematical ideas and techniques inevitably grows because they often provide the only tools by which one can probe further into the unknown. This is particularly true for domains involving complexity of organization or nonlinearity of interaction, the future frontier of the major themes of scientific advance. Although they may become the subject matter of major scientific disciplines in their own right, I doubt that this will lead to the disappearance of professional differences between specialists in various disciplines in attacking these scientific problems. The difference between specialties has a positive function as well as negative consequences. Specialists can rely upon the intellectual traditions and resources of their scientific specialty, and this applies with the greatest force to the mathematician. We can ask for a broader and more effective effort at communication, however, among those concerned with common problems, and we can cultivate an active interest in and sympathy with the thematic concerns of other specialties than our own.

Notes

1. In *The Works of the Mind*, ed. Heywood and Nef (Chicago: University of Chicago Press, 1945); reprinted in *The World of Mathematics*, ed. J. Newman, vol. 4 (New York: Simon and Schuster, 1956), pp. 2053-63.

2. In *Semi-Centennial Addresses of the American Mathematical Society*, 1938.

Mathematical Naturalism

Virtually all the discussion of the "philosophy of mathematics" in our century has been concerned with the enterprise of providing a foundation for mathematics. There is no doubt that this enterprise has often been mathematically fruitful. Indeed, the growth of logic as an important field within mathematics owes much to the pioneering work of scholars who hoped to exhibit the foundations of mathematics. Yet it should be almost equally obvious that the major foundational programs have not achieved their main goals. The mathematical results that they have brought forth seem more of a piece with the rest of mathematics than first points from which the entire edifice of mathematics can be built.

Under these circumstances, it is tempting to echo a title question of Dedekind's and to ask what the philosophy of mathematics is and what it ought to be. Many practicing mathematicians and historians of mathematics will have a brusque reply to the first part of the question: a subject noted as much for its irrelevance as for its vaunted rigor, carried out with minute attention to a small number of atypical parts of mathematics and with enormous neglect of what most mathematicians spend most of their time doing. My aim in the present essay is to offer an answer to the second half of the question, leaving others to quarrel about the proper response to the first. I shall argue that there are good reasons to reject the presuppositions of the foundationalist enterprises and that, by doing so, we obtain a picture of mathematics that raises different—and, to my mind, more interesting—philosophical problems. By trying to formulate these problems clearly, I shall try to draw up an agenda for a naturalistic philosophy of mathematics that will, I hope, have a more obvious relevance for mathematicians and historians of mathematics.

1. Epistemological Commitments

Seeking a foundation for a part of mathematics can make exactly the same sense as looking for a foundation for some problematic piece of scientific theory. At some times in the history of mathematics, practitioners have self-consciously set themselves the task of clarifying concepts whose antecedent use skirted paradox or of systematizing results whose connections were previously only dimly perceived. Weierstrass's efforts with concepts of convergence and Lagrange's explanation of the successes of techniques for solving cubic and quartic equations are prime examples of both forms of activity. But the grand foundational programs move beyond these local projects of intellectual slum clearance.[1]

Foundationalist philosophies of mathematics bear a tacit commitment to apriorist epistemology. If mathematics were not taken to be a priori, then the foundational programs would have point only insofar as they responded to some particular difficulty internal to a field of mathematics. Subtract the apriorist commitments and there is no motivation for thinking that there must be some first mathematics, some special discipline from which all the rest must be built.

Mathematical apriorism has traditionally been popular, so popular that there has seemed little reason to articulate and defend it, because it has been opposed to the most simplistic versions of empiricism. Apriorists come in two varieties. Conservative apriorists claim that there is no possibility of obtaining mathematical *knowledge* without the use of certain special procedures: one does not know a theorem unless one has carried out the appropriate procedures for gaining knowledge of the axioms (enlightenment by Platonic intuition, construction in pure intuition, stipulative fixing of the meanings of terms, or whatever) and has followed a gapless chain of inferences leading from axioms to theorem. Frege, at his most militant, is an example of a conservative apriorist. By contrast, liberals do not insist on the strict impossibility of knowing a mathematical truth without appealing to the favored procedures. Their suggestion is that any knowledge so obtained can ultimately be generated through the use of a prior procedures and that mathematical knowledge is ultimately improved through the production of genuine proofs.

Empiricists and naturalists[2] dissent from both versions of apriorism by questioning the existence or the power of the alleged special procedures. Insofar as we can make sense of the procedures to which apriorist epistemologies make their dim appeals, those procedures will not generate knowledge that is independent of our experience. Platonic or constructivist intuition, stipulative definition, yield knowledge—to the extent that they function at all—only against the background of a kindly experience that underwrites their deliverances. From the naturalistic perspective, apriorists have misjudged the epistemological status of the features they invoke, supposing that processes that are analogous to the heuristic arguments or *Gedankenexperimente* of the natural scientist are able to warrant belief come what may.

The contrast may be sharpened by considering Zermelo's introduction of his set-theoretic axioms.[3] For the conservative apriorist, there was no set-theoretic knowledge prior to 1905 and, in consequence, no knowledge of analysis, arithmetic, or any other part of mathematics before the pure cumulative hierarchy beamed in on Zermelo's consciousness. Liberals may relax this harsh judgment about the pre-Zermelian ignoramuses, but they remain committed to the view that mathematical knowledge underwent a dramatic transformation when the "merely empirical" justifications of the Greeks, of Fermat, Newton, Euler, Gauss, Cauchy, and Weierstrass finally gave way, in the first decade of our century, to genuine proof.

Mathematical naturalism opposes to both positions a different philosophical picture. Zermelo's knowledge of the axioms that he introduced was based on his recognition of the possibility of systematizing the prior corpus of claims about sets, claims that had been tacitly or explicitly employed in reasoning about real numbers. Zermelo proposed that these antecedently accepted claims could be derived from the principles he selected as basic. The justification is exactly analogous to that of a scientist who introduces a novel collection of theoretical principles on the grounds that they can explain the results achieved by previous workers in the field.

Strictly speaking, one might accept this picture of Zermelo's knowledge without subscribing to mathematical naturalism. Perhaps there are some who believe that *Zermelo* may have known the axioms of his set theory by recognizing their ability to systematize previous mathematical work but that his successors did better. Their consciousnesses have been illuminated by the beauties of the cumulative hierarchy. Mathematical naturalists take this prospect to be illusory. Not only is Zermelo's knowledge to be understood in the way that I have suggested but—in a fashion that I shall describe in more detail below—the same type of justification is inherited by those of us who come after him.

Yet even this much may be conceded without adopting naturalism. There is a hybrid position, with which Russell and Whitehead seem briefly to have flirted,[4] according to which our knowledge of the principles of set theory is based on our knowledge of arithmetic, even though the principles of arithmetic are themselves taken to be a priori. Mathematical intuition, stipulative definition, or whatever, is supposed to give us a priori knowledge of the Peano postulates and of all the theorems we can deduce from them; however, we know the axioms of set theory by seeing that they suffice to systematize the Peano postulates.

Mathematical naturalism embodies a ruthless consistency. Having set its face against the procedures to which a priori epistemologies for mathematics have appealed, it will not allow that those procedures operate by themselves to justify any mathematical beliefs possessed by any mathematician, past or present. I have tried elsewhere to give reasons for this principled objection against the varieties of intuition and apprehension of meanings on which apriorist accounts have traditionally relied—usually without much explicit epistemological commentary.[5]

Here I shall only offer the sketch of a motivating argument. Many people will have no hesitation in supposing that mathematicians know the statements they record on blackboards and in books and articles. Perhaps there are some who think of the practice of inscribing mathematical symbols as a contentless formal game, and who feel squeamish in talking about mathematical "knowledge." Yet, even for those who adopt this position, there must be an analog to the epistemic concept of justification. Inscriptional practices can be performed well or badly, usefully or pointlessly, and our common view is that the mathematical community has some justification for proceeding in the way that it does.

Now it seems that there can be a mathematical analog of the distinction between knowledge and conjecture (or ungrounded belief). Sometimes a professional mathematician begins with a conjecture and concludes with a piece of knowledge. Sometimes completely uninitiated people guess correctly. If two friends who know barely enough of number theory to understand some outstanding problems decide to divide up the alternatives, one claiming that Goldbach's conjecture is true, the other that it is false, and so forth, then there will not be a sudden advance in mathematical knowledge. Mathematical knowledge, like knowledge generally, requires more than believing the truth. (Similar distinctions can be drawn from the perspective of the approach that takes mathematical statements to be contentless. I shall henceforth ignore this approach and leave it to the interested reader to see how the argument would be developed from the preferred point of view.)

The obvious way to distinguish mathematical knowledge from mere true belief is to suggest that a person only knows a mathematical statement when that person has evidence for the truth of the statement—typically, though not invariably, what mathematicians count as a proof.[6] But that evidence must begin somewhere, and an epistemology for mathematics ought to tell us where. If we trace the evidence for the statement back to the acknowledged axioms for some part of mathematics, then we can ask how the person knows those axioms.

In almost all cases, there will be a straightforward answer to the question of how the person learned the axioms. They were displayed on a blackboard or discovered in a book, endorsed by the appropriate authorities, and committed to the learner's memory. But nonnaturalistic epistemologies of mathematics deny that the axioms are known because they were acquired in this way. Apriorists offer us the picture of individuals throwing away the props that they originally used to obtain their belief in the axioms and coming to know those axioms in special ways. At this point we should ask a series of questions. What are these special ways of knowing? How do they function? Are they able to produce knowledge that is independent of the processes through which the beliefs were originally acquired? One line of naturalistic argument consists in examining the possibilities and showing that the questions cannot be answered in ways that are consistent with apriorism.[7]

There is a second, simpler way to argue for a naturalistic epistemology for mathematics. Consider the special cases, the episodes in which a new axiom or concept is introduced and accepted by the mathematical community. Naturalists regard such episodes as involving the assembly of evidence to show that the modification of mathematics through the adoption of the new axiom or concept would bring some advance in mathematical knowledge. Often the arguments involved will be complex—in the way that scientific arguments in behalf of a new theoretical idea are complex—and the pages in which the innovators argue for the merits of their proposal will not simply consist in epistemologically superfluous rhetoric. On the rival picture, the history of mathematics is punctuated by events in which individuals are illuminated by new insights *that bear no particular relation to the antecedent state of the discipline.*

To appreciate the difference, and the merits of naturalism, consider Cauchy's introduction of the algebraic concept of limit in the definitions of the concepts of convergence, continuity, and derivative. If we suppose that Cauchy's claims were backed by some special a priori insight, then it is appropriate to ask why this insight was unavailable to his predecessors, many of whom had considered the possibility of employing the concept of limit to reconstruct fundamental arguments in the differential calculus. We seem compelled to develop an analogy with superior powers of visual discrimination: Cauchy just had better powers of mathematical intuition, so he saw what Lacroix, Lagrange, l'Huilier, d'Alembert, Euler, Maclaurian, Leibniz, and Newton had all missed. This far-fetched story is not only unnecessary, it also fails to do justice to the argumentative structure of Cauchy's work on the calculus. The *Cours d'Analyse* displays how a thorough use of the algebraic limit concept can be employed to reformulate problematic reasoning in the calculus and thus to prepare the way for the resolution of outstanding problems. Cauchy does not invite his fellow mathematicians to intuit the correctness of his new claims. He shows, at some length, how they are useful in continuing the practice of mathematics.[8]

This second line of argument anticipates a theme that I shall elaborate in the next section. Mathematical naturalism offers an account of mathematical knowledge that does justice to the historical development of mathematics. Apriorism has seemed attractive because the only available rival seemed to be the simplistic empiricism of John Stuart Mill.[9] I claim that we can transcend a pair of inadequate alternatives by recognizing that mathematical knowledge is a historical product.

2. A Role for History

Mathematical knowledge is not built from the beginning in each generation. During the course of their education, young mathematicians absorb the ideas accepted by the previous generation. If they go on to creative work in mathematics, they may alter that body of ideas in ways that are reflected in the training of their successors. Like any other part of science, mathematics builds new knowledge on what has already been achieved. For the epistemologist of mathematics, as for the epistemologist of science, a crucial task is to identify those modifications of the corpus of knowledge that can yield a new corpus of knowledge.

We can now outline a naturalist account of mathematical knowledge.

Our present body of mathematical beliefs is justified in virtue of its relation to a prior body of beliefs; that prior body of beliefs is justified in virtue of its relation to a yet earlier corpus; and so it goes. Somewhere, of course, the chain must be grounded. Here, perhaps, we discover a type of mathematics about which Mill was right, a state of rudimentary mathematical knowledge in which people are justified through their perceptual experiences in situations where they manipulate their environments (for example, by shuffling small groups of objects). What naturalism has to show is that contemporary mathematical knowledge results from this primitive state through a sequence of rational transitions.[10]

A preliminary task is to replace the vague talk of "states of mathematical knowledge" with a more precise account of the units of mathematical change. The problem is formally parallel to that which confronts the philosopher concerned with the growth of scientific knowledge, and the solutions are also analogous. In both cases, I suggest, we should understand the growth of knowledge in terms of changes in a multidimensional unit, a *practice*, that consists of several different components.[11] Each generation transmits to its successor its own practice. In each generation, the practice is modified by the creative workers in the field. If the result is knowledge, then the new practice emerged from the old by a *rational interpractice transition*.

As a first analysis, I propose that a mathematical practice has five components: a language employed by the mathematicians whose practice it is, a set of statements accepted by those mathematicians, a set of questions that they regard as important and as currently unsolved, a set of reasonings that they use to justify the statements they accept, and a set of mathematical views embodying their ideas about how mathematics should be done, the ordering of mathematical disciplines, and so forth. I claim that we can regard the history of mathematics as a sequence of changes in mathematical practices, that most of these change are rational,[12] and that contemporary mathematical practice can be connected with the primitive, empirically grounded practice through a chain of interpractice transitions, all of which are rational.[13]

It is probably easier to discern the various components in a mathematical practice by looking at a community of mathematicians other than our own. As an illustration of my skeletal account, let us consider the practice of the British community of mathematicians around 1700, shortly after Newton's calculus had become publicly available.[14] This community adopted a language that not only lacks concepts belonging to contemporary mathematics but also contains concepts we no longer use: Newtonian concepts of fluxion, number, series, and function are, I think, different from anything that has survived to the present, although are contemporary concepts that are *similar* to these in certain respects. The British mathematicians of the time accepted a wide variety of claims about the properties of tangents to curves, areas under curves, motions of bodies, and sums of infinite series. Many of these statements can be translated into results that we would now accept, although some of the suggestions about infinite series might give us pause. The mathematicians posed for themselves the problems of achieving analogous claims for a broader class of curves and motions; they did not endorse (as their Leibnizian contemporaries did) the general question of finding canonical algebraic representations for the integrals of arbitrary functions or for the sums of infinite series; problems in algebra arose out of, and were to be interpreted in terms of, problems in geometry and kinematics. Members of the community were prepared to justify some of their claims by offering geometric proofs—which they took to be strict synthetic demonstrations in the style of traditional geometry—but, in some cases, they were forced to rely on reasoning that appealed to infinitesimals. Because of their background metamathematical views, reasoning of this kind appeared less than ideal. On the Newtonian conception of mathematics, the fundamental mathematical disciplines are geometry and kinematics, and there is a serious foundational problem of showing how infinitesimalist justifications can either be recast as or replaced by reasonings in proper geometric style.

The central part of a naturalistic account of mathematical knowledge will consist in specifying the conditions under which transitions between practices preserve justification. As I have tried to show elsewhere,[15] the Newtonian practice just described can be connected with the practice of the late-nineteenth-century community of analysts by a series of transitions that are recognizably rational. But it is important to acknowledge that my previous efforts at articulating naturalism with respect to this example fall short of providing a full account of rational interpractice transitions in mathematics. We can appreciate rationality in the growth of mathematical knowledge by describing episodes in the history of mathematics in a way that makes clear their affinity with transitions in science (and in our everyday modification of our beliefs) that we count as rational. A *theory* of rationality in mathematics (or, more precisely, in the growth of mathematics) would go further by providing principles that underlie such intuitive judgments, making clear the fundamental factors on which the rationality of the various transitions rests.

By emphasizing the connection between transitions in the history of mathematics and transitions in the history of science that we view as rational, it is possible to argue for a reduction of the epistemology of mathematics to the epistemology of science. In both cases it will be important to distinguish two main types of transition. The first type justifies some change in some component (or components) of a prior practice on the basis of the state of that practice alone. I shall call such transitions *internal transitions*. The modifications of the second type, the *external transitions*, owe their rationality to something outside the prior state of mathematics, some recent experience on the part of the those who make the change or some feature of another area of inquiry. It is tempting to believe that the natural sciences are altered through external transitions, and through these alone, and that mathematics grows solely through internal transitions. The temptation should be resisted. Theoretical sciences are often changed in dramatic ways through attempts to resolve tensions in the prevailing practice—think of Copernicus's original efforts to resolve the problem of the planets or the development of population genetics in the early 1930s. By the same token, the development of mathematics is sometimes affected by the state of other sciences, or even by ordinary experience. To cite just one example, the pursuit of analysis in the early nineteenth century was profoundly modified through the study of problems in theoretical physics. I shall consider the importance of external transitions in mathematics in greater detail below.

The insight that underlies the tempting—but oversimple—thesis is that there is a continuum of cases. Some areas of inquiry develop primarily through external transitions, others grow mainly through internal transitions. What is usually called "pure" mathematics lies at one end of the continuum; the applied sciences (for example, metallurgy) lie at the other. We can appreciate the fact that there are differences of degree without making the mistake of believing that there are differences in kind.

Because internal transitions are obviously important in mathematics, it is useful to consider some main patterns of mathematical change, to consider how they fit into the scheme that I have outlined, and to see how they are exemplified in the history of mathematics. Unresolved problems that have emerged as significant frequently provide the spur for the further articulation of a branch of mathematics. New language, new statements, and new forms of reasoning are introduced to solve them. In at least some cases, the newly introduced language is initially ill understood, the workings of the new reasonings may be mysterious, and there may be legitimate doubts about the truth of the new statements. However, when things turn out well, the unclarities are ultimately resolved and the collection of answers is eventually systematized by principles that accord with prevailing metamathematical criteria. Moreover, we can easily understand prominent ways in which new questions are generated in mathematics: as new language is introduced or old problems are resolved, it becomes rational to pose new, often more general, questions.

Consider many famous examples, each of which counts as an internal interpractice transition. Descartes's introduction of the concepts of analytic geometry was motivated by the ability of the new concepts and modes of reasoning to yield answers to questions about locus problems, answers that were recognizably correct and recognizably more general than those attained by his predecessors. Mysterious though the language of the Leibnizian calculus may have been, its defenders could point out, as the Marquis de l'Hospital did, that it enabled them to solve problems that "previously no one had dared to attempt." Lagrange's use of concepts that we see as precursors of group-theoretic ideas was justified through its ability to bring order to the ramshackle collection of methods for solving cubic and quartic equations that had been assembled by earlier mathematicians.

In each case, the new extension generated new problems. After Descartes, it was rational to seek techniques for associating recalcitrant curves with algebraic equations and for computing the "functions" of curves.[16] Through their use of infinite series, Newton and Leibniz generated new families of mathematical questions: What is the series expansion of a given function? What is the sum of a given infinite series? This process of question generation illustrates a general pattern that has been repeated countless times in the history of function theory: given two schemes for representing functions, one asks how the two are coordinated, typically by seeking canonical representations in well-established terms for functions that are defined in some novel fashion. Similarly, the combination of two discoveries in the theory of equations—the proof of the insolubility of the quintic and Gauss's demonstration that the cyclotomic equation, $x^p - 1 = 0$, is soluble when p is prime—posed the question Galois answered: Under what conditions is an equation soluble in radicals?

All these episodes can be represented within the framework I have sketched, and when they are so represented we see interpractice transitions that are recognizably rational. So much suffices for the reduction of the epistemology of mathematics to the epistemology of science. But it is only the first step toward a complete naturalistic epistemology. Eventually our confident judgments of rationality should be subsumed under principles that explain the rationality of the transitions we applaud. Instead of responding to a skeptic by pointing to the details of the episode, saying "Look! Don't you see that what is going on when Galois poses this problem or when Leibniz introduces this reasoning is just like what goes on in any number of scientific situations?", we would strive to identify the crucial epistemic features of the situation and to incorporate them within a general theory of rational interpractice transitions, a theory that would cover both the mathematical and the scientific cases.

When we see the problem, and our current ability to tackle it, in this way, we should realize that there is presently no prospect of resolving hard cases. Consider the current situation in the foundations of set theory. (Set theory deserves to count as *a* subject for philosophers of mathematics to explore, even if it is not the only such subject.) There are several different proposals for extending the standard collection of set-theoretic axioms. Ideally, a naturalistic epistemology of mathematics would enable us to adjudicate the situation, to decide which, if any, of the alternatives should be incorporated into set theory.[17] If my naturalistic reduction of the philosophy of mathematics is correct, then the problem has been properly defined. But it has not been solved.[18]

Mathematical naturalism thus identifies a program for philosophical research. Apriorists will not find the program worth pursuing because they will suppose that the genuine sources of mathematical knowledge lie elsewhere. (But, with luck, the naturalistic challenge may prompt them to be more forthcoming about exactly where.) However, even those who join me in rejecting apriorism may maintain that naturalism is wrongheaded. For they may perceive a deep difficulty in understanding the rationality of mathematics—either because they are skeptical about the rationality of science or because they think that the conditions that allow the rationality of science are absent in the mathematical case. The objector speaks:

Naturalists take the reconstruction of our mathematical knowledge to consist in a selective narration of parts of the history of mathematics, a narration designed to give us comfortable feelings about the innovative ideas of the great mathematicians. But why should we count the end product as knowledge? What is it about these transitions that makes us dignify them with the name *rational*? Even if we had a clear set of principles that subsumed the patterns of interpractice transition to which the label is attached, it would still be incumbent on the naturalist to explain why they are to be given a preferred epistemic status.

There is a tangle of problems here, among which is a cousin of Humean worries about the grounds of methodological principles. In the rest of this essay, I shall try to confront the problems, showing that they may help us to clarify the naturalistic picture and that some interesting and unanticipated conclusions emerge.

3. Varieties of Rationality

Those who offer theories of the growth of scientific knowledge must answer two main problems. The first of these, the problem of progress, requires us to specify the conditions under which fields of science make progress. The second, the problem of rationality, requires us to specify the conditions under which fields of science proceed rationally. Everyone ought to agree that the two problems are closely connected. For a field to proceed rationally, the transitions between states of the field at different times must offer those who make the transitions the best available strategies for making progress. Rationality, as countless philosophers have remarked, consists in adjustment of means to ends. Our typical judgments of rationality in discussions about scientific change or the growth of knowledge tacitly assume that the ends are epistemic. When we attribute rationality to a past community of scientists, we consider how people in their position would best act to achieve those ends that direct all inquiry, and we recognize a fit between what was actually done and our envisaged ideal.

So I start with a general thesis, one that applies equally to mathematics and to the sciences. Interpractice transitions count as rational insofar as they maximize the chances of attaining the ends of inquiry. The notion of rationality with which I am concerned is an absolute one. I suppose that there are some goals that dominate the context of inquiry, that are not goals simply because they would serve as stepping stones to yet further ends. The assumption is tricky. What are these goals? Who (or what) is it that stands a chance of attaining them? In what ways should the chances be maximized? Finally, are there special mathematical goals, or does mathematics serve us in our attempts to achieve more general ends?

Following Kant, I take it that there are ends of rational inquiry and that to be a rational inquirer is to be a being who strives to achieve these ends.[19] The ends I identify as ultimate are the achieving of truth and the attainment of understanding. If we ask a mathematician why a certain style of proof has been widely adopted, or why a particular systematization of a body of mathematical knowledge has been proposed and accepted, then we shall expect to be told that the proof-pattern is likely to issue in the acceptance of true conclusions and that the favored axiomatization yields understanding of the body of mathematical theory that it systematizes. Similar responses are anticipated if we raise questions about the value of certain mathematical concepts or the urgency of some mathematical questions: the former help us to appreciate the interrelationships among mathematical claims, the latter signal gaps in our current understanding. But, if we continue by asking why we should be interested in true conclusions or in understanding, we can only answer that these are the ends of inquiry.

The ends of rational inquiry are not our only ends, and, in consequence, they do not completely dictate the course of rational development of mathematics. Our aims also include the goals of providing for the welfare of present and future members of our species (and perhaps members of other species as well), of securing free and just social arrangements, and so forth. The relations between the growth of science and these practical ends are typically tenuous and indirect, and the connection with mathematics is even more attenuated. Nevertheless, if asked to justify the pursuit of a particular collection of mathematical problems, we may reply that these problems arise within the context of a particular scientific inquiry. That scientific inquiry, in its turn, may be motivated by the desire for rational ends of scientific inquiry—greater understanding of some facet of the universe, say of the structure of matter or of the springs of animal behavior. Or it may be justified through its contribution to some practical project, the securing of a steady supply of food for a group of people or the improvement of transmission of information or resources around the world. Finally, that practical project, in its turn, may connect directly with our ultimate practical ends.

My division between ends of rational inquiry and practical ends obviously relates to the previous distinction between internal and external rational interpractice transitions. More exactly, an internal interpractice transition is rational in virtue of its advancement of the ends of rational inquiry *in mathematics*. An external interpractice transition is rational in virtue of its advancement either of the ends of rational inquiry in some other branch of knowledge or of some practical ends.[20]

When we consider the rationality of mathematics and science as a problem about the adjusting of means to ends, then it becomes evident that judgments of rationality are highly ambiguous. I want to distinguish four main senses:

(1) Individual epistemic rationality: an individual is epistemically rational when he or she adjusts practice so as to maximize the probability that he or she will attain his or her epistemic ends.

(2) Individual overall rationality: an individual is rational overall when that individual adjusts practice so as to maximize the probability that he or she will attain his or her total set of ends.

(3) Community epistemic rationality: a community is epistemically rational when the distribution of practices within it maximizes the probability that the community will ultimately attain its epistemic ends.

(4) Community overall rationality: a community is rational overall when the distribution of practices within it maximizes the probability that the community will ultimately attain its total set of ends.

I claim that these notions of rationality are quite distinct and that it is important to know which we have in mind before we inquire about the rationality of mathematics or of some other science.[21]

To see the difference between the individual and community perspectives, consider the following kind of example. Workers in a scientific field seek the answer to a certain question, with respect to which various methods are available. Considering the decision for each individual scientist, we quickly reach the conclusion that, if there is one method that is recognizably more likely to lead to the solution, then each scientist ought to elect to pursue that method. However, if all the methods are roughly comparable, then this is a bad bargain from the point of view of the community. Each individual ought to prefer that he or she belong to a community in which the full range of methods is employed. Hence, from the community perspective, a rational transition from the initial situation would be one in which the community splits into subgroups, each homogeneous with respect to the pursuit of method. The intuition is that the ends of the community can be better advanced if some of the members act against their own epistemic interests.[22]

Although the points I have been making may seem to have little to do with the growth of mathematical knowledge, they enable us to resolve an important type of disagreement among historians and philosophers of mathematics. Consider the different research strategies pursued by British and Continental mathematicians in the years after the priority dispute between Newton and Leibniz.[23] Leibnizians confidently set about using new algebraic techniques, vastly increased the set of problems in analysis, and postponed the task of attempting to provide a rigorous account of their concepts and reasonings. Their attitude is not only made explicit in Leibniz's exhortations to his followers to extend the scope of his methods, without worrying too much about what the more mysterious algebraic maneuvers might mean, but also in the acceptance of results about infinite series sums that their successors would abandon as wrongheaded.[24] Insofar as they were concerned to articulate the foundations of the new mathematics, the Leibnizians seem to have thought that the proper way to clarify their concepts and reasonings would emerge from the vigorous pursuit of the new techniques. In retrospect, we can say that their confidence was justified.

By contrast, Newton's successors were deeply worried about the significance of the symbols that they employed in solving geometric and kinematic problems. They refused to admit into their mathematical work questions or modes of reasoning that could not be construed in geometric terms, and they lavished attention on the problem of giving clear and convincing demonstrations of elementary rules for differentiating and integrating (to speak somewhat anachronistically). Some of the ideas that emerged from their work were ultimately made central to analysis in the reconstruction offered by Cauchy.[25]

Now, we ask, "Which, if either, of these developments was rational?" Before we answer the question, it is important to be aware of the distinctions that I have drawn above. When we demand an individualistic assessment of the rationality of the rival interpractice transitions, we seem compelled to engage in a difficult cost-benefit analysis. To a first approximation, we can view both groups of mathematicians as aiming at two kinds of epistemic ends: proliferation of answers to problems and increased understanding of those answers, the concepts they contain, and the methods by which they are generated. Newtonians emphasize the need for clarity first, holding that secure results in problem solving will be found once the concepts and reasonings are well understood. Leibnizians suggest that the means of clarification will emerge once a wide variety of problems have been tackled. Each tradition is gambling. What are the expected costs and benefits?

Those who are uneasy, as I am, about attributing rationality to one tradition and denying it to the other have an obvious first response to the situation. They may propose that the alternative strategies were initially so close in terms of their objective merits that it would have been perfectly reasonable to pursue either of them. Hence there can be no condemnation of either the early Newtonians or the early Leibnizians. Neither group behaved irrationally. Yet this proposal is unsatisfactory as it stands. For it is hard to extend it to account for the continued pursuit of the Newtonian tradition once it had become apparent that the Continental approach stemming from Leibniz and the Bernoullis was achieving vast numbers of solutions to problems that the Newtonians also viewed as significant (problems that could readily be interpreted in geometric or kinematic terms, even though the techniques used to solve them could not). Moreover, historians who like to emphasize the role of social factors in the development of science will note, quite correctly, that the national pride of the British mathematicians and the legacy of the dispute over priority in the elaboration of the calculus both played an important role in the continued opposition to Continental mathematics.

I suggest that we can understand why British mathematicians doggedly persisted in offering clumsy and often opaque geometric arguments if we recognize the broader set of goals that they struggled to attain. Among these goals was that of establishing the eminence of indigenous British mathematics, and we can imagine that this end became especially important after the Hanoverian succession and after Berkeley's clever challenge to the credentials of the Newtonian calculus.[26] Moreover, before we deplore the fact that some of Newton's successors in the 1740s and 1750s may have been moved by such nonepistemic interests as national pride, we should also appreciate the possibility that the maintenance of a variety of points of view (which chauvinism may sometimes achieve) can advance the epistemic ends of the community. The goals of promoting acceptance of truth and understanding in the total mathematical community were ultimately achieved (in the nineteenth century) because both traditions were kept alive through the eighteenth century. If, as I suspect, one of the traditions was maintained because some mathematicians were motivated by nonepistemic interests, then perhaps we should envisage the possibility that the deviation from individual epistemic rationality signals the presence of an institution in the community of knowers that promotes community epistemic rationality. A rational community of knowers will find ways to exploit individual overall rationality in the interests of maximizing the chances that the community will attain its epistemic ends.[27]

Only if we restrict ourselves to the notion of individual epistemic rationality and seek to find this everywhere in the history of mathematics (or of science generally) does the search for rationality in that history commit us to a Whiggish enterprise of distributing gold stars and black marks. The general form of an interpractice transition is more complex than we might have supposed, and the historiography of mathematics should reflect the added complexity. We are to imagine that the community of mathematicians is initially divided into a number of homogeneous groups that pursue different practices. In some cases, the claims made by members of the groups will be incompatible (Newtonians versus Leibnizians, Kronecker and his disciples against Dedekind and Cantor, militant constructivists against classical mathematicians). In other cases, they will just be different. An interpractice transition may modify the particular practices, the group structure, or both. Such interpractice transitions may be viewed from the perspective of considering whether (a) they maximize the chances that the individuals who participate in them will achieve their individual epistemic ends, (b) they maximize the chances that those individuals will attain their total set of ends with some epistemic ends being sacrificed to nonepistemic ends, (c) they maximize the chances that the community will attain its epistemic ends, or (d) they maximize the chances that the community will attain its total set of ends, with epistemic ends being sacrificed to nonepistemic ends. Moreover, with respect to each case we may focus on epistemic ends that are internal to mathematics or we may look to see whether the ends of other areas of inquiry are also involved, and, where (b) and (c) both obtain, we may look for institutions within the total mathematical community that promote the attainment of epistemic ends by the community at cost to the individual.

I have been concerned to stress the broad variety of questions that arise for the history of mathematics once we adopt the naturalistic perspective I have outlined and once we have extended it by differentiating notions of rationality. To the best of my knowledge, these questions remain virtually unanswered for virtually all of the major transitions in the history of mathematics.

4. The Ends of Inquiry

So far I have remained vague about the epistemic ends at which I take inquiry to be directed. The vagueness should provoke questions. Are there epistemic ends intrinsic to mathematics? If so, what are they? In other words, what counts as *mathematical* progress? How are the epistemic ends of mathematics, if there are any, to be balanced against the epistemic ends of other branches of inquiry or against nonepistemic (practical) ends? These are important issues for a naturalistic philosophy of mathematics to address. I shall try to show that they connect with questions that should concern historians of mathematics and even professional mathematicians.

Compare paradigm cases of internal and external interpractice transitions. For the former, imagine that a group of mathematicians introduces new concepts or new axioms for the sole purpose of improving their understanding of prior mathematical claims. For the latter, suppose that they decide to lavish great attention on methods for solution to complicated differential equations with important applications, or to develop broadly applicable techniques of approximation that allow practical problems to be solved with assignable error. In either case, we may applaud the transition for its contribution to some end. Yet the applause is consistent with a more global criticism. Perhaps the transition may be felt to strike the wrong balance between our epistemic and practical (or, more precisely, our extramathematical) ends. The gain in understanding was achieved by sacrificing opportunities to pursue a more significant goal, or the concern with the practical applications overshadowed a more important epistemic interest.

The recent history of mathematics has shown that the criticism can run either way. Critics of post-Weierstrassian analysis (and of the educational program that it influenced in the Cambridge mathematical Tripos of most decades of our century) may suggest that the emphasis on solving recondite problems involving special functions (originally introduced into mathematics because of their significance for certain problems in physics)[28] detracted from the achievement of insight into fundamental theorems. Detractors of contemporary mathematics might concede that the habit of discerning maximum generality, and the embedding of classical results in the broadest possible context, has yielded a collection of sterile investigations that should be abandoned in favor of projects more closely connected with the projects of the sciences and with everyday life.

To formulate these issues in a clear way we need a solution to the problem of mathematical progress. We need to know the epistemic ends at which mathematics aims. Let us begin with a liberal conception of progress. According to this conception, we must distinguish between epistemic and pragmatic appraisal. We make progress so long as we add to the store of knowledge in *any* way. Perhaps some additions are more profitable in advancing us toward the totality of our goals, but that is a purely pragmatic matter.

The liberal conception suggests that our applause for both our paradigm cases should not be tempered with criticism. To see if so ecumenical an attitude can be sustained, we need to consider what theses about the ontology of mathematics can be integrated with a naturalistic approach. There is a variant of ontological Platonism that is compatible with almost all my claims about mathematical knowledge. According to this variant, mathematical knowledge begins, in prehistoric times, with the apprehension of those structures that are instantiated in everyday physical phenomena. On the simplest version, we perceive the properties of small concrete sets (that is, sets whose members are physical objects).[29] Mathematics proceeds by systematically investigating the abstract realm, to which our rudimentary perceptual experiences give us initial access. That investigation is guided by further perception, by the uses of mathematics in the physical sciences, and by attempts to explain and to systematize the body of results that has so far been acquired. Platonists can simply take over my stories about rational interpractice transitions, regarding those transitions as issuing in the recognition of further aspects of the realm of abstract objects.

Mathematical naturalists who are also Platonists can readily draw the distinction between pragmatic appraisal and epistemic appraisal. The mathematician's task is to draw a map of Platonic heaven, and the acquisition of any "geographical" information constitutes progress.[30] In some cases (the embedding of classical results in an abstract and highly general context), we may learn about abstract features of that heaven that are of little practical value in steering ourselves around the earth. In other instances (the discovery of techniques for solving very special classes of differential equations), we may come to know details of considerable utility but of little general import. So long as we continue to amass truths, there is no question about making progress.

Now I think that it is possible to attack this liberal view of progress even within the Platonistic perspective. Those who see the natural science as aiming at truth do not have to suppose that *any* accumulation of truths constitutes scientific progress: insignificant truth is typically not hard to come by.[31] Thus my point is not that a naturalistic Platonist is compelled to adopt the liberal conception of progress, but that Platonism is one way of elaborating the liberal conception. However, I want to press a more radical challenge. I claim that we should reject the Platonist's view of the ontology of mathematics, substituting for it a picture that does not allow for the distinction between pragmatic assessment and epistemic assessment on which the liberal conception builds.

Many of the reasons for worrying about Platonism are familiar. How is reference to mathematical objects to be secured? How is it that perception provides us with knowledge about such objects? Even though these questions may be blunted by claiming that we initially refer to and know about *concrete* sets, it still remains mysterious how we are ultimately able to refer to and know about *abstract* sets.[32] Other anxieties arise in the particular context of a naturalistic approach to mathematical knowledge. Like other theoretical realists, Platonists must explain why our ability to systematize a body of results provides a basis for belief in the existence of antecedently unrecognized entities. It is not easy to understand how Lagrange's insights into the possibilities of leaving certain expressions invariant through the permutation of roots should constitute recognition of the existence of hitherto unappreciated mathematical objects—to wit, groups.

The general problem of understanding the historical transitions that have occurred in mathematics as revelations of the inhabitants of Platonic heaven arises with particular force when we consider a particular type of interpractice transition that is common in mathematics: the resolution of apparent incompatibility by reinterpreting alternatives. When mathematicians discovered that non-Euclidean geometries were consistent, the traditional vocabulary of geometry was reconstrued. On the Platonist's account, new objects—non-Euclidean spaces and their constituents—were discovered. By contrast, the resolution of incompatibilities in the sciences often proceeds by dismissing previously countenanced entities. The chemical revolution did not conclude by allowing for both Lavoisierian and non-Lavoisierian combustibles.

Focus on this last example may prompt the conventionalist reaction that mathematics simply consists in exploring the consequences of arbitrarily selected conventions and that the history of mathematics reveals the emergence of the conventions that have happened to interest past mathematicians. This is an overreaction to the pitfalls of Platonism. Interpractice transitions typically involve a nonarbitrary modification of antecedent concepts, reasonings, problems, and statements. Thus our task is to find an account of the ontology of mathematics that will avoid supposing that practices are altered in accordance with the whim of the moment, without lapsing into the idea that the alterations issue in the disclosure of truths about Platonic heaven.

I have suggested elsewhere that the problem can be solved if we treat mathematics as an idealized science of human operations.[33] The ultimate subject matter of mathematics is the way in which human beings structure the world, either through performing crude physical manipulations or through operations of thought. We idealize the science of human physical and mental operations by considering all the ways in which we could collect and order the constituents of our world if we were freed from various limitations of time, energy, and ability. One way to articulate the content of the science is to conceive of mathematics as a collection of stories about the performances of an ideal subject to whom we attribute powers in the hope of illuminating the abilities we have to structure our environment.

This proposal goes beyond conventionalism in placing restrictions on the stories we tell when we make progress in mathematics. Some stories, the stories of elementary mathematics, achieve the epistemic end of directly systematizing the operations we find ourselves able to perform on physical objects (and on mental representations of such objects). Others achieve the epistemic end of answering questions that arise from stories that achieve an epistemic end, or of systematizing the results obtained within stories that achieve an epistemic end. Ultimately, there must be a link, however indirect, to operations on our environment. Nevertheless, it would be wrong to think of the entire structure of mathematics as an attempt to systematize and illuminate the elementary operations that are described in the rudimentary portions of the subject. For, in an important sense, *mathematics generates its own content.* The new forms of mathematical notation that we introduce not only enable us to systematize and extend the mathematics that has already been achieved, but also to perform new operations or to appreciate the possibility that beings released from certain physical limitations could perform such operations. Such extensions of our repertoire seem to me to have occurred with the development of notation for representing morphisms on groups and for constructing sets of sets. In both cases, the notation is a vehicle for iterating operations that we would not be able to perform without it.[34]

This proposal (which I shall call *naturalistic constructivism*) resolves the difficulties that beset Platonism. There is no question of securing reference to or gaining perceptual knowledge of anything other than unproblematic entities—to wit, operations that we perform and recognize ourselves as performing. The transitions that occur in the history of mathematics are taken at face value: they consist in introducing concepts, statements, problems, and reasonings as parts of stories that help us understand the operations we are able to perform on our environments; in some cases, the activity of elaborating these stories itself generates new kinds of operations for later mathematics to consider. Finally, the instances in which an apparent disagreement is resolved through reinterpretation are cases in which the illumination of what has so far been achieved is obtained by embedding what appeared to be a single story within a range of alternatives.

Naturalistic constructivism collapses the notions of justification and truth in an interesting way. To say that a mathematical statement is true is to make a claim about the powers that are properly attributed to the ideal subject (or, more generally, to make a claim to the effect that the statement figures in a story that is properly told). What "properly" means here is that, in the limit of the development of rational mathematical inquiry, our mathematical practice contains that statement. Truth is what rational inquiry will produce, in the long run.

We can now see why naturalistic constructivism undercuts the distinction between pragmatic and epistemic appraisal. What we mean by "the limit of the development of rational mathematical inquiry" is the state to which mathematics will tend if we allow our entire body of investigations to run their course and to be guided by procedures designed to maximize our chances of attaining our ends. Since there is no independent notion of mathematical truth, the only epistemic end in the case of mathematics is the understanding of the mathematical results so far achieved. Mathematics proceeds autonomously through attempting to systematize whatever claims have previously been made about our powers to order and collect—including those that have been made possible for us through the activity of mathematics itself—and is inevitably dependent on other sciences and on practical concerns for the material on which it will work. Alternatively, we can say that the goal of mathematics is to bring system and understanding to the physical and mental operations *we find it worth performing* on the objects of our world, so that the shape and content of mathematics are ultimately dictated by our practical interests and the epistemic goals of other sciences.

If this is correct, then epistemic appraisal in mathematics inevitably involves pragmatic considerations. It is not only legitimate to reform a body of mathematics on the grounds that it fails to answer to our pragmatic needs, but also senseless to defend the mathematics so reformed on the grounds that it is "useless knowledge." There is room for just one kind of useless knowledge in mathematics: claims that have in themselves no practical implications but serve to enhance our understanding of results that are practically significant. This does not mean that Hardy was wrong when he gloried in the uselessness of the results of number theory. But, if he was right, then there must be a chain of mathematical practices culminating in the refined abstractions of number theory and beginning with material of genuine practical significance, such that each member of the chain illuminates its predecessor.

Mathematical progress, in a nutshell, consists in constructing a systematic and idealized account of the operations that humans find it profitable to perform in organizing their experience. Some of these operations are the primitive manipulations with which elementary arithmetic and elementary geometry begin. Others are first performed by us through the development of mathematical notation that is then employed in the sciences as a vehicle for the scientific organization of some area of experience. But there is no independent notion of mathematical truth and mathematical progress that stands apart from the rational conduct of inquiry and our pursuit of nonmathematical ends, both epistemic and nonepistemic.

I draw a radical conclusion. *Epistemic* justification of a body of mathematics must show that the corpus we have obtained contributes either to the aims of science or to our practical goals. If parts can be excised without loss of understanding or of fruitfulness, then we have no *epistemic* warrant for retaining them. If there is a distinction between mathematics as art and mathematics as cognitive endeavor, it is here that it must be drawn.[35] Drawing it must wait on the development of a full theory of rational interpractice transitions, both in mathematics and in the sciences.

5. An Agenda

In *The Nature of Mathematical Knowledge*, I focused on internal transitions and on individual epistemic rationality. My project was to show that there have been important and unrecognized inferences in the history of mathematics and that similar inferences underlie our knowledge of those statements philosophers have often taken to be the foundations of our mathematical knowledge. I now think that the position I took was too conservative in several different ways.

First, external interpractice transitions need more emphasis. As I have suggested above, mathematics is dependent on other sciences and on our practical interests for the concepts that are employed in the spinning of our mathematical stories. It is shortsighted to think that the systematization of a branch of mathematics can proceed in neglect of the ways in which adjacent fields are responding to external demands. Hamilton thought that generalization of claims about complex numbers would necessarily be a fruitful project. Ultimately, the field in which he labored was changed decisively through attempts to come to terms with problems in mathematical physics. Vector algebra and analysis offered a perspective from which the lengthy derivations of recondite properties of quaternions look beside the point.[36]

Second, the history of mathematics, like the history of other areas of science, needs to be approached from both the individual and the community perspectives. We should ask not only about the reasons people have for changing their minds, but also about the fashion in which the community takes advantage of our idiosyncrasies to guide us toward ends we might otherwise have missed. As I have already noted, the pursuit of more than one research program may often advance the community's epistemic projects—despite the fact that it will require of some members of the community that they act against their individual epistemic interests. We can properly ask whether there are enough incentives in mathematics for the encouragement of diversity.

Third, there are serious questions about the balance between the pursuit of epistemic ends and the pursuit of nonepistemic ends. I have been arguing that it is difficult to distinguish questions of "science policy" from questions of epistemology when the science under study is mathematics. An important part of an epistemology for mathematics ought to be the consideration of the relative importance of mathematical understanding and the articulation of methods that will promote our extramathematical projects (including our practical projects).

These conclusions are not likely to be popular, for they contradict an image of purity that has dominated much contemporary thinking about mathematics and much past philosophy of mathematics. I hold them because I take them to follow from a thoroughly naturalistic approach to mathematical knowledge, and because I believe that there is no plausible alternative to a naturalistic mathematical epistemology. These conclusions should be corrected, refined, or discarded by undertaking a pair of studies.

The first of these studies is the philosophical enterprise of giving a precise account of rationality and progress in the sciences. The discussions of the last sections need to give way to a detailed account of our epistemic ends, of the ways in which interpractice transitions contribute to these ends, and of the kinds of institutions that can play a role in shaping the community pursuit of the ends. If my claims above are correct, then the problems of epistemology of mathematics reduce to questions in the philosophy of science, questions that I have tried to formulate in a preliminary fashion here.

The second project is more historical. As I have already suggested, virtually all the major questions about the growth of mathematical knowledge remain unanswered for almost every major transition in the history of mathematics. I want to conclude by mentioning two examples that seem to me to illustrate some of the historical issues that I am recommending and that point toward distinctions that have been underemphasized in the previous discussion.

Philosophers of mathematics routinely concentrate their attention on the emergence of set theory and of modern logic in the early decades of our century. That historical episode is important, but, to my mind, it is far less significant than a contemporaneous development. *Part* (but only part) of that development begins in papers by Dedekind (specifically the supplements to Dirichlet's lectures on number theory), is pursued by Emmy Noether and her colleagues, and is completed with the publication of van der Waerden's *Moderne Algebra*. In the process, the language of mathematics was radically changed and the problems of mathematics were transformed. If we are to assess the way in which the mathematics of our century has been driven by internal changes, then there can be no better model for our study than the evolution in which the lineage begun by Dedekind is one significant strand.

The example is not only important because it focuses on a transition in which an entirely new ideal of mathematical understanding was fashioned—an ideal that sees our ability to identify particular results about numbers, spaces, functions, and so forth, as special cases of claims about very abstract structure. Insight into this transition is essential if the naturalistic approach I espouse is to be defended against a commonly heard objection. "Naturalism," say the critics, "is all very well for pre-twentieth-century mathematics, but in our own times the subject has come of age and has been transformed."[37] Now I shall assume that the historical reconstructions that I have given in the case of the emergence of the abstract group concept and (in far more detail) in the emergence of the main concepts of nineteenth-century analysis are satisfactory. If that is so, then it is possible to show how major modifications of mathematical practice were achieved on the basis of tensions within prior practice—how such tensions led to the the introduction of new problems, how the new problems were solved by employing new concepts and methods of reasoning, how ill-understood reasonings and concepts were finally systematized with axioms and definitions. There is no obvious reason why the development of contemporary mathematics should have gone differently, and it is natural to suggest that every successful generation of inquirers takes itself to have brought the discipline to maturity. Nonetheless, the critics make a telling point when they emphasize the extent of the differences between current investigations and nineteenth-century mathematics, and it is a legitimate challenge to ask whether the transition can be understood in the same way as the examples of group theory and classical analysis.

Here is a simple version of a fragment of the story. In the tenth supplement to Dirichlet's lectures, Dedekind undertakes to introduce a new, general, and perspicuous way of reformulating some of Kummer's results on "ideal numbers." Part of his strategy is to consider collections of elements in the number domain and to define an analog of multiplication when one of the multiplicands is not an element of the domain but a certain kind of collection. So we might see set-theoretic constructions as admitted into mathematics because of their ability to systematize a class of results that had already been assembled in tackling significant problems in number theory.

But the simple version is far too simple. As Harold Edwards has argued, there was a serious debate about whether or not Dedekind's method offered an explanatory systematization (and extension) of the theory of ideal numbers.[38] Kronecker offered a rival version, bereft of set-theoretic constructions but also avoiding a detour that Dedekind's reasoning was forced to undertake.[39] So why did Dedekind's proposals win adherents and, eventually, success? Here we must look at the manifold ways in which thinking about various types of abstract structures was beginning to prove useful in late-nineteenth-century mathematics. I conjecture that a more realistic version of the rational acceptance of the concept of an ideal (or, more generally, of the idea of defining analogs of operations on elements on collection of such elements) will need to consider the ways in which related notions proved useful in a number of quite different mathematical subdisciplines.[40]

This is *only* a conjecture. I offer it merely to turn back the charge that it is quite incomprehensible how the naturalistic approach that I have used in understanding the transitions that took place in analysis from Newton and Leibniz to the end of the nineteenth century could achieve similar success in understanding the emergence of contemporary mathematics. Critics who contend that it is *impossible* that twentieth-century algebra (for example) could have been the product of a sequence of interpractice transitions of the kinds I describe are mistaken. Whether any of the sequences of rational interpractice transitions that my scheme allows corresponds to the actual historical developments is another matter—a matter for detailed historical research.

Contrast this first example with cases suggested by an obvious question. There is little doubt that the practice of some areas of pre-twentieth-century mathematics was decisively affected by developments in the natural sciences (particularly in physics). Has this process continued into our own times? Three obvious possible positive instances come to mind. The first is the development of catastrophe theory, with its roots in the attempts to study aspects of biological and social systems. Second is the work of Traub and his associates on identifying the reliability of error-prone algorithms. Third, and perhaps most exciting, is the investigation of periodic equilibria and "chaos," pioneered by Feigenbaum and culminating in the identification of new "fundamental constants."[41] In each of these instances, mathematics is being used in novel ways to address practical and scientific problems. The interesting issue is whether the new work provides a basis for modifying mathematical practice.

Recall a paradigm from nineteenth-century analysis. Fourier's studies of the diffusion equation led to problems that could not be resolved within the framework of the analysis of the time. In Cauchy's treatment, the central concepts of analysis were redefined and clarified so as to provide means of answering the new questions. Conservatives may declare that the contemporary applications to which I have alluded are quite different from the paradigm. All that Thom, Traub, and Feigenbaum have done is to call to our attention hitherto unanticipated ways of developing received ideas. Others may see in these (or in different) contemporary developments the basis for introducing new concepts that will radically alter mathematical practice.

Who is right? Again, the question can only be resolved by detailed investigation. But it is important to see that there is a live issue here, an issue related to the exchanges in which pessimistic mathematicians engage over the coffeepot at numerous institutions all over the world. Is the field going to the dogs because it has lost itself in arid abstractions that serve no cognitive purpose? Or is it in decline because mathematics is no longer queen, but very much the servant, of science (or, worse, engineering)? On my view of mathematical knowledge, these common complaints pose central—but neglected—problems in the philosophy of mathematics: What is the right balance between the epistemic end of understanding the mathematics already achieved and the ends set for us by the sciences and our practical needs? Does contemporary practice strike that balance? I have tried to construct a framework within which historians, philosophers, and mathematicians can collaborate to find answers.

Notes

1. The transition is evident in the work of Frege, who notes in the Introduction to the *Grundlagen (The Foundations of Arithmetic* [Oxford: Blackwell, 1959]) that he has "felt bound to go back rather further into the general logical foundations of our science than perhaps most mathematicians will consider necessary" (p. x). His stated reason is that, without the successful completion of the project he undertakes, mathematics has no more than "an empirical certainty" (ibid.). Later, he suggests that the achievements of the nineteenth century in defining the main concepts of analysis point inexorably to an analogous clarification of the concept of natural number. To those who ask why, Frege offers the same epistemological contrast, pointing out that mathematics prefers "proof, where proof is possible, to any confirmation by induction" (pp. 1-2). For a more detailed investigation of Frege's

epistemological motives and of his transformation of the philosophy of mathematics, see my "Frege's Epistomology," *Philosophical Review* 88 (1979): 235-62, and "Frege, Dedekind, and the Philosophy of Mathematics" (to appear in *Synthesizing Frege*, ed. L. Haaparanta and J. Hintikka).

2. In *The Nature of Mathematical Knowledge* (New York: Oxford University Press, 1983), I developed an anti-apriorist philosophy of mathematics that I called *empiricism*. However, that position differs from most approaches that have called themselves *empiricism* in several important ways, and I now prefer the name *naturalism*, both for the view I defended in the book and the refinement of it presented in this essay. I hope that the change of labels will help some of my mathematical readers to avoid the provocations of my earlier terminology. I am grateful to Felix Browder for suggesting that the choice of that earlier terminology was "a philosophical fetish."

3. This example was used to similar ends by Hilary Putnam in "What Is Mathematical Truth?" in his *Collected Papers*, vol. *Mathematics, Matter, and Method* (Cambridge: Cambridge University Press, 1976).

4. *Principia Mathematica*, vol. 1 (Cambridge: Cambridge University Press, 1910), p. x.

5. See *Nature of Mathematical Knowledge*, chap. 1-4. One important part of my argument is that the notion of a priori knowledge stands in need of preliminary clarification. Perhaps the thesis that mathematical knowledge is a priori obtains its total credibility from the fact that the crucial notion of apriority is usually left so vague that it is possible to mistake heuristic devices for vehicles of a priori knowledge.

6. Not, of course, what logicians count as a proof. Complete derivations with all steps made explicit are not available except in the case of extremely elementary parts of mathematics. Moreover, it is a serious epistemological question to ask what such derivations would do for us if we had them.

7. Again, see *Nature of Mathematical Knowledge*, chap. 1-4. I think that the best reply for the apriorist is to contend that the conditions I take to be necessary for a priori knowledge are too stringent, and this reply has been offered by a number of people, most lucidly by Charles Parsons (review of *The Nature of Mathematical Knowledge*, in *Philosophical Review* 95 [1986]: 129-37). As Parsons correctly points out, my arguments against apriorism depend on a condition to the effect that the procedures that give a priori justificatory function no matter what our experience. So apriorism could be salvaged by dropping that condition. But it seems to me that the suggestion is vulnerable in two different ways. First, if "a priori" procedures could be undermined by recalcitrant experiences, then it would appear that experience is doing some positive work when the procedures actually function to justify our beliefs. Thus our "a priori" knowledge would be dependent on our experience. To overcome this difficulty, it would seem necessary to argue for an asymmetry: recalcitrant experience can play a negative role, but kindly experience does nothing positive. I do not at present see how such an asymmetry could be articulated and defended. Second, there seem to be many procedures in the natural sciences—the thought experiments of Galileo and Einstein, for example—that also seem to fit the notion of "a priori knowledge" if one abandons the condition that generates trouble for the apriorist. Hence, I do not think there is much point in defending "apriorism" by dropping that condition.

8. For an account with much more detail, see chap. 10 of *Nature of Mathematical Knowledge*. Other significant features of the episode are presented in Judith Grabiner's *The Origins of Cauchy's Rigorous Calculus* (Cambridge, Mass.: MIT Press, 1981).

9. However, Mill was not as muddled as his critics (notably Frege) have often made him out to be. For a more charitable assessment of his views, see Glenn Kessler's "Frege, Mill, and the Foundations of Arithmetic," *Journal of Philosophy* 77 (1980): 65-79; and my paper "Arithmetic for the Millian," *Philosophical Studies* 37 (1980): 215-36.

10. Two important points need to be noted here. First, because the chain is so long it seems misleading to emphasize the *empirical* character of the foundation. Indeed, it seems to me to be possible that the roots of primitive mathematical knowledge may lie so deep in prehistory that our first mathematical knowledge may be coeval with our first propositional knowledge of any kind. Thus, as we envision the evolution of human thought (or

of hominid thought, or of primate thought) from a state in which there is no propositional knowledge to a state in which some of our ancestors know some propositions, elements of mathematical knowledge may emerge with the first elements of the system of representation. Of course, this is extremely speculative, but it should serve as a reminder that the main thrust of the naturalistic approach to mathematical knowledge is to understand *changes* in mathematical knowledge; although a naturalist contends that mathematical knowledge originated in some kind of responses to the environment, it is eminently reasonable to propose that there are a number of possibilities and that this aspect of the naturalistic theory of knowledge is (for the moment, and perhaps permanently) less accessible to elaboration. (I am indebted to Thomas Kuhn for an illuminating discussion of this point.)

Second, it is not necessary for a naturalist to believe that *all* the transitions that have occurred in the history of mathematics were rational. There may be temporary fallacies in the arguments that mathematicians initially give to introduce new ideas, or the reasons they present may be inadequate. Such lapses are of no account provided that good reasons are later supplied, for, in such cases, we may see the change as an episode in which mathematical practice develops for a while in an unjustified fashion before securing proper justification that then prepares the way for later transitions. Hence it is possible for the chain of rational interpractice transitions to diverge from the actual course of events. It seems to me likely that this occurred in the case of the introduction of complex numbers, where the initial reasons for extending the language of mathematics were not very strong. However, Euler's demonstrations of the fruitfulness of the new language eventually provided compelling reasons for the modification, and the "rational reconstruction" favored by the naturalistic epistemological will depart from the chronology by treating the Eulerian transition as the pertinent link in the justificatory chain and by ignoring the arguments originally proposed by Bombelli.

11. The same approach is also useful in understanding issues in the growth of scientific knowledge. I have discussed two particularly important cases in "1953 and All That: A Tale of Two Sciences," *Philosophical Review* 93 (1984): 335-73; and "Darwin's Achievement," in *Reason and Rationality in Science*, ed. N. Rescher (Washington, D.C.: University Press of America, 1985), pp. 127-89.

12. Strictly speaking, it is not necessary to assume the rationality of the majority—or even of *any*—of the major interpractice transitions that have actually occurred in the history of mathematics. All that is needed is to suppose that unjustified leaps are later made good through the provision of reasons that support the practices that emerged from the leaps. As a matter of fact, I think that modifications of practice in the history of mathematics are usually made for good reason, and hence I offer the formulation of the text. However, because the point has so frequently been misunderstood, it seems worth reemphasizing one conclusion of note 10: the naturalism I have proposed allows for a distinction between discovery and justification, and it does not commit the genetic fallacy.

13. One part of this work is done in chap. 10 of *Nature of Mathematical Knowledge*.

14. For an illuminating account of how Newton's mathematical ideas became available to his contemporaries, see R. S. Westfall's *Never at Rest* (Cambridge: Cambridge University Press, 1980).

15. *Nature of Mathematical Knowledge*, chap. 10.

16. The original notion of function is geometric. The functions of curves are such things as subtangents, subnormals, radii of curvature, and so forth.

17. It is possible that mathematics should rationally proceed by adopting more than one, so that there would be alternative set theories as there are alternative geometries.

18. Penelope Maddy has argued that a naturalistic account of mathematical knowledge ought to enable us to resolve current disputes in the foundations of set theory (review of *Nature of Mathematical Knowledge*, in *Philosophy of Science* 52 [1985]: 312-14). Here, I think she expects too much. The initial task for naturalistic epistemology, the task undertaken in *Nature of Mathematical Knowledge*, is to integrate mathematical knowledge into the naturalistic framework. A subsequent project is to give a sufficiently detailed account of scientific methodology to allow for the resolution of hard cases. The latter is the vast problem of characterizing scientific rationality.

19. Here I draw on an interpretation of Kant that I have developed in "Kant's Philosophy of Science," in *Self and Nature in Kant's Philosophy*, ed. A. Wood (Ithaca, N.Y.: Cornell University Press, 1984), pp. 185-215; and "Projecting the Order of Nature" (to appear in *Kant's Philosophy of Physical Science*, ed. R. E. Butts). I should note that my interpretation is heretical in cutting away the apriorist strands in Kant's thought.

20. Of course, external transitions may create new branches of mathematics that are then subject to internal interpractice transitions. If we credit popular anecdotes about Pascal and Euler, then the fields of probability theory and topology may have originated in this way.

21. A large number of further distinctions may obviously be drawn here, for we may take very different approaches to the question of how the maximization is to be done. This is especially clear in cases where our ends admit of degrees, so that we may contrast maximizing the expected value with minimizing the risk of failing to obtain a certain value, and so forth. I ignore such niceties for the purposes of present discussion.

22. I think that it can often be shown in cases of this type that consideration of the practical interests of individuals reveals that the *community* optimum is more likely to result if the individuals are motivated by nonepistemic factors. In other words, a sine qua non for community epistemic rationality may be the abandonment by some individuals of individual epistemic rationality. However, it is possible that those individuals are *overall* rational. Something like this has been suggested by Kuhn (see, for example, his "Objectivity, Value Judgment, and Theory Choice," in *The Essential Tension* [Chicago: University of Chicago Press, 1977], pp. 320-39).

23. For a brief account, see chap. 10 of *Nature of Mathematical Knowledge*, and, for more detail about the Leibnizians, Ivor Grattan-Guinness, *The Development of the Foundations of Analysis from Euler to Riemann* (Cambridge, Mass.: MIT Press, 1970).

24. An especially clear example is furnished by the discussion among the Leibnizians of the "result" that $1 - 1 + 1 - 1 + \ldots = 1/2$. See Leibniz's *Mathematische Schriften*, ed. Gerhardt, 5 vols. (Halle, 1849-63), vol. 5, pp. 382ff., and vol. 4, p. 388. Euler was extremely dubious about the conclusions favored by Leibniz and Varignon. Nevertheless, his own writings are full of inspired attempts to assign sums to divergent series that such later writers as Abel would find appalling.

25. Important figures in the sequence are Benjamin Robins, Colin Maclaurin, and Simon l'Huilier. The case of Maclaurin offers a clear contrast with the Continental tradition. When *Treatise on Fluxions* is compared with any volume of Euler's works in analysis, one sees two talented (though not *equally* talented mathematicians) proceeding by working on very different problems. Maclaurin turns again and again to the question of finding an explanation of the basic rules of the Newtonian calculus. Euler builds up a wealth of results about integrals, series, maximization problems, and so forth, and is almost perfunctory about the basic algorithms for differentiating and integrating.

26. In *The Analyst* (reprinted in *The Works of George Berkeley*, vol. 4, A. Luce and T. Jessop, eds. [London: Nelson, 1950]). Berkeley's challenge provoked a number of responses, some fairly inept (the essays of James Jurin, for example), others that helped elucidate some important Newtonian ideas (the work of Maclaurian and, even more, the papers of Benjamin Robins).

27. Plainly, this is simply part of a long and complicated story. The purpose of telling it here is to show that a simplistic historiography is not forced on us by thinking about the rationality of mathematical change. (I am grateful to Lorraine Daston for some penetrating remarks that raised for me the issue of whether my ascriptions of rationality to past mathematicians commit me to Whig history. See her review of *Nature of Mathematical Knowledge*, in *Isis* 75 [1984]: 717-21).

28. See Roger Cooke, *The Mathematics of Sonya Kovaleskaya* (New York: Springer, 1984), for some beautiful examples of the influence of physical problems on late-nineteenth-century analysis.

29. See Penelope Maddy's "Perception and Mathematical Intuition," *Philosophical Review* 89, (1980): 163-96. Related views have been elaborated by Michael Resnik in his "Mathematics As a Science of Patterns: Ontology," *Nous* 15 (1981): 529-50.

30. The geographical analogy stems from Frege; see *Grundlagen*, p. 108.

31. This is a point that has been emphasized by Karl Popper (see, for example, *The Logic of Scientific Discovery* [London: Hutchinson, 1959], pp. 27-145); in the Popperian tradition, it leads to the notorious problems of constructing measures of verisimilitude (for reviews, see I. Niiniluoto's "Scientific Progress," *Synthese* 45 [1980]: 427-62; and W. Newton-Smith's *The Rationality of Science* [London: Routledge and Kegan Paul, 1981], chap. 2 and 8). I believe that the problem can be overcome if we break the spell of the idea that the search for the significant is always the search for the general, but this is a long story for another occasion.

32. Thus, for example, the account offered by Maddy in her "Perception and Mathematical Intuition" seems at best to reveal how we are able to refer and to know about concrete sets. It is not at all clear how this knowledge is supposed to provide us with a basis for reference to and knowledge of abstract objects, where we are no longer in causal interaction with the supposed objects. So even if we grant that our causal relation to an object provides us with a basis for knowledge about the set whose sole member is that object, it is hard to see how we obtain a similar basis when the sets under discussion do not have concrete objects as members.

33. *Nature of Mathematical Knowledge*, chap. 6. This chapter has often been misunderstood, and I have been taken to substitute one kind of abstract object (ideal agents) for another (sets). But, as I took some pains to emphasize, there are no more any ideal agents than there are such things as ideal gases. In both ideal gas theory and in mathematics, we tell stories—stories designed to highlight salient features of a messy reality. I hope that my present stress on storytelling will forestall any further misconceptions on this point.

34. See *Nature of Mathematical Knowledge*, pp. 128-29. It is crucial to appreciate that some forms of human constructive activity consist in achieving representations of objects—as when, paradigmatically, we cluster objects in thought. My claim is that the use of various kinds of mathematical notation—designed to describe the properties of various constructions—makes possible new constructive activity. Thus we have constructive operations that are iterated, sometimes to quite dizzying complexity, through the use of notation. Mathematics does not describe the notation but does provide (idealized) descriptions of the constructive acts that we can carry out with the help of the notation.

35. Mathematicians sometimes toy with the idea that mathematics is art—or is like art. One consequence of my naturalistic epistemology for mathematics is that it enables us to see what this idea might amount to and how it might apply to various parts of mathematics.

36. Since quaternions are now coming back into fashion, it may appear that the example does not support my claims. However, what is now being done with quaternions is quite distinct from what Hamilton did. For Hamilton, quaternions were to be treated in just the ways that real and complex numbers had previously been treated. So, to cite only one example, Hamilton set himself the task of defining the logarithm of a quaternion. So far as I know, that perspective is a long way from the context of present discussion.

37. This is a common response from mathematicians who have read *Nature of Mathematical Knowledge*. As I shall argue below in the text, the complaint seems to me a very important one, and its justice can only be resolved by combining sophisticated understanding of contemporary mathematics with sophisticated understanding of the philosophical and historical issues. Here, I think, collaboration is clearly required

38. See H. M. Edwards, "The Genesis of Ideal Theory," *Archive for the History of the Exact Sciences* 23 (1980): 321-78.

39. For Dedekind's argument, see his essay *Sur la theorie des nombres entiers algebriques*, in *Gesammelte Mathematische Werke*, vol. 3, ed. E. Noether and O. Ore (Braunschweigh: Vieweg, 1932). The crucial passage occurs on pp. 268-69.

40. In his presentation in Minneapolis, Garrett Birkhoff stressed the intertwining of threads in "the tapestry of mathematics." It seems to me that there are numerous occasions in the history of mathematics in which one area of mathematical practice is modified in response to the state of others, and that concentration on the development of a mathematical field can blind one to the ways in which fields emerge, modify one another, and are fused. For

some stimulating attempts to reveal these processes in concrete cases, see Emily Grosholz's papers "Descartes' Unification of Algebra and Geometry," in *Descartes, Mathematics and Physics*, ed. S. Gaukroger (Hassocks: Harvester Press, 1980); and "The Unification of Logic and Topology," *British Journal for the Philosophy of Science* 36 (1985): 147-57.

41. See R. Thom, *Structural Stability and Morphogenesis*, trans. D. Fowler (New York: Benjamin, 1975); J. F. Traub and H. Wozniakowksi, Information and Computation," *Advances in Computers* 23 (1984): 23-92; and M. Feigenbaum, "Universal Behavior in Nonlinear Systems," *Los Alamos Science* (Summer 1980), pp. 3-27.

IV. THE SOCIAL CONTEXT OF MODERN MATHEMATICS

Partisans and Critics of a New Science: The Case of Artificial Intelligence and Some Historical Parallels

Everywhere in our society—at the supermarket, in the stock market, in mathematics courses—we see the presence of the computer. We are told that we are entering a new age, that of the Computer Revolution, and that the field known as Artificial Intelligence is at the forefront of that revolution. Its practitioners have proclaimed that they will solve the age-old problems of the nature of human thought and of using technology to build a peaceful and prosperous world. These claims have provoked considerable controversy.

The field known as AI is, of course, vast, but the subject matters that have fueled the most debate are these: computer systems that exhibit behaviors that, if exhibited by people, should be called "intelligent"; and "expert" computer systems that, though not exhibiting a broad range of such behaviors, can, because of speed and size of memory and built-in knowledge, outperform humans at specific tasks. The chief implications of research in these areas, according to its advocates, and the claims most viewed with alarm by the research's critics are that AI has a new way of understanding the nature of human beings and their place in the universe and that the "technology" of task-oriented AI—sometimes called *expert systems*, sometimes *knowledge engineering*—will be able to solve the outstanding problems facing man and society.

Critics of AI enthusiasts perceive these claims as wrong: as based, first, on assumptions about the nature of man and society that are inadequate at best and dangerously misleading at worst; and, second, as overly optimistic about the feasibility of finding technical solutions to human problems. In the first case, the critics say, advocates of AI reify a mechanical model of man and of society, one that is not only insufficient but that encourages the treatment of individuals mechanically and as parts of a larger machine. In the second case, they propose technical solutions to what may not be technical problems, and they underestimate the costs of technical innovations while overestimating the benefits.

Upon listening to a debate like this, a historian might well ask whether the claims of the partisans of the science—and of their critics—are solely reactions to the current and probable future state of their science, or whether they are drawing instead on the rhetoric and ideas of other debates about somewhat different topics. If the latter, how did those earlier debates—on which more of the returns are in—actually come out?

In this paper, I hope to provide some historical perspective on the implications of Artificial Intelligence by describing a pattern followed by similar debates about the implications of scientific breakthroughs earlier in history. I will present several examples to show the frequency with which this pattern has been followed. But the purpose of giving these examples is not only to demonstrate that this pattern has occurred in the past. I want also to suggest that causes similar to those that operated in the past are working now; that some of the content of past debates has helped shape the current one; and, finally, that recognizing the pattern will help clarify what is really at issue now. In short, I hope to illuminate the present debate over AI by showing that it shares many characteristics with past debates between partisans of new scientific approaches and critics of those approaches.

The pattern followed is roughly this: (1) The new methods solve some outstanding set of problems despaired over in the past, and these successes are impressive to outsiders and inspiring to those who have achieved them. (I call this first stage *technical success*.) (2) The practitioners extrapolate their successes to other fields (they have found the language of the book of nature, or the secret of life); at the very least, they argue that those who use their methods in other areas will succeed by these new means where earlier investigators have failed. (I call this second stage *extrapolation*.) (3) In reaction, antiscientific currents that always surround the sciences— currents hostile even to successes of the delimited kind first mentioned— are intensified by the extreme claims made in the second stage. (4) Finally, more serious critics enter the fray—critics often from within the scientific community itself, who perceive unfounded enthusiastic claims, claims whose consequences—with no apparent scientific warrant—threaten cherished values.

There are three other striking phenomena. First, the promises of the first stage—successes in the original area and its nearby neighbors—are in fact fulfilled, but the claims of the second stage largely are not; and there are costs even of the first. Second, the mature science growing out of the initial success consolidates the earlier gains and continues. Third, the enthusiasm and inspiration of the second stage seem valuable in consolidating the gains of the first stage of initial success, even though their wide-ranging predictions are not fulfilled.

The debates also make reference to historical analogies taken from similar debates in the past, and in so doing, they often blur the distinction between the stage of initial success and the second, more enthusiastic stage. Critics opposing the second-stage extrapolation are often accused of opposing the first, and are called Luddites or persecutors of Galileo. Keeping the distinctions between the four stages clear, even though there may be some blurring near the boundaries, would improve the current debates over the support that society should give to fifth-generation AI research, the human tasks that should be entrusted to computers, and the information that artificial intelligence might give us about the nature of man. It is important that these debates not be carried out with arguments like "They laughed at Fulton" on the one hand and "AI is the new alchemy" on the other. Of course, history can help, but one must first understand that history.

To support these general points, I shall outline a set of examples from the history of science so that we can then describe recent debates over AI with this historical perspective. I will in particular discuss the methodological revolution in seventeenth-century science; the eighteenth-century "spirit of systems" and the visions of society stemming from it; the Industrial Revolution; and Darwinian evolution and the extrapolations—social Darwinism, racism, atheistic materialism—from it. Finally, I will take up the key ideas at the early stages of AI in the work of Alan Turing; the claims of AI pioneers like Herbert Simon, Alan Newell, John McCarthy, and Marvin Minsky, and some of their successors; some reactions to the overly enthusiastic claims; and, finally, the judicious, historically informed criticism of AI by computer scientist Joseph Weizenbaum.

The first example is to be found in the period of the scientific revolution—in particular, the methodological revolution of seventeenth-century science. Let us begin by measuring the first stage: its successes in its own sphere. As recently as the sixteenth century, Michel de Montaigne could speak of the futility of learning and could use the Copernican system and the fact that it had ancient predecessors as an example of the way ideas about nature came and went as did other fashions (*Defense of Raymond Sebond*, 86-87). According to Montaigne, the sciences had established no real knowledge. But to Francis Bacon, this lack of past success called not for despair but for a new approach: "Things which have never yet been done can be done [only] by methods which have never yet been tried" (*Novum Organum*, aphorism vi). And, indeed, the mechanical philosophy did things that had never been done. Using the Cartesian method of "analysis"—studying macroscopic phenomena by resolving them into their component parts of bits of matter in motion—men like Pascal, Torricelli, and Boyle were able to explain, by mechanical means, a whole range of phenomena previously requiring notions like the abhorrence of a vacuum (Westfall 1971, 44-49). William Harvey, treating the heart as if it were a pump and the blood as though it were an ordinary fluid, demonstrated the circulation of the blood. Descartes saw Harvey's work—though Harvey himself was far from being a mechanist—as a triumph of the mechanical and analytic methods (Westfall 1971, 93; *Discourse on Method*, 47-55). Cartesians saw the method as opening all of nature to being reduced to Descartes's principles, with Descartes himself believing that he had exhibited the mechanism of the solar system. Some of his followers took the mechanical philosophy even further: LaMettrie in 1747 spoke of man as a machine, and Hobbes (unlike Descartes) saw even thought as a purely mechanical process (*Leviathan*; in Burtt 1939, 143-44).

But the outstanding predictive success of the mechanical philosophy when applied to the vacuum or (with Borrelli) to the mechanical advantage of muscles did not extend to psychology, or even to all of physics. Descartes's prediction that a science of nature on his principles was almost with his reach (*Discourse on Method*, 68) was an example of extrapolation beyond the domain of success. So were the views that man was wholly mechanical, and so were the views that the Cartesian principles sufficed to explain the phenomena of the solar system. As is well known, reactions even against the initial stage of predictive successes of science occurred in the seventeenth century, from the fabled few who refused to look through Galileo's telescope to the poets who grieved that "new philosophy calls all in doubt" (Donne, "Anatomy of the World"; quoted by Koyré 1957, 29). But there were other opponents of the new philosophy who appreciated what the scientists had accomplished but who saw no reason to believe that it extended as far as the scientists claimed. Blaise

Pascal made clear that, though scientific reasoning had its place, there was much in the situation of man in the world, and man's understanding of himself, that was not amenable to reason. Pascal's *esprit de finesse* recognized intuition and tacit ways of knowing as just as valid in their sphere as mathematical reasoning is in its proper sphere (Pensées 1).

As for Newton and the Newtonians, they found the Cartesian insistence that causes must be mechanical as the same sort of excess. Perhaps the cause of gravity was not known; this did *not* mean that it was reducible to mechanism, any more than the fact that a clock *could* be run by a spring meant that a *given* clock was not actually moved by a weight. (For this Turing test for pendulum clocks, see Roger Cotes's Preface to Newton's *Principia* [pp. xxvii-xxviii].) Besides, as Newton had shown, Cartesian mechanism had not been able to account for all the phenomena of the solar system (notably Kepler's laws), whereas gravity seemed really to exist, though its cause was unknown, and gravity could explain the motions of the heavenly bodies, the fall of terrestrial objects, and the interaction between them in the movement of the tides (Newton's *Principia*, 547). Thus the methodological imperialism of the Cartesians had not in fact produced the promised understanding of the universe, much less of man. That critics like Pascal and the Newtonians waged part of their battle on behalf of their views of religion does not invalidate their conclusions about the limited scope of the successes of the mechanical philosophy.

My second example is found in the "spirit of systems" characteristic of such thinkers of the Enlightenment as Voltaire, Condorcet, Comte, and St.-Simon. Buoyed up by the successful application of reason in the realm of nature, and by the analysis of macroscopic phenomena in terms of simple statements about the elements, they hoped to achieve in the same way a science of society capable of remaking the world in a rational way. As one opponent, A. Thibaudet, put it, they "ventured to create a religion as one learns at the Ecole [polytechnique] to build a bridge or a road" (Hayek 1955, 113). They had, however, left out of their calculations the passions unleashed by, for instance, the French Revolution, and the tenacious holding power of old institutions and ideas and the powerful people who wanted those preserved. In particular, Voltaire's faith in the power of reasonable arguments to persuade anybody who heard them, based on the success of Newtonian science (Voltaire 1764, article "Sect"), was not supported by events. The institutions of the past, as Burke pointed out, and the forces preserving them, as Marx pointed out, had too much staying power. The philosophes, to be sure, left a lasting legacy of new political and social institutions, but things did not work out as they had predicted and hoped. Representing the scientists of our fourth stage, A.-L. Cauchy in 1821 said to his colleagues: "Let us enthusiastically cultivate the mathematical sciences, without trying to extend them beyond their domain; and let us not imagine that one can attack history with formulas, nor give as sanctions to morality theorems of algebra or of integral calculus" (*Cours d'analyse*, vii).

Another example is to be found in the Industrial Revolution. Indeed, the parallel between the Industrial Revolution and the "Information Revolution" we are now supposedly experiencing is made repeatedly in the AI community. The Industrial Revolution in Britain, whose history has been thoroughly written (see, e.g., Deane 1979), was billed as arising from science (though there was more rationalism of the Cartesian variety and analysis of systems using the concept of "division of labor" than much new science, at least before the 1850s). Nineteenth-century partisans of the Industrial Revolution, drawing on Baconian prophecies of the benevolence of applied science and science's ability to improve the lot of all mankind, theorized far beyond their data. Industrialization was certainly bringing about new jobs and providing new products and services, which on balance raised the average standard of living (Deane 1979, 272-95). But it also altered people's lives, took employment from some workers, and produced a new type of job and a new relation between employer and employee. There was incredible urban filth as people were crowded together without adequate water and sewers (Deane 1979, 260-61), and the worker worked at the rhythm of the factory, which led to feelings of alienation. Charles Dickens reflects these widespread feelings in his novel *Hard Times*, when the worker Stephen Blackpool, whose trade union expelled him because he wanted to go his own way and who later died after falling into an old mineshaft that it had not been cost-effective to make safe, expressed his view of the inhumanity of the system: It is all a "muddle." And, Adam Smith to the contrary, it will not fix itself: "Nor yet lettin alone will never do 't. . . . Rating 'em [workers] as so much Power, and reg'latin 'em as if they was figures in a soom [sum], or machines, wi'out loves and likens, wi'out memories and inclinations, wi'out souls to weary and souls to hope . . . this will never do 't" (pp. 115-16). The enthusiasts of the Industrial Revolution consistently underestimated these costs; rather than working to minimize the effect of the costs, they often justified them as part of the workings of the system of efficient organization of labor, whose ultimate tendency, they thought, was better (at least statistically) for everyone.

These tendencies continued into the twentieth century, notably with the work of F. W. Taylor on what is called "scientific management." The organization of production into separate tasks, each one optimized, is the key to efficiency, said Taylor in 1911: "In the past the man has been first; in the future the system must be first" (Pursell 1969, 246). Listen to Taylor instructing a workman to listen to the scientific manager: "A high-priced man has to do exactly as he's told from morning till night. . . . You will do exactly as this man tells you. . . . When he tells you to pick up a [piece of pig iron] and walk, you pick it up and you walk, and when he tells you to sit down and rest, you sit down. . . . And what's more, no back talk" (Pursell 1969, 246). This is *not* Dickensian satire.

Even enemies of Taylorism often conceded that, like it or not, it was "the most efficient means of production" (Keir 1917; in Pursell 1969, 255). However, the neglect of the human costs of doing a mechanized, though optimized, set of tasks surely is one of the reasons workers feel alienated, and many labor historians hold this feeling to be at the root of inefficiencies ranging from goldbricking to labor strife. The famous Hawthorne experiments of the late 1920s, in fact, showed that workers worked better when management paid more attention to them (Noble 1977, 319). This research gave rise to a set of new disciplines, like industrial psychology, and showed that Taylorism, even from the standpoint of maximizing worker efficiency, was inadequate. Similarly, it has recently been suggested that cooperative models, as used in Japan, are more efficient than the manufacturer-initiated, top-down organization characteristic of nineteenth-century European industry. One more point needs to be made to underline the applicability of our pattern to this example: it is true that some critics of industrialism wanted to reverse the whole process, but others, like George Orwell—who certainly had no love for mechanization (Orwell 1937, chap. 12, esp. p 160)—stressed that technological advances had now made it at least possible to raise everyone's standard of living "if we chose to set our minds to it" (Orwell 1943, 213) and criticized industrialism principally for its failure to provide the predicted benefits to all.

My last example is Darwinism. Charles Darwin and Alfred Russel Wallace proposed the idea of evolution by natural selection to explain the origin of biological species, their geographic and fossil distribution, and their adaptation to the conditions of life. Some of their followers used the removal of God from the creation of species to advocate a materialistic biology from which all nonmechanical causes were to be banished. T. H. Huxley went so far as to predict that, just as there was a mechanical equivalent of heat, so we would find a "mechanical equivalent of consciousness" (Huxley 1870, 191). Others used evolution to argue that the survivors in economic competition were the "fittest," or to justify the racial superiority of Caucasians, which, after Darwin, could no longer be scientifically justified by the special creation of the white race (Stanton 1960; Hofstadter 1955). The religious attacks on evolution were strongly reinforced because of these extensions—with ominous consequences—of the doctrine, as the examples of the anti-slavery Bishop Wilberforce and the populist and anti-imperialist William Jennings Bryan show. The eugenics movement in the United States provides another example of the unwarranted extension of evolutionary ideas. During the debate in Congress over immigration restriction in 1924, the eugenicists made the selective restriction against southern and eastern European immigration "seem like a biological imperative" (Ludmerer 1972, 113).

Critical responses from within the scientific community to these extrapolations of evolution ultimately were to prevail. Thus Asa Gray in the nineteenth century (Dupree 1959, 375-77) showed that an evolutionist could be a Christian and not a reductionist, saying that what "science removes from the supernatural to the natural [is not therefore] lost to theism." Similarly, Kirtley Mather in the twentieth century, testifying at the Scopes trial as a geologist and a prominent Baptist layman, showed by example as well as precept that one could be a professing Christian as well as an evolutionist (Ginger 1969). Again, the geneticist Herbert Spencer Jennings (the only scientist to testify against immigration restriction before Congress in 1924) and the anthropologist Arthur Kroeber argued in the 1920s that the eugenicists' arguments went far beyond anything justified by genetic science (Ludmerer 1972, 113; Cravens 1978, 105). By the 1930s, most American geneticists had repudiated eugenics, attributing to it, in the words of geneticist Edward M. East, "too great a tendency to push ahead of the facts" (Ludmerer 1972, 123).

The examples I have cited could be multiplied many times, from the Pythagorean idea that all of nature could be reduced to simple mathematics to the Skinnerian idea of learning by reinforcement and the resulting predictions of the imminent successful introduction of teaching machines.

(See also Lorraine Daston's paper on the probability of judgments in the present volume.) The specific examples I have chosen to discuss are chosen not only for their importance, but because they have a direct bearing on the current debates over AI. Not only is the structure of the debates similar, but the same themes—rationality and mechanism as a complete picture of reality, technological development as the solution to social problems—have been revived and redeveloped in the present debates. Let us now look at the recent history of Artificial Intelligence: at some successes, at some claims made, and at the types of critical responses that have been produced.

Borrowing the idea of a completely mechanized thought process with no room for free will from the mechanical world views of the seventeenth and eighteenth centuries, and drawing also on logic and on his own formal analysis of computability, Alan Turing saw no reason to conclude that a machine could not think, if by *think* one meant to imitate a thinking human being to the point where the machine could fool a human conversational partner into believing it to be another human (the Turing test; Turing 1956). The mathematical abstraction he had created, the Turing machine, enables one to prove many results about what can and cannot be computed.

> Given the initial state of the machine and the input signals, it is always possible to predict all future states. This is reminiscent of Laplace's view that from the complete state of the universe at one moment of time...it should be possible to predict all future states. The prediction we are considering is, however, rather nearer to practicability than that considered by Laplace. (Turing 1956, 2105)

Turing suggested that a computer with 10^9 memory locations would in about 1990 have a 70 percent chance of passing the Turing test (2107) and thus a 70 percent chance of being able to think. Responding to the possible objection that thinking requires a soul, Turing said that "we should not be irreverently [usurping] God's power of creating souls, any more than we are in the procreation of children; rather we are, in either case, instruments of His will providing mansions for the souls that he creates" (2108). But this is all beside the point: "I am not very impressed with theological arguments whatever they may be used to support," he says, dismissing them all with a reference to the case of Galileo.

The mind, then, seems for Turing to be a Turing machine. The adult mind might be hard to imitate, Turing continues, since the processes that have brought it to the "state that it is in" include "the initial state of the mind, say at birth," education, and other experiences. Education and experience are not provided in the most efficient possible manner; this could be done more rationally if we started with the initial state: "Presumably the child-brain [has] so little mechanism that something like it can be easily programmed" (2119).

Turing's views on computability and thought as both being formally describable, rule-based processes appeared to some philosophers, notably J. R. Lucas, to be similar to the old mechanistic ideas, though in a new dress. Even if these new mechanists do not assert that minds come out of matter, said Lucas, they do claim that minds "operate according to 'mechanical principles'"—that is, think that the whole is just the sum of the operations of the separate parts and that the operation of each part is determined by previous states or by random choices between states (Lucas 1961, 126). I have discussed this controversy elsewhere (Grabiner 1984, 1986). For our present purposes, the key point is that Lucas's opposition to Turing's ideas came from Lucas's desire to refute both "traditional materialism and its modern mechanist version" (Lucas 1968, 156), a "mechanism" that is not implied by, but extrapolated from, Turing's research in computability theory.

Beginning in 1956 with the Dartmouth conference that launched AI as an organized discipline in the United States many more arguments characteristic of the "extrapolation" stage began to appear. The original proposal to the Rockefeller Foundation for the conference said: "The study is to proceed on the basis of the conjecture that every aspect of learning or any other feature of intelligence can in principle be so precisely described that a machine can be made to simulate it" (McCorduck 1979, 93). The key ideas used to "precisely describe" intelligence are that the human mind is a system, either like a machine or like Laplace's world-system, hierarchically organized and programmed by its environment; thought takes place by the rule-generated manipulation of formal symbols. The mind's organization, like the Tables of Discovery of Francis Bacon (*Novum Organum*, aphorism cii), provides for empty slots into which pieces of information from the world (disentangled from their context) can be placed. For instance, John McCarthy: "One of the basic entities in our theory is the situation. Intuitively, a situation is the complete state of affairs at some instant in time. The laws of motion of a system determine all future situations from a given situation" (McCorduck 1979, 213).

Similarly, Simon and Newell state, "The free behavior of a reasonably intelligent human can be understood as the product of a complex but finite and determinate set of laws" (quoted by Dreyfus 1979, 170). And, in an often-quoted passage: "A man, viewed as a behaving system, is quite simple. The apparent complexity of his behavior over time is largely a reflection of the complexity of the environment in which he finds himself.... I myself believe that the hypothesis holds *even for the whole man*" (Simon 1981, 65; my emphasis).

We should acknowledge the initial success of early AI programs in imitating some aspects of human thought: solving mathematical and logical problems; "memorizing" nonsense syllables in ways similar to those psychologists find in human subjects; conducting dialogues about limited worlds; answering questions about brief, formula-based stories; and playing checkers and chess. All this was accompanied by enthusiastic predictions. Herbert Simon in 1960 proclaimed that in ten years a computer would be world chess champion, discover and prove an important new mathematical theorem, and translate natural languages, and also that most psychological theories would be embodied in computer models (quoted by Dreyfus 1979, 81-82). Marvin Minsky proposed that computers would write good music and draw significant pictures as soon as aesthetic analysts do something about "the traditionally disgraceful state of analytic criticism in the arts" (Minsky 1968, 12; quoted by Weizenbaum 1976, 157). In words deliberately (and only too successfully) provocative, John McCarthy asked, "What do judges know that we cannot eventually tell a computer?" (McCorduck 1979, 324), and Minsky called the brain "a meat machine" (McCorduck 1979, 170).

Let us now turn to some of the prominent critics. An attack was made on the entire AI program by Hubert Dreyfus, who claimed that the "successes" were trivial and that predictions about uses in the future would never be fulfilled. His weapons were chiefly Wittgensteinian language theory and a phenomenological view of the world. John Searle attacked AI enthusiasts on other grounds: he argued that even the successful passing of the Turing test by a formal system—that is, a formal system capable of making people believe it was thinking—would not mean that the machine really understood anything. He asks us to imagine a human being "simulating" natural-language understanding by simply matching symbols according to a book of rules, without ever knowing what the symbols mean. The person would still not understand. This rejection of behavioral equivalence on Searle's part is not an attack on the actual accomplishments of AI research, but on the interpretation the researchers place on that work (Searle 1980; Grabiner 1984, 474-76).

The work of MIT computer scientist Joseph Weizenbaum itself produced a capsule version of our pattern, Weizenbaum in 1966 had published a paper on a program called ELIZA that illustrates its ability to match patterns in English sentences by mimicking the "mirroring of statements" of a therapist of the school of Carl Rogers (Weizenbaum 1966). Others, not Weizenbaum, did the enthusiastic extrapolation of his work: that it had passed the Turing test, that it meant that automated psychiatry was just around the corner (Grabiner 1984, 1986; Weizenbaum 1976, 180). Weizenbaum was appalled. He has since devoted much of his professional life to warning against extrapolating what he calls the "computer metaphor" to areas of human life to which, he argued, it does not pertain. Objecting to the contention that the mind is just a computer operating on formal symbols, Weizenbaum argued instead that humans use language always in context—notably "in a context of experiences like love and trust that machines cannot share" (Weizenbaum 1976, 208-9). And Weizenbaum, being himself involved in the field, made a criticism of the basis for AI's extrapolations that the philosophers cited earlier had been unable to make: the most successful AI programs, he said, lacked strong theory-based models of human intelligence, being instead largely collections of ad hoc programming tricks (Weizenbaum 1976, 232); therefore, such programs, however well they seemed to imitate certain human behaviors, did not in any sense "explain" the nature of human thought. I will return to Weizenbaum's critique of AI later on.

As AI has developed, what some people have characterized as a paradigm change has come about (Duda and Shortliffe 1983). There has, in particular, been a realization that building programs that can learn the way a child does and that can proceed with common sense in everyday situations is very hard indeed. However, programs embodying human expertise in specific areas, with the knowledge organized according to AI-based theories of knowledge representation and using the AI techniques of heuristic search, have been developed that perform impressively. DENDRAL in chemical structure, MYCIN in diagnosing bacterial infections, and PROSPECTOR in mineral exploration are well-known examples (Duda and Shortliffe 1983; Hayes-Roth et al. 1983; Feigenbaum and McCorduck 1983). These examples of applied AI, which are called expert systems or knowledge engineering, somewhat resemble the way the manufacturers of the Industrial Revolution applied some physical principles and the idea of systems; but the amount of science involved here is considerably greater. Many people working in expert systems have drawn back from the claims made earlier for AI and have conceded that whether computers can imitate the full range of human intelligent behavior is "unknown and perhaps . . . forever unknowable" (Albus 1981, 299; cf. Hayes-Roth et al. 1983, 3).

But this modesty of theoretical goals has not precluded another set of enthusiastic predictions, this time reminding us of the prophets of the Industrial Revolution: that expert systems provide an "unprecedented opportunity" (the subtitle of the recent book *Intelligent Systems* (1983) by J. E. Hayes and Donald Michie) to solve the world's problems. Thus intelligent computers will revolutionize education (Papert 1980; Schanck 1984), will solve world problems like terrorism as they read and analyze news stories (Schanck 1984, 220). Edward Feigenbaum and Pamela McCorduck say that, like Adam Smith, they "also write inspired by a machine . . . . An intelligent machine [is] the engine that will produce the new wealth of nations" (1983, 17). The Japanese futurist Yoneji Masuda predicts that technical advances in computing mean that "as the 21st century approaches . . . the possibilities of a universally opulent society being realized have appeared in the sense that Smith envisioned it . . . a society that actually moves toward a universal society of plenty" (Masuda 1985, 625). Thus the claims about what computer technology will make possible are expanded past the actual capabilities of today's computers, and the nontechnological difficulties that will certainly arise are minimized. In an odd use of history, potential ill effects are dealt with by detaching them from computers. Thus it is claimed that there has already been considerable unemployment and dehumanization without computers (e.g., Schanck 1984, 235), so there is no reason to fear that computers will make things worse, and they might help. This follows the same pattern that we saw in the Industrial Revolution, in the Enlightenment's desire to build a commonwealth based on reason, and in Francis Bacon's view that the fruits of science could only help, not hurt mankind (*Novum Organum* aphorism cxxix). It is hard, after Auschwitz and Hiroshima, to accept this kind of optimism in exactly this form; there are, at the very least, costs that will need to be considered in assessing the proposed benefits.

The sociologist Sherry Turkle has characterized the "extrapolation" stage of AI research in words that equally describe the mechanical philosophers of the seventeenth century, the social theorists of the age of the French Revolution and of Taylorism, and the apostles of social Darwinism and eugenics—in terms of the concept of a *colonizing discipline*. "Being in a colonizing discipline," she writes, "first demands and then encourages an attitude [of] intellectual hubris . . . , intellectual principles . . . universal enough to give you the feeling that you have something to say about everything . . . [and] confidence that your knowledge makes the 'traditional wisdom' unworthy of serious consideration." This "so fundamentally changes the rules of every game in the social and behavioral sciences that everything that came before is relegated to a period of intellectual immaturity." And "you have to feel that nothing is beyond your intellectual reach if you are smart enough" (Turkle 1984, 251-52; cf. Descartes's *Discourse on Method*, 68).

Joseph Weizenbaum, as his continuing critique of AI and expert systems shows, is aware of the historical pattern I have been discussing. The introduction of computers into our society, says Weizenbaum, "merely reinforced and amplified those antecedent pressures that have driven man to an ever more highly rationalistic view of his society and an ever more mechanistic image of himself" (1976, 11). And this matters because of the way these views lead people to treat their fellows and to view themselves as programmed and thus not responsible for their actions (1976, chap. 10).

Responding to the methodological universalism of AI, Weizenbaum argues that what AI has done so far "is build a machine that performs certain specific tasks, just as, say, 17th-century artisans built machines that kept time," fired cannon balls, etc. They would have been mistaken, he continues, had they concluded that "because their machines worked, they had validated the idea that the laws of the universe are formalizable" in the same terms. But AI people, he concludes, advance "exactly these mistaken ideas about the machines [they have] succeeded in building" (1976, 196).

Weizenbaum would doubtless agree with Pascal with the *esprit géometrique* has its proper realm: "Nothing is wrong with viewing man as an information processor," writes Weizenbaum, "providing, however, that we never act as though any single perspective can comprehend the whole man" (1976, 140). "Physicists, mathematicians, and philosophers" all accept that "there are limits to how far the world can be comprehended" in what he calls Leibnizian terms, where knowing the initial conditions lets us fully predict the future (1976, 221). But, as Aldous Huxley warned, "The popular Weltanschauung of our times contains a large element of what may be called 'nothing-but' thinking. Human beings . . . are nothing but bodies, animals, even machines . . . . Mental happenings are nothing but epiphenomena" (quoted by Weizenbaum 1976, 129). Max Horkheimer sums this all up to Weizenbaum's satisfaction: too many people believe that there is just one authority, namely science, and, furthermore, a science conceived merely as "the classification of facts and the calculation of probabilities" (Horkheimer 1947; quoted by Weizenbaum 1976, 252). Weizenbaum summarizes his own position: "The computer is a powerful new metaphor for helping us to understand many aspects of the world, but . . . it enslaves the mind that has no other metaphors and few other resources to call on." The computer scientist "must teach the limitations of his tools as well as their power" (1976, 277).

Before ending our discussion of Weizenbaum's ideas, let me remind the reader that there is nothing inevitable about the timing in the pattern I have described. In particular, the fourth-stage critics may be rare, and they do not necessarily win. As H. S. Jennings was unsuccessful in his lone opposition to immigration restriction in 1924, so Weizenbaum now is almost alone in the computer science community.

What, then, are we to conclude from the pattern common to all these examples from the seventeenth-century mechanists to AI? It is possible, as Max Weber has said, that scientific research must be a vocation and that the researcher must be passionately committed to the belief that the outcome of his next piece of research is the most important thing in the world. But as the rest of us try to understand the implications of artificial intelligence for our view of ourselves and for the future of our complex world, we should remember that, though the sciences have over time greatly enlarged the domain in which they can successfully predict and explain phenomena, the enthusiastic predictions about extending methods successful in one area to solve society's problems and to understand the world have generally not come to pass.

References

Albus, James. 1981. *Brains, Behavior, and Robotics.* Peterborough, N.H.: Byte Books.

Bacon, Francis. 1620. *Novum Organum.* Reprinted in (Burtt 1939).

Burtt, E., ed. 1939. *The English Philosophers from Bacon to Mill.* New York: Modern Library.

Cauchy, A.-L. 1821. *Cours d'analyse de l'école royale polytechnique.* Part 1. *Analyse algébrique.* In *Oeuvres complètes.* Paris: Gauthier-Villars, ser. 2, vol. 3, 1897.

Cravens, H. 1978. *The Triumph of Evolution*. Philadelphia: University of Pennsylvania Press.

Deane, Phyllis. 1979. *The First Industrial Revolution*, 2d ed. Cambridge: Cambridge University Press.

Descartes, Rene. [1637] 1956. *A Discourse on Method*. Reprinted (with page numbers from Adam-Tannery edition) New York: Bobbs-Merrill.

Dickens, Charles. [1854] 1966. *Hard Times*. Reprinted New York: W. W. Norton.

Dreyfus, H. 1979. *What Computers Can't Do: A Critique of Artificial Reason*, 2d ed. New York: Harper and Row.

Duda, R., and Shortliffe, E. 1983. Expert Systems Research. *Science* 220: 261-68.

Dupree, A. Hunter. 1959. *Asa Gray*. New York: Atheneum.

Feigenbaum, E., and McCorduck, P. 1983. *The Fifth Generation: Artificial Intelligence and Japan's Computer Challenge to the World*. Reading, Mass.: Addison-Wesley.

Forester, Tom, ed. 1985. *The Information Technology Revolution*. Cambridge, Mass.: MIT Press.

Ginger, Ray. 1969. *Six Days or Forever*. New York: New American Library.

Grabiner, Judith. 1984. Artificial Intelligence: Debates about Its Use and Abuse. *Historia Mathematica* 11: 471-80.

———. 1986. Computers and the Nature of Man: A Historian's Perspective on Controversies about Artificial Intelligence. *Bulletin of the American Mathematical Society* 15: 113-26.

Hayek, F. 1955. The Source of the Scientistic Hubris: l'Ecole Polytechnique. In *The Counter-Revolution of Science: Studies on the Abuse of Reason*. Glencoe, Ill.: Free Press.

Hayes, J. E., and Michie, D. 1983. *Intelligent Systems: The Unprecedented Opportunity*. Chichester, England: Ellis Horwood.

Hayes-Roth, F., Waterman, D., and Lenat, D. B. eds. 1983. *Building Expert Systems*. Reading, Mass.: Addison-Wesley.

Hobbes, Thomas. 1651. *Leviathan, or the Matter, Form, and Power of a Commonwealth Ecclesiastical and Civil*. Reprinted in (Burtt 1939).

Hofstadter, Richard. 1955. *Social Darwinism in American Thought*. Boston: Beacon Press.

Horkheimer, M. 1947. *Eclipse of Reason*. Oxford: Oxford University Press. Reprinted New York: Seabury, 1974. Cited in (Weizenbaum 1974), p. 252.

Huxley, Thomas Henry. 1870. On Descartes' "Discourse Touching the Method of Using One's Reason Rightly and of Seeking Scientific Truth." Reprinted in (Huxley 1898), pp. 166-98.

———. 1874. On the Hypothesis That Animals Are Automata, and Its History. Reprinted in (Huxley 1898), pp. 199-250.

———. 1898. *Method and Results*. New York: Appleton.

Keir, M. 1917. Scientific Management and Socialism. *Scientific Monthly* 5: 359-67. Reprinted in (Pursell 1969), pp. 248-55.

Koyré, Alexandre. 1957. *From the Closed World to the Infinite Universe*. Baltimore: Johns Hopkins.

Lucas, J. R. 1961. Minds, machines, and Gödel. *Philosophy* 36: 112-27.

———. 1968. Human and Machine Logic: A Rejoinder. *British Journal for the Philosophy of Science* 19: 155-56.

Ludmerer, Kenneth M. 1972. *Genetics and American Society: A Historical Appraisal*. Baltimore: Johns Hopkins.

Masuda, Y. 1985. Computopia. In (Forester 1985), pp. 620-34.

McCorduck, Pamela. 1979. *Machines Who Think. A Personal Inquiry Into the History and Prospects of Artificial Intelligence*. San Francisco: Freeman.

Minsky, Marvin, ed. 1968. *Semantic Information Processing*. Cambridge, Mass.: MIT Press.

Montaigne, Michel de. [1580] 1959. The Futility of Learning. In *In Defense of Raymond Sebond*, chap 3. Trans. A. H. Beattie. New York: F. Ungar.

Newton, Isaac. [1713] 1934. *Mathematical Principles of Natural Philosophy*, 2d ed. Trans. F. Cajori. Reprinted Berkeley: University of California Press.

Noble, David. 1977. *America by Design: Science, Technology, and the Rise of Corporate Capitalism*. Oxford: Oxford University Press.

Orwell, George, 1937. *The Road to Wigan Pier*. Reprinted New York: Berkley, 1961.
——. 1943. Looking Back on the Spanish War. Reprinted in *A Collection of Essays*. Garden City, N.Y.: Doubleday, 1957, pp. 193-214.
Papert, Seymour. 1980. *Mindstorms: Children, Computers, and Powerful Ideas*. New York: Basic Books.
Pascal, Blaise. [ca. 1660] 1958. *Pensées*. Reprinted New York: F. Dutton. (Numbering follows the L. Brunschvicg edition).
Pursell, Carroll W., Jr., ed. 1969. *Readings in Technology and American Life*. New York: Oxford University Press.
Schanck, Roger C. 1984. *The Cognitive Computer: On Language, Learning, and Artificial Intelligence*. Reading, Mass.: Addison-Wesley.
Searle, John. 1980. Minds, Brains, and Programs; Author's Response. *Behavioral and Brain Sciences* 3: 417-24, 450-57.
Simon, Herbert A. 1981. The Psychology of Thinking: Embedding Artifice in Nature. In *The Sciences of the Artificial*, 2d ed. Cambridge, Mass.: MIT Press.
Stanton, William. 1960. *The Leopard's Spots*. Chicago: University of Chicago Press.
Taylor, F. W. 1911. *Scientific Management*. New York: Harper. Excerpts from pp. 5-7, 15-16, 38-47, and 116-19 in (Pursell 1969), pp. 240-48.
Turing, Alan. 1956. Can a Machine Think? Reprinted from *Mind* (1950) in *The World of Mathematics*, vol. 4, ed. J. R. Newman. New York: Simon and Schuster, pp. 2099-2123.
Turkle, Sherry. 1984. *The Second Self: Computers and the Human Spirit*. New York: Simon and Schuster.
Voltaire. 1764. Sect. *Dictionnaire philosophique*. Reprinted in *Portable Voltaire*. New York: Viking Press, 1949, pp. 195-99.
Weizenbaum, Joseph. 1966. ELIZA—A Computer Program for the Study of Natural Language Communication between Man and Machine. *Communications of the ACM* 9: 36-45.
——. 1976. *Computer Power and Human Reason: From Judgment to Calculation*. San Francisco: Freeman.
Westfall, Richard S. 1971. *The Construction of Modern Science: Mechanisms and Mechanics*. New York: Wiley.

The Emergence of Princeton as a
World Center for
Mathematical Research, 1896-1939

1. Introduction

In 1896 the College of New Jersey changed its name to Princeton University, reflecting its ambitions for graduate education and research. At the time, Princeton, like other American universities, was primarily a teaching institution that made few significant contributions to mathematics. Just four decades later, by the mid-1930s, Princeton had become a world center for mathematical research and advanced education.[1] This paper reviews some social and institutional factors significant in this rapid rise to excellence.[2]

The decade of the 1930s was a critical period for American research mathematics generally, and for Princeton in particular. The charter of the Institute for Advanced Study in 1930 and the completion of a university mathematics building (Fine Hall) in 1931 frame the opening of the period in Princeton; the completion of separate quarters (Fuld Hall) for the institute mathematicians in 1939 and the entrance of the United States into World War II effectively close it. During this decade, Princeton had the unique atmosphere of an exclusive and highly productive mathematical club. This social environment changed after the war with the increase in university personnel and the move of the institute to separate quarters, and the uniqueness was challenged by the improvement of mathematical research and advanced education at other American institutions.

I appreciate the useful comments and corrections of Richard Askey, Saunders MacLane, and V. Frederick Rickey to the conference presentation and of Garrett Birkhoff and Albert W. Tucker to later drafts of the manuscript. The writing of this paper was stimulated by my participation in the Princeton Mathematical Community in the 1930s Oral History Project, which resulted from the efforts of Charles Gillespie, Frederik Nebeker, and especially Albert Tucker and from the financial support of the Alfred P. Sloan Foundation.

2. A Fine Start

Efforts to establish a research program in mathematics at Princeton University began in the first decade of the twentieth century at the hands of Henry Burchard Fine. Fine had completed an undergraduate major in classics at Princeton in 1880 and remained until 1884, first as a fellow in experimental physics and then as a tutor in mathematics. The latter position brought him into contact with George Bruce Halsted, a mathematics instructor fresh from his dissertation work under J. J. Sylvester at Johns Hopkins University.[3] Fine wrote of their relationship in a testimonial:

> I am glad of this opportunity of acknowledging my obligations to Dr. Halsted. Though all my early prejudices and previous training had been in favor of classical study, through his influence I was turned from the Classics to Mathematics, and under his instruction or direction almost all of my mathematical training had been acquired. (Eisenhart 1950, 31-32)

On Halsted's advice, Fine traveled to Leipzig in the spring of 1884 to study with Felix Klein. Halsted's ability to inspire proved greater than his ability to teach Fine mathematics, for Klein found Fine to know no German and little mathematics. Nevertheless, Fine was encouraged to attend lectures. He progressed quickly and was awarded the Ph.D. after only a year for his solution to a problem in algebraic geometry. In the summer of 1885, and again in 1891, Fine visited Berlin to study with Leopold Kronecker. Fine's first book and several of his papers are testimony to the profound influence of Kronecker (Fine 1891, 1892, 1914).

Fine returned to Princeton in the fall of 1885 as an assistant professor of mathematics with an admiration for the German system, which provided opportunities for young mathematicians to work closely with established researchers. He progressed steadily through the ranks. In 1889 he was promoted to professor and in 1898 to Dod Professor; by 1900, he was the senior member of the department. During Woodrow Wilson's tenure as university president, from 1903 to 1911, Fine's career and his influence on Princeton mathematics advanced most rapidly.[4] Fine was appointed chairman of mathematics (1904-28), dean of the faculty (1903-12), and dean of the science departments (1909-28). When Wilson resigned to run for governor of New Jersey in 1911, Fine served as acting president of the university until John Grier Hibben was appointed president.

Fine published several research papers in geometry and numerical analysis, but he was most prominent as a textbook writer (Fine 1905, 1927; Fine and Thompson 1909) and an institution builder. In the latter capacity, he was one of two professors of mathematics to support Thomas Fiske's 1888 plan to found the New York Mathematical Society, which became the American Mathematical Society (AMS) after an international meeting at the time of the Chicago World's Fair in 1893. In 1911 and 1912, Fine served as president of the society.

When Wilson was called to the Princeton presidency in 1903, his first priority was to match the quality of the educational program to the upgraded status of a university. At Wilson's instigation, the preceptorial system was introduced in 1905 to provide smaller classes and more personalized instruction. Fine was a strong proponent of the system, and he recognized the opportunity to strengthen the mathematics program through the new appointments the preceptorial system required.[5]

At the time of Wilson's appointment, the mathematics department numbered eight members—none more distinguished a researcher than Fine. Undergraduate teaching loads were heavy, salaries low, and opportunities for research limited. The department had an office in the library (East Pyne Hall), but most individual faculty members had to work at home.

Fine planned to build a strong research program in mathematics slowly through appointments to young mathematicians with research promise. In 1905 he appointed the young American mathematicians Gilbert Bliss, Luther Eisenhart, Oswald Veblen, and John Wesley Young to preceptorships. In a move that was uncharacteristic of American mathematical institutions, Fine also sought to hire English-speaking European mathematicians.[6] That same year he hired James Jeans to a professorship in applied mathematics and offered a senior position to Jeans's fellow Englishman, Arthur Eddington, who declined it in favor of a post at the Greenwich Observatory. When Bliss left for Chicago and Young for Illinois in 1908,[7] Fine replaced them with the promising American mathematician G. D. Birkhoff and the young Scottish algebraist J. H. M. Wedderburn. Birkhoff left for Harvard in 1912 and was replaced in 1913 by the Parisian Pierre Boutroux. Fine added the Swede Thomas Gronwall in 1913, Princeton-born and -educated James Alexander in 1915, another young Swede, Einar Hille, in 1922, and Paris-trained Solomon Lefschetz in 1924. Thus, between 1905 and 1925 many of the young mathematicians who were to become leaders in American mathematics were members of the Princeton faculty.[8] Princeton was beginning to collect mathematical talent that rivaled that of the established world centers: Göttingen, Berlin, Paris, Cambridge, Harvard, and Chicago.

3. The 1920s

Although individual members of the mathematics faculty carried on intensive research activities, Princeton remained principally a teaching institution in the 1920s. As at most American universities during those years, the Princeton faculty was saddled with heavy undergraduate teaching loads and had little money to improve facilities or research opportunities. The European mathematicians who came to Princeton recognized this clearly. As Einar Hille remembers his first year there, in 1922-23: "Princeton was somewhat of a disappointment. There were in power old undergraduate teachers Gillespie, McInnes, Thompson. I think that during my first term there I had two divisions of trigonometry with endless homework" (Hille 1962). Solomon Lefschetz confirmed this situation:

> When I came in 1924 there were only seven men there engaged in mathematical research.[9] These were Fine, Eisenhart, Veblen, Wedderburn, Alexander, Einar Hille and myself. In the beginning we had no quarters. Everyone worked at home. Two rooms in Palmer [Laboratory of Physics] had been assigned to us. One was used as a library, and the other for everything else! Only three members of the department had offices. Fine and Eisenhart [as administrators] had offices in Nassau Hall, and Veblen had an office in Palmer. (Bienen 1970, 18-19)

The situation began to change around 1924 when an effort was made to raise funds to support mathematical research. With the turnover in the preceptorial rank and the disinterest of Wedderburn and others in institutional matters, the responsibility for building the Princeton research program devolved to Fine, Eisenhart, and Veblen. The first step was taken by Veblen during his term in 1923-24 as president of the American Mathematical Society.[10] In an effort to improve American mathematics nationally, he arranged for mathematicians to be included in the National Research Council fellowship program already established for physicists and chemists. He also established an endowment fund for the AMS and raised funds to subsidize its publications.

Within Princeton, the move to improve the research environment was spearheaded by Fine with the assistance of Eisenhart and Veblen. As dean of the sciences, Fine assumed responsibility for helping Princeton President Hibben to raise and allocate funds for research in the sciences. In a fund-raising document of 1926, Fine outlined the "means to the full realization for the purposes of the Mathematics Department":

(1) Endowment for Research Professorships.
(2) Improvement and increase of personnel with schedules compatible with better teaching and more research.
(3) A departmental research fund to meet changing conditions.
(4) A visiting professorship which might well bear the name of Boutroux [in memorium].
(5) A group of offices and other rooms for mathematical work, both undergraduate and advanced.
(6) Continued financial support for the *Annals of Mathematics*.
(7) A number of graduate scholarships. (Fine 1926)

It is instructive to compare this list of objectives outlined by Fine to a plan for an Institute for Mathematical Research proposed by Veblen in the period of 1924-26 to both the National Research Council and the General Education Board of the Rockefeller Foundation. Veblen's plan not only amplifies on the reasoning behind Fine's list, but also illustrates the greater vision of Veblen—realized in the 1930s with the founding of the Institute for Advanced Study.

Veblen's argument began with the premise that "the surest way of promoting such research [in pure science] is to provide the opportunities for competent men to devote themselves to it" (letter to H. J. Thorkelson, 21 October 1925, Veblen Papers). According to the American system, Veblen noted, this is a "by-product of teaching. The consequence has been that although our country has produced a great many men of high abilities, very few of them have an output which corresponds to their native gifts." Playing to the desire of funding organizations to build strong American research institutions, Veblen added that "men of considerably less ability have been able to do greater things in the European environment . . . [because] their time and energy have been free for the prosecution of their research." Elaborating this argument elsewhere, Veblen noted that his American colleagues taught nine to fifteen hours a week as compared with three hours by a mathematician in the College de France;[11] and that the American mathematician's primary task was the teaching of elementary subjects to freshmen and sophomores. These subjects were taught in the lycées and Gymnasia of Europe, and university research mathematicians there could concentrate on the teaching of more advanced subjects (letter to Vernon Kellogg, 10 June 1924, Veblen Papers).

The simplest solution, to Veblen's mind, was to provide research positions for mathematicians in which teaching duties would be limited or not required. But how was this to be accomplished? Veblen rejected the idea of "distinguished service professorships," which he was skeptical in general "would be held by men of high distinction, but who often would have passed the most active stage of research" ("Institute for Mathematical Research at Princeton"; undated, unsigned proposal, Veblen Papers).

Instead, he proposed an institute consisting of "a balanced group of first rank productive mathematicians who have opportunities for mathematical research comparable with opportunities ordinarily given those who conduct research, and train research workers, in the laboratory sciences" (document labeled "C. Mathematical Research," probably prepared for the Princeton Scientific Fund proposal to the General Education Board; undated, unsigned, Veblen Papers).

The institute he envisioned would consist of four or five senior mathematicians and an equal number of junior colleagues. The senior men would devote themselves "entirely to research and to the guidance of the research of younger men," though all institute members should "be free to offer occasional courses for advanced students" (letter to Vernon Kellogg, 10 June 1924, Veblen Papers). Beyond salaries, professorial needs were not very great: a library, a few offices, lecture rooms, a few computing machines, and money for stenographers and (human) computers. The institute could operate successfully, he maintained, either in conjunction with a university or independently.

Veblen's preference for an institute over individual research professorships was based on his and Eisenhart's assessment of the success of Göttingen over other German universities as a mathematical research center.

In those cases where a good scientific tradition has been established and has subsequently broken down, it will be found that the organization was such as to depend on a single leader. The break in the tradition came when the leader died. But if instead of having a single outstanding figure, you have a group of men of different ages who are working together so that the replacements which take place are gradual, then if you have made a good start, the conservative forces inherent in such a group tend to maintain it. A good illustration of this is to be found in the mathematical tradition of Göttingen. While there have often been men of the first magnitude at Göttingen, there has always been a large group gathered together which has maintained itself so well that the prestige of the Mathematical Institute at Göttingen is, if possible, greater now than it ever has been. During the same period of time the other German universities, which have depended for their eminence on particular individuals, have had vicissitudes of all kinds. The preeminence of Göttingen is due to the laws of statistics and the power of tradition. (Letter to H. J. Thorkelson, 21 October 1925, Veblen Papers)

As a further incentive for funding, Veblen sought to dedicate the institute to applied mathematics, a discipline he regarded as underrepresented in the United States. He pointed out that through the work of Eisenhart, Veblen, and Tracy Thomas in topology and differential geometry, Princeton had already initiated "a very definite programme" in this direction. "This programme embraces studies in the geometry of paths and analysis situs which are becoming more and more clearly the foundations of dynamics and the quantum theory" ("Institute for Mathematical Research at Princeton"; undated, unsigned proposal, Veblen Papers).

For whatever reasons, Veblen's plan for an institute was not realized at this time. No record of the National Research Council's response has been found. His proposal to the General Education Board was included as part of Princeton's general campaign to raise money to support fundamental work in the sciences, a campaign that Fine directed for President Hibben. This proposal did call for support for, among projects in astronomy, physics, chemistry, and biology, a mathematical institute with focus on applied mathematics; and it specifically referred to Veblen's earlier contacts with the board ("Memorandum for Dr. Wickliffe Rose . . . for Endowment of Research in the Fundamental Sciences"; undated, Veblen Papers). Although this grant was made, the money was not used to form the institute Veblen desired. One reason may have been the board's concern about the long-term productivity of research mathematicians:

(1) that one cannot be absolutely sure that a man who is appointed to a research position will continue for the rest of his life to do research of a high grade, and (2) that supposing your first appointments to be of the right quality, it is not certain that this quality will be maintained through the long future. (As repeated by Veblen in letter to H. J. Thorkelson, 21 October 1925, Veblen Papers)

Although the institute was not funded, the umbrella grant for fundamental scientific research was. The board awarded Princeton $1 million on the condition that it raise twice that amount. By 1928 the university had raised the $2 million through alumni gifts, and one-fifth of the total amount ($600,000) was made available to the mathematics department. It was used to buy library materials, support the journal *Annals of Mathematics*,[12] reduce teaching loads, and pay salaries of visiting mathematicians.[13]

Most of the other objectives on Fine's list were also met. Soon after Fine became involved with the Fund Campaign Committee in 1926, he approached Thomas Jones, an old friend and former Princeton classmate who had made a fortune through a Chicago law practice and his presidency of the Mineral Point Zinc Company. Jones endowed the Fine Professorship, the most distinguished chair in American mathematics at the time. Together with his niece Gwenthalyn, Jones also provided $500,000 to the research fund and endowed three chairs, including the Jones Chair in Mathematical Physics, which was first held by Hermann Weyl in 1928-29.

Princeton was able to provide good financial support for doctoral and postdoctoral mathematicians in the late 1920s and the 1930s. It attracted more National Research Council fellows than any other U.S. university.[14] British and French students were supported by the Commonwealth and Procter Fellowship programs and American graduate students by university funds.

Of the items on Fine's list, an endowment for research professorships, a departmental research fund, a visiting professorship, support for *Annals of Mathematics*, and graduate scholarships were all met. Only two items caused difficulty: increase in personnel and housing. Both needs were met in the early 1930s.

4. Fine Hall

In the late 1920s, the University of Chicago began construction of Eckart Hall for its mathematics department.[15] Veblen, a Chicago Ph.D. with continuing ties to his alma mater, kept closely informed about the new mathematics building.[16] He recognized its potential value in nurturing a mathematics community and "so he worked on Dean Fine to have this as a goal in connection with the Scientific Research Fund" (private communication from A. W. Tucker, 1985). Although Fine understood the need for adequate space (item 6 on his list), he resisted Veblen's exhortations because he knew that money was not available for similar buildings for the other sciences. Psychology, in particular, Fine regarded as having a greater space need than mathematics.

Curiously, Fine's accidental death in 1928 made possible the realization of Veblen's plan. Within a few weeks after Fine's death, the Jones family offered funds for the construction and maintenance of a mathematics building in memory of Dean Fine. Veblen and Wedderburn were given responsibility for designing the new building. Veblen took charge and designed a building that, as Jones said, "any mathematician would be loath to leave" (*Princeton University Alumni Weekly* 1931, 113). The building was constructed of red brick and limestone in the "college Gothic" style of the universities of Paris and Oxford that Veblen so admired.[17] Fine Hall was situated adjacent to the Palmer Laboratory of Physics, with a connecting corridor to enable the physicists easy access to the library and Common Room.

Veblen attended to every last detail in the design and finishing of the building. He worked closely with a high-quality decorating firm from New York on the furnishings and insisted on extensive sound testing of the classrooms. All design features were carefully chosen to promote a research environment and communal interaction. As Veblen observed:

> The modern American university is a complicated organism devoted to a variety of purposes among which creative scholarship is sometimes overlooked. Those universities which do recognize it as one of their purposes are beginning to feel the necessity of providing centers about which people of like intellectual interests can group themselves for mutual encouragement and support, and where the young recruit and the old campaigner can have those informal and easy contacts that are so important to each of them. (*Princeton University Alumni Weekly* 1931, 112)

The top floor housed the library for mathematics and physics. An open central court provided natural lighting and quiet space with carrels for each graduate student and postdoctoral visitor. Conversation rooms with blackboards were available in each corner. Eisenhart convinced the university administration to transfer there from the main library all books of research value to mathematicians and physicists, and all researchers had 24-hour access to the collection. Subscriptions of *Annals of Mathematics* were traded to other institutions in order to build up a complete collection of research journals, including all of the major foreign journals.

On the second floor the faculty members had "studies"—not offices— some of which were large rooms lavishly appointed with fireplaces, carved oak paneling, leather sofas, oriental rugs, concealed blackboards and coat

closets, and leaded windows with mathematical designs. To promote continued close ties with physics, the mathematical physicists were also assigned studies there.[18]

The second floor also housed a Common Room and a Professor's
Lounge, following a tradition Veblen had admired at Oxford. Princeton
faculty proved to be less interested than their Oxford colleagues in having a place to retire from their students, and the rather formal Professor's
Lounge was seldom used. However, at most any time of day or night one
could find graduate students, faculty members, or visitors in the Common Room discussing mathematics, playing Kriegspiel, Go, or chess, or
sleeping. The social and perhaps intellectual zenith was attained each week-
day afternoon when the mathematics and physics communities would
gather for tea. Here twenty or thirty would meet to socialize and discuss
their craft.

The first floor housed additional studies, the chairman's (Eisenhart's)
office, and various classrooms sized to accommodate seminars or lectures,
all with ample blackboard space, good acoustics, and carefully determined
dimensions. Completing the club atmosphere was a locker room with
showers to facilitate short breaks by faculty (especially Alexander) on the
nearby tennis or squash courts.

Fine Hall succeeded in promoting a community atmosphere for the
mathematical researchers. Because of the Depression and the rule against
marriage for graduate students, many of these poor, single, male students
rented small furnished rooms in town and ate meals in restaurants. The
Common Room was their main living space. The many foreign faculty
members, students, and visitors also congregated in Fine Hall, which they
found to be a place of congeniality.[19] Faculty members regularly used their
studies, which contributed to the close ties between students and staff.
Between 1933 and 1939, when Fuld Hall was completed for the institute,
Fine Hall also accommodated the institute's faculty and many visitors.

Today, many departments have facilities similar to, if more modest
than, Fine Hall. But this environment, unusual for its time, promoted a
sense of community that was impossible to foster at many universities,
like Harvard, where the faculty was scattered across the campus without
a common meeting place.[20] Albert Tucker, a graduate student and later
a faculty member at Princeton in the 1930s, describes the importance of
Fine Hall to his career:[21]

It was the amenities of Fine Hall that certainly caused me to come back to Princeton in 1933, after the year that I had my National Research Council Fellowship. I could have had a second year as a Fellow, just for the asking; but if I had, I would be supposed to spend it somewhere else than Fine Hall because that was where I had gotten my doctor's degree. The experiences that I had had at Cambridge, England, but particularly at Harvard and Chicago, made me long for the comforts, the social atmosphere, the library convenience of Fine Hall. Of course there were other things, Lefschetz, Eisenhart, and so on. But at the time, [when] Marston Morse told me I was a fool not to take the instructorship at Harvard . . . it was the opportunity to be in the Fine Hall community [that mattered]. (Oral history, 11 April 1984; The Princeton Mathematical Community in the 1930s Oral History Project, Princeton University Archives)

Between the time of the Jones gift in 1928 and the completion of Fine Hall in the fall of 1931, mathematics buildings opened at Chicago, Paris, Göttingen, and Jena. Yet, as one of the earliest and certainly the most successful building for this purpose, Fine Hall served frequently as an architectural model.[22] For example, Dartmouth, Wisconsin, Arizona State, and Western Australia modeled their mathematics buildings after Fine Hall.

5. The Institute for Advanced Study

By the time the department moved into Fine Hall in 1931, the conditions were favorable for research. Of Fine's original list of objectives only item 2, improvement and increase of personnel with schedules compatible with better teaching and more research, was unfulfilled. This was soon to be met by the founding of the Institute for Advanced Study.

The story of the institute is well known and need not be told in detail here.[23] Money from the Bamberger and Fuld family fortunes was donated in 1930 to endow an institute for advanced research to be situated in the state of New Jersey or a contiguous area. Abraham Flexner, retired from the General Education Board and the original proponent of this advanced research institute, was appointed as director. Flexner chose Princeton as the site because he believed its rural environment was suitable to pure scholarly endeavor and because the university possessed a good research library. He decided to focus the institute's activities initially in a single area, and he chose mathematics as that area for three reasons:

(1) It was fundamental.
(2) It required the least investment in plant or books.

(3) It had become obvious to me [Flexner] that I could secure greater agreement upon personnel in the field of mathematics than in any other subject. (Flexner 1940, 235)

Between 1930 and 1932, Flexner toured Europe and North America discussing his plans with leading scholars. Veblen impressed Flexner with his counsel,[24] and, once mathematics was chosen in 1932 as the institute's first mission, Veblen was given the first faculty appointment. On an earlier trip to Europe, Flexner had discussed with Albert Einstein and Hermann Weyl the prospect of their joining the institute faculty. Veblen assumed responsibility for selecting the other original faculty members of the institute's School of Mathematics. James Alexander, John von Neumann, Einstein, and Weyl joined the institute in 1933 and Marston Morse in 1935.

Flexner and Veblen assembled an impressive international group of research mathematicians. Alexander was a distinguished topologist. Einstein was already world-renowned for his contributions to theoretical physics. Morse was an accomplished Harvard mathematician known for his research in "analysis in the large." Veblen was the senior American geometer. Von Neumann was a brilliant young mathematician who had already made major contributions to logic, quantum mechanics, and analysis. Weyl, who was regarded as having the widest range of mathematical knowledge since Poincaré, was Hilbert's successor in the mathematics chair at Göttingen. The institute had many visitors each year and a few research associates (known as "permanent members") like Kurt Gödel, but these six constituted the regular faculty of the School of Mathematics for about ten years.[25]

Although the research environment was undoubtedly an attraction, social factors also contributed to their decisions to accept appointments. Alexander, whose family wealth freed him from the need to work, was relieved at the release from teaching responsibility that the institute position offered. The Nazis drove Einstein, von Neumann, and Weyl from Europe: von Neumann lost his position at Berlin for being Jewish,[26] and Weyl feared for his Jewish wife and found the Nazi interference in the affairs of the mathematics institute at Göttingen intolerable. Morse had a better opportunity at the institute to forward his research program through the availability of funds for visiting postdoctoral researchers, and it must also have been some comfort to escape from the Harvard environment where his wife had recently divorced him to marry another of the Harvard mathematicians.

During the 1932-33 academic year, Veblen rented office space for the institute in a commercial building near the Princeton campus while retaining his research study in Fine Hall. President Hibben was sympathetic to the institute's goals and offered a five-year lease on space in Fine Hall, which had been built with expansion of the department in mind and therefore exceeded current space needs. The offer was gratefully accepted and the institute settled into Fine Hall the following year.

In the early years, it was uncertain what direction the institute might take. The charter provided for it to become an educational institution and grant doctoral degrees—in competition with the university. Others may have envisioned it as an "Ivory Tower" where distinguished mathematicians could pursue their research without distraction. Veblen steered a third course, however, emphasizing both postdoctoral education and research. The institute's School of Mathematics realized the plans for an Institute for Mathematical Research that Veblen had outlined in 1925, with the exception that applied mathematics was only the principal, not the exclusive, subject of research.

Veblen envisioned a place where promising young Ph.D.s and established research mathematicians could interact and pursue common mathematical research interests without interruption by undergraduate teaching or other routine faculty duties. To that end, he arranged for adequate funds for visitors, including funds for many young Ph.D.s to work as research assistants of the permanent institute faculty.

There was free interchange between institute and university personnel. Following Veblen's example, institute faculty members offered advanced seminars open to both institute and university faculty members, visitors, and students. For example, when Gödel presented his famous institute lectures on incompleteness in 1934, graduate students Stephen Kleene and J. Barkley Rosser prepared the course notes for distribution. Institute members were commonly employed as informal advisors of doctoral dissertations, and during this period, when both university and institute mathematicians had offices in Fine Hall, it was difficult to determine the official affiliation of visitors.

6. Growth of the Department

The founding of the institute could have seriously weakened the university's mathematics program. With Veblen, Alexander, and von Neumann accepting positions at the institute and Hille leaving for a professorship at Yale in 1933, representing over a third of the faculty and a greater percentage of those productive in research, the university's strength in mathematics was seriously threatened.[27] Both the institute and the university recognized the advantages of cooperation. Both gained from having a larger community of permanent and visiting mathematicians. The university retained the services of Alexander, Veblen, and von Neumann at least to the degree that they were available to consult, help direct theses, referee articles for *Annals of Mathematics*, and present advanced seminars.[28] The institute gained in return a good research library and excellent physical facilities.

Nonetheless, the department confronted a major rebuilding project. The able leadership of Eisenhart and Lefschetz carried the department through this tumultuous period. As department chairman and dean of the faculty, Eisenhart was in a position to facilitate close ties with the institute, which he supported as favorable to the university. When a new dean of the faculty was appointed, the position of dean of the graduate school was assigned to Eisenhart.[29] Although his greatest interest had always been in the undergraduate mathematics program, he devoted himself to his new duties and was able to use his position within the university to secure adequate resources for the graduate program in mathematics.

Lefschetz was appointed to the Fine Research Professorship vacated by Veblen. Although this position did not require teaching, Lefschetz followed Veblen's example and taught a research seminar every semester, which was well attended by the graduate students. Lefschetz was also departmental representative to the Committee on the Graduate School, which enabled him to further protect the interests of the graduate mathematics program. It also stimulated him to take a personal interest in the graduate students: he meticulously inspected graduate applications to select the ten to twenty percent that he believed could handle the demanding but unstructured graduate program. He went to great lengths to meet each graduate student and monitor his progress, and he was adept at identifying dissertation problems of appropriate content and difficulty. He directed many dissertations, including those of Hugh Dowker, William Flexner, Ralph Fox, Paul Smith, Norman Steenrod, Albert Tucker, John Tukey, Robert Walker, and Henry Wallman in the period 1925-39. While Eisenhart administered the department and the relations with the rest of the university, Lefschetz built up the research and graduate programs.

After considering senior appointments to outsiders, the decision was made to continue the Fine approach and replace the departing senior mathematicians with promising junior researchers.

> It is the desire of this department to maintain its distinction by giving constant attention to the question of its personnel. Because we publish the *Annals of Mathematics*, we know rather well who the good men of the country are. Furthermore, because of the presence here of the Institute for Advanced Study, we have an additional means of keeping in contact with this situation.

Salomon Bochner, H. F. Bohnenblust, E. J. McShane, Albert Tucker, and Samuel Wilks were given junior appointments in 1933, as were Norman Steenrod, Walter Strodt, E. W. Titt, and C. B. Tompkins later in the decade.

The department continued to concentrate in a few select fields and did not try to provide uniform coverage of all areas of mathematics. Research in geometry and topology was carried out by Eisenhart, Lefschetz, Tracy Thomas, Morris Knebelman, Tucker, and Steenrod in the university and by Veblen and Alexander in the institute; mathematical physics by Eugene Wigner and H. P. Robertson in the university and by Weyl, von Neumann, and Einstein in the institute; mathematical logic by Church in the university and Gödel in the institute; and analysis by Bohnenblust, Bochner, and McShane in the university and by von Neumann and Weyl in the institute. A new area of research was initiated with the hiring of S. S. Wilks, one of the pioneers in mathematical statistics.

Measured by number or productivity of postgraduate mathematicians, the Princeton mathematical community excelled in the 1930s. Between 1930 and 1939 the university produced 39 mathematics Ph.D.s, including (with others already mentioned) Carl Allendoerfer, John Bardeen, J. L. Barnes, A. L. Foster, Wallace Givens, Robert Greenwood, Israel Halperin, Banesh Hoffmann, Nathan Jacobson, Malcolm Robertson, Ernst Snapper, Abraham Taub, Alan Turing, J. L. Vanderslice, J. H. C. Whitehead, and Shaun Wylie. Between 1923 and 1941, the years of the NRC Fellowship program in mathematics, 59 mathematicians visited Princeton, including (among others already mentioned) A. A. Albert, Gustav Hedlund, Derrick Lehmer, Neal McCoy, Deane Montgomery, Charles Morrey, Hassler Whitney, and Leo Zippin. One hundred eighty-nine mathematicians were visitors to the institute between 1933 and 1939, notably Reinhold Baer, Valentine Bargmann, Paul Bernays, Garrett Birkhoff, Eduard Cech, A. H. Clifford, P. A. M. Dirac, Witold Hurewicz, Deane Montgomery, Marshall Stone, Stanislaw Ulam, Andre Weil, and Oscar Zariski.

7. Conclusion

Institutional factors clearly helped to shape the development of the mathematics programs in Princeton in the 1930s. Careful planning by Fine, Eisenhart, and Veblen over the preceding quarter-century placed Princeton in a position to establish a world-class center of mathematics once funds started to become available in the late 1920s.

Funding for these purposes was fairly easily acquired. It may have been fortunate that the Fuld and Bamberger families were willing to endow the Institute in 1930, but it was no accident that Veblen was ready with a strong plan to build it in Princeton and to devote it to mathematical research. It is clear that the trend in the 1920s of the great foundations to support American scholarship benefited Princeton mathematics. However, it must be remembered that the bulk of the money for the university's program in mathematics came through alumni gifts. The ease at raising these matching funds is perhaps indicative of the general wealth of the United States in the late 1920s.

The existing strength of the department and a workable plan for the future were undoubtedly strong factors in attracting financial support. Several principles were consistently applied by Fine, Veblen, and others over the first forty years of the century to build excellence in the department and later in the institute. Foremost was the emphasis on research, as demonstrated in appointments and promotions, training offered to graduate students, teaching loads, and many other ways. This ran counter to the well-established tradition of American colleges as undergraduate teaching institutions. Second was the attempt to build a community of mathematical researchers so that the "old campaigner" and the "young recruit" could exchange ideas. Third was the concentration on a few areas (topology, differential geometry, mathematical physics, and logic), instead of attempting to provide uniform coverage across all of mathematics. Fourth was the adoption of an international perspective. More than any other American university in the period 1905-40, Princeton sought out students, visitors, and faculty from around the world. When the Nazi peril disrupted European mathematics, Veblen and Weyl led the way in placing émigré mathematicians in American institutions, including some in Princeton (Reingold 1981). Fifth was the decision to build up a research community through the cultivation of young mathematical talent. Although the institute took great advantage of the Nazi situation in attracting Einstein, von Neumann, and Weyl to its faculty, most of the Princeton staff was hired at the junior level and promoted from within.

Finally, great attention was given to environmental factors that would affect the research community. Prominent among these was Fine Hall. It is hard to overemphasize the importance Princeton mathematicians of the 1930s attached to their physical quarters. Soon universities throughout the world came to recognize the value of a place where their mathematicians could gather to discuss mathematics, with excellent support facilities. Another factor was the editing of professional journals (*Annals of Mathematics* and *Studies, Journal of Symbolic Logic*, and *Annals of Mathematical Statistics*) at Princeton. They provided the faculty and their students with an outlet for research and gave the faculty some control over the direction of American research. These journals also provided extensive contacts with the wider mathematical community and a vehicle for scouting new talent for appointments. Financial support for graduate students and visitors and for reduced teaching loads of staff also promoted the growth of a large community focused on mathematical research.

The success in Princeton is even more remarkable when it is considered that it occurred at the same time as the Great Depression and the growth of Nazism. General economic circumstances severely depressed academic salaries, limited funds for graduate and postdoctoral support, and restricted job placement for Princeton Ph.D.s and junior faculty. The political disruption of European academics resulted in an influx of European mathematicians into the United States, further straining the appointment and promotion of American-bred and -trained mathematicians.

Notes

1. Others were even more lavish in their praise of Princeton: R. C. Archibald wrote of Princeton as "the greatest center of mathematical activity in this country" (Archibald 1938, 169); the Danish mathematician Harald Bohr referred to Princeton as "the mathematical center of the world" when addressing an international scientific audience in 1936 (Chaplin 1958).

2. This paper about Princeton tells part of a larger story of the emergence of mathematics research in U.S. institutions in the period 1875-1940. The full story involves the rise of mathematics at Brown, Chicago, Clark, Johns Hopkins, Harvard, and Yale universities in the first half of this period and at Berkeley, MIT, Michigan, Stanford, and Wisconsin near the end. Harvard and Chicago, in particular, have many parallels with Princeton. Some information on this topic can be found in (Archibald 1938; Birkhoff 1977; Bocher et al. 1911; Lewis 1976; Reid 1976). Dr. Uta Merzbach of the Smithsonian Institution is preparing a history of American mathematics and mathematical institutions.

In this paper, I have focused on social and institutional issues. I am planning additional papers on the contributions in the 1930s of Princeton mathematicians to topology and logic.

3. Halsted was widely influential in the early development of American mathematics, e.g. inspiring Leonard Dickson and R. L. Moore as well as Fine (Birkhoff 1976; Lewis 1976).

4. Fine and Wilson were lifelong friends, having first become acquainted while working on the editorial board of the student newspaper, the *Princetonian*.

5. Harvard adopted the preceptorial system in 1910, using it to finance graduate students and train them to teach, rather than hiring additional junior faculty members.

6. Other distinguished European research mathematicians that came to the United States in its formative period include J. J. Sylvester at Johns Hopkins, R. Perrault at Johns Hopkins and Clark, and Oskar Bolza and Heinrich Maschke at Clark and Chicago. The number of foreign mathematicians willing to accept appointments in the United States was small. Many of the senior American mathematicians in this period were not distinguished researchers. Thus, Fine's appointment strategy appears sound.

7. Bliss became a leading figure in the Chicago department and in American mathematics generally. After two years at the University of Illinois and a year as mathematics department chairman at the University of Kansas, Young devoted many years to building up the mathematics program at Dartmouth.

8. Note the interesting, but perhaps coincidental, pattern of junior appointments at the university: 1905-10, American, trained elsewhere; 1910-25, mostly European; 1925-40, mostly American, several having been trained at Princeton. Princeton was not the first experience with U.S. institutions for some of these young European mathematicians. For example, Wedderburn first spent a year at Chicago and Hille a year at Harvard.

9. By the standards of the major European centers of the time, or of major American universities when Lefschetz made this comment in 1970, seven was not a large number of research mathematicians for an institution. But few other American universities, if any, had that large a number in 1924.

10. Veblen's interest in fund-raising dates from after World War I, perhaps stemming from his wartime administrative experience at Aberdeen Proving Grounds in Maryland.

11. At most American universities in the 1920s, mathematicians taught twelve or more hours per week. According to Garrett Birkhoff, it was considered a great coup at Harvard in 1928 when the weekly load for a mathematician was reduced to 4½ hours of lectures, 1½ hours of theoretical tutoring, and 3 hours of graduate student supervision (private correspondence, 4 October 1985).

12. Princeton had assumed editorial responsibility for *Annals of Mathematics* in 1911. Previously it was edited at Harvard, and before that at the University of Virginia.

13. Long-term visitors to the department in the late 1920s and 1930s included: Paul Alexandroff and Heinz Hopf (1927-28); G. H. Hardy (1928-29); Thornton Fry, John von Neumann, and Eugene Wigner (1929-30, the last two returning in subsequent years); J. H. Roberts and J. H. van Vleck (1937-38); and C. Chevalley (1939-40).

14. Veblen was one of three members of the NRC Fellowship selection committee for mathematics and was thus positioned to assist Princeton mathematics. This arrangement is characteristic of the organization of American mathematics in the 1920s and 1930s, where the power was concentrated in a small number of individuals, including G. D. Birkhoff of Harvard, G. A. Bliss of Chicago, R. G. D. Richardson of Brown, Veblen, and perhaps a few others.

15. Increased wealth in the United States in the latter 1920s redounded on the universities. Chicago's ability to construct Eckart Hall, Harvard's reduction of the teaching load, and Princeton's ease at matching the Rockefeller grant through alumni contributions all indicate this improved economic condition.

16. Veblen was at the University of Chicago from 1900 to 1905, receiving his Ph.D. in 1903 under the direction of E. H. Moore. Veblen's drive to build up American mathematics may have been stimulated by his experience at the University of Chicago, where Moore had built a strong program that produced many prominent research mathematicians, including L. E. Dickson, G. A. Bliss, G. D. Birkhoff, and Veblen, as well as supplying several major midwestern universities with their mathematics chairmen.

17. Veblen drew many ideas for Fine Hall from a visit to Oxford in 1928-29. As Savilian Professor, G. H. Hardy was expected to lecture occasionally on geometry. To avoid this responsibility, he exchanged positions that year with Veblen, whose principal mathematical interest was geometry.

18. Physics and mathematics had shared quarters earlier in Palmer Laboratory. Pro-

fessor Condon of physics was a supporter and advisor on the planning of Fine Hall; there are many references to Condon's role in the Oswald Veblen Papers.

19. Weyl reported to his former colleagues in Göttingen that German was spoken as much as English in the institute, then located in Fine Hall (Reid 1976, 157). Garrett Birkhoff reports that in the latter 1930s the official language of the institute was jokingly said to be "broken English" (private communication, 24 October 1985).

20. Harvard did have a common room, Room O, in Widener Library, but it did not have the facilities or receive the use that Fine Hall did.

21. Although Tucker may attach more significance to the environment of Fine Hall than others might care to, more than twenty mathematicians commented on the amenities of Fine Hall in the Princeton Mathematics Community of the 1930s Oral History Project.

22. The Conference Board of the Mathematical Sciences used Fine Hall as a model in a book on mathematical facilities (Frame 1963).

23. The most accessible account is in (Flexner 1940, chap. 27 and 28).

24. Flexner and Veblen held similar views about education and research. Both were enamored with the European university systems (see Flexner 1930), and both felt a need "to emphasize scholarship and the capacity for severe intellectual efforts" (Flexner 1927, 10). Both saw a need for an environment where the research faculty would be free from "routine duties . . . —from administrative burdens, from secondary instruction, from distracting tasks undertaken to piece out a livelihood" (Flexner 1930, 10-11). Both saw the research institute as a "specialized and advanced university laboratory" (Flexner 1930, 35). However, they disagreed on the addition of schools of study to the institute, support of European mathematicians, a new institute building off the university campus, administrative responsibilities for institute faculty, and other matters (see Reingold and Reingold 1982, chap. 13).

25. There is little firm evidence of who else was considered or offered an original appointment at the institute. An offer to G. D. Birkhoff was declined. Harvard countered the offer by making Birkhoff a Cabot Fellow. The address pages of Veblen's 1932 diary list five groupings of names that may have been candidates for institute positions or people whose advice Veblen sought about candidates during a 1932 tour of Europe and North America. The names (separated by semicolons the way Veblen grouped them) are as follows: Dirac, Artin, Lefschetz, Morse; Alexandroff, Wiener, Kolmogorov, von Neumann; Albert, R. Brauer, Gödel, Douglas; Bernays, Peterson, Kloosterman, Heyting, Chapin; Deuring, McShane, Whitney, Mahler.

26. Von Neumann was teaching half-time at Berlin and half-time at Princeton. After wavering for a considerable time, the university offered him a full-time position. He chose the appointment at the institute instead.

27. The 1933-34 directory of members of the mathematics department lists professors L. P. Eisenhart, W. Gillespie, S. Lefschetz, J. H. M. Wedderburn, and E. P. Wigner; associate professors H. P. Robertson and T. Y. Thomas (on leave); assistant professors H. F. Bohnenblust, A. Church, and M. S. Knebelman; and instructors E. G. McShane, J. Singer, A. W. Tucker, and S. S. Wilks. There were others listed as part-time instructors, research assistants, and advanced fellows.

28. Additional information about mathematics journals edited at Princeton in the 1930s and their contributions to the research community can be found in an oral history I conducted with Albert Tucker on 13 April 1984 (Princeton Mathematical Community in the 1930s, Oral History PMC 34).

29. In 1932-33, Eisenhart headed the Princeton administration as dean of the faculty— between the death of President Hibben and the appointment of President Dodds. Dodds appointed Eisenhart to replace the physicist Augustus Trowbridge as dean of the Graduate School and made Robert Root the new dean of the faculty.

Bibliography

The following bibliography includes only published and widely distributed printed materials. Biographies of the many distinguished Princeton mathematicians and physicists discussed herein are excluded, except those that have been quoted. *Bulletin of the American Mathematical Society*, *Dictionary of Scientific Biography*, *Proceedings of the National Academy of Sciences*, and *Princeton Alumni Weekly* have carried biographies of many of these mathematicians.

Two fertile archival sources for this research are the Oswald Veblen Papers and the Princeton Mathematical Community in the 1930s Oral History Project. The Veblen Papers are held by the Manuscripts Division of the Library of Congress, Washington, D.C. The Oral History Project records are held by the Seeley G. Mudd Manuscript Library of Princeton University. Copies are on deposit at the American Philosophical Society in Philadelphia and the Charles Babbage Institute, University of Minnesota, Minneapolis.

Archibald, R. C. 1938. *A Semicentennial History of the American Mathematical Society: 1888-1938*. Semicentennial Publications, vol. 1. Providence, R. I.: American Mathematical Society.

Bienen, Leigh Buchanan. 1970. Notes Found in a Klein Bottle. *Princeton Alumni Weekly* (April 21), pp. 17-20.

Birkhoff, Garrett. 1977. Some Leaders in American Mathematics: 1891-1941. In *The Bicentennial Tribute to American Mathematics 1776-1976*. ed. J. Dalton Tarwater. Providence, R. I.: American Mathematical Society.

Bocher, M., Curtiss, D. R., Smith, P. F., and Van Vleck, E. B. 1911. Graduate Work in Mathematics. *Bulletin of the American Mathematical Society* 18: 122-37.

Chaplin, Virginia. 1958. A History of Mathematics at Princeton. *Princeton Alumni Weekly* (May 9).

Eisenhart, Katherine. 1950. Brief Biographies of Princeton Mathematicians and Physicists. Unpublished, Princeton.

Eisenhart, L. P. 1931. The Progress of Science: Henry Burchard Fine and the Fine Memorial Hall. *Scientific Monthly* 33: 565-68.

Fine, H. B. 1891. *The Number System of Algebra Treated Theoretically and Historically*. Boston: Lench, Showall, and Sanborn.

——. 1892. Kronecker and His Arithmetical Theory of the Algebraic Equation. *Bulletin of the New York Mathematical Society* 1: 173-84.

——. 1905. *A College Algebra*. Boston: Ginn.

——. 1914. An Unpublished Theorem of Kronecker Respecting Numerical Equations. *Bulletin of the American Mathematical Society* 20: 339-58.

——. 1926. The Role of Mathematics. Limited distribution pamphlet (unsigned). The Princeton Fund Committee, Princeton University Press.

——. 1927. *Calculus*. New York: Macmillan.

Fine, H. B., and Thompson, Henry Dallas. 1909. *Coordinate Geometry*. New York: Macmillan.

Flexner, Abraham. 1927. *Do Americans Really Value Education?* The Inglis Lecture. Cambridge, Mass.: Harvard University Press.

——. 1930. *Universities: American, English, German*. New York: Oxford University Press.

——. 1940. *I Remember: The Autobiography of Abraham Flexner*. New York: Simon and Schuster.

Frame, J. Sutherland. 1963. *Buildings and Facilities for the Mathematical Sciences*. Washington, D.C.: Conference Board of the Mathematical Sciences.

Grabiner, Judith V. 1977. Mathematics in America: The First Hundred Years. In *The Bicentennial Tribute to American Mathematics 1776-1976*, ed. J. Dalton Tarwater. Providence, R.I.: American Mathematical Society. 9-25.

Hille, Einar. 1962. In Retrospect. Unpublished notes dated 16 May, Yale University Mathematics Colloquium.

Lewis, Albert. 1976. George Bruce Halsted and the Development of American Mathematics. In *Men and Institutions in American Mathematics*, ed. J. Dalton Tarwater, John T. White, and John D. Miller. Lubbock, Tex.: Texas Tech Press, Graduate Studies 13, pp. 123-30.

Montgomery, Deane. 1963. Oswald Veblen. *Bulletin of the American Mathematical Society* 69: 26-36.

Princeton University Alumni Weekly. 1931. A Memorial to a Scholar-Teacher. 30 October, pp. 111-13.

Reid, Constance. 1976. *Courant in Göttigen and New York*. New York: Springer-Verlag.

Reingold, Nathan. 1981. Refugee Mathematicians in the United States of America, 1933-1941: Reception and Reaction. *Annals of Science* 38: 313-38.

Reingold, Nathan, and Reingold, Ida H., eds. 1982. *Science in America: A Documentary History, 1900-1939*. Chicago: University of Chicago Press.

Richardson, R. G. D. 1936. The Ph.D. Degree and Mathematical Research. *American Mathematical Monthly* (April), pp. 199-204.

Stone, Marshall H. 1976. Men and Institutions in American Mathematics. In *Men and Institutions in American Mathematics*, ed. J. Dalton Tarwater, John T. White, and John D. Miller. Lubbock, Tex.: Texas Tech Press, Graduate Studies 13, pp. 7-30.

Veblen, Oswald. 1929. Henry Burchard Fine—In Memoriam. *Bulletin of the American Mathematical Society* 35: 726-30.

Contributors

Richard Askey, Professor of Mathematics, University of Wisconsin at Madison

William Aspray, Associate Director, Charles Babbage Institute for the History of Information Processing, University of Minnesota

M. K. Bennett, Professor of Mathematics, University of Massachusetts at Amherst

Garrett Birkhoff, Professor of Mathematics, Harvard University

Felix E. Browder, Dean, Rutgers University

Michael J. Crowe, Professor of the History of Science, University of Notre Dame

Lorraine J. Daston, Associate Professor of History, Brandeis University

Joseph W. Dauben, Professor of History, Lehmann College of the City University of New York

Harold Edwards, Professor of Mathematics, New York University

Michael Friedman, Professor of Philosophy, University of Illinois at Chicago

Warren Goldfarb, Professor of Philosophy, Harvard University

Judith V. Grabiner, Professor of Mathematics and History, Pitzer College

Philip Kitcher, Professor of Philosophy, University of California, San Diego

Gregory H. Moore, Associate Professor of Mathematics, McMaster University

Howard Stein, Professor of Philosophy, University of Chicago

INDEX

Index

Abel, Niels Henrik, 240
Abelian integrals, 147
Aberdeen Proving Grounds, 363n.10
Ackermann, Wilhelm, 114-16, 118, 119, 128
Acta Mathematica, 161
Affine group, 151
Aim and Structure of Physical Theory (Duhem), 267, 271, 273
Albert, A. A., 360, 364n.25
Alexander, James W., 348, 349, 355, 357-60 *passim*
Alexandroff, Paul S., 363n.13, 364n.25
Algebra, 240-44 *passim*, 270, 287, 300, 307
Algebraic geometry, 287
Algebraic number theory, 284
Allan, F., 214, 215
Allendoerfer, Carl, 360
American Academy of Arts and Sciences, 187
American Academy of Arts and Sciences workshop (Boston 1974), 192
American Mathematical Monthly, 190
American Mathematical Society: Dartmouth conference, 338; meetings, 178, 214; members and officers, 163, 164, 349; New York Mathematical Society, 348; R. C. Archibald's history of, 30
American Postulate Theorists, 108
Ampère, André-Marie, 264
Analysis: Cauchy's theorem, 36, 37; vs. deduction, 289; foundations of, 240, 242, 243, 251; nineteenth-century, 300, 301, 316, 318-20; nonstandard, 269; at Princeton, 348, 352, 357, 360; real and functional, 287; of sensation, 222
Analytic geometry, 272, 289

Analytic number theory, 242
Analytic truth, 82, 84, 94
Analytic-in-L, 87, 88, 89, 92, 93
Analyticity, 33, 84, 85, 86, 89
Andrews, George, 207, 211
Annals of Mathematical Statistics, 362
Annals of Mathematics, 350, 353-62 *passim*, 363n.12
Annals of the History of Computing, 26
Anschauliche Geometrie (Cohn-Vossen), 168
Antipsychologism, 70, 91, 93
Antrittsrede (Klein), 165
Apollonius, 266
Applied mathematics: vs. mixed mathematics, 221, 222, 231, 236; at Princeton, 352, 358
Apriorism, 44, 294-98, 303
Archibald, R. C., 21, 30, 362n.1
Archimedean axiom, 105, 106, 181
Archimedes, 238, 266, 289
Archive for History of Exact Science, 26
Archives of American Mathematics, 26
Aristotle: deduction, 288; description of world, 289; *Metaphysics*, 222; and mixed mathematics, 221, 223; vs. nineteenth-century mathematics, 238, 242; *Posterior Analytics*, 265; syllogism, 96, 97
Arithmetic, 84, 242, 243, 245, 270
Arizona State University, 356
Ars conjectandi (Jakob Bernoulli), 226
Art of Computer Programming, The (Knuth), 206
Artificial intelligence: computer science, 285; historical parallels, 329-31, 337-40, 343
Artin, Emil, 364n.25

371

William Aspray is associate director of the
Charles Babbage Institute for the History of
Information Processing at the University of
Minnesota. Aspray received his Ph.D. from the
University of Wisconsin-Madison in the history
of science in 1980 and then taught mathematical
science at Williams College and history of
science at Harvard University. He is co-editor of
*John von Neumann's Papers on Computers &
Computer Theory* and a regular contributor to
the *Annals of the History of Computing*.

Currently a professor of philosophy at the University of California, San Diego, **Philip Kitcher** has
also taught at Vassar College, the University of
Vermont, and the University of Minnesota,
where he served as director of the Minnesota
Center for Philosophy of Science from 1984 until mid-1986. In 1974 he obtained a doctorate in
the history and philosophy of science from
Princeton University. Kitcher is the author of
*Abusing Science: The Case Against Creationism,
The Nature of Mathematical Knowledge,* and
*Vaulting Ambition: Sociobiology and the Quest
for Human Nature.* He has contributed articles
to numerous journals including *Philosophical
Review, Philosophy of Science, Journal of
Philosophy* and *British Journal for the
Philosophy of Science.*